D1443194

basics of
structural
steel design

second edition

basics of structural steel design

Samuel H. Marcus, P.E.

Structural Engineer

Reston Publishing Company, Inc.
A Prentice-Hall Company
Reston, Virginia

Library of Congress Cataloging in Publication Data

Marcus, Samuel H.
    Basics of structural steel design.

    Includes bibliographical references and index.
    1. Building, Iron and steel. 2. Steel, Structural.
I. Title.
TA684.M35 1981   624.1′821   81–2666
ISBN 0–8359–0419–9   AACR2

Editorial/production supervision by
    Norma M. Karlin
Manufacturing buyer: Ron Chapman

© 1981 by Reston Publishing Company, Inc.
A Prentice-Hall Company
Reston, Virginia 22090

PRINTED IN THE UNITED STATES OF AMERICA

To my mother and father, who instilled the basic will to achieve.

To my wife, who developed and furthered this will.

To my children, who make it all worthwhile.

And finally to the AISC, who made it all possible.

contents

preface

This book is written specifically for the two-year college and technology student with the intention of fully acquainting him with all the facets of structural steel and its design. The book is of equal value both to the four-year engineering or architectural student, as well as the practicing engineer or architect, because it incorporates the practical aspects of design that were gained by the author in his years of exposure to engineering and architectural practice. In addition, the author has attempted to have the book closely parallel the AISC Manual of Steel Construction and the Specification for the Design, Fabrication and Erection of Structural Steel for Buildings, giving his interpretation of the intent behind many of the specifications, tables, and design aids.

The book is intended for use as a student text and reference text. Although for the most part, this book discusses the design of structural steel for buildings, it will also provide a basic understanding of how and why structural steel behaves as it does. Thus, the reader can easily adopt the content to other structure types, using specifications other than the AISC Specification.

The presentation commences by assuming that the reader has little familiarity with the subject of structural steel and its design. It therefore starts with a discussion of the material, plans, drawings, nomenclature, shapes of sections, loads and forces, types of framing, types of fasteners, and grades of structural steel, and continues with a discussion of tension members, bending members, axially loaded columns, beam columns, buckling of bending members, composite design, and connections. It is assumed that the reader has first studied elementary courses in statics and/or dynamics, basic strength of materials, and structural analysis.

Upon completion of the study of this textbook, the reader should have a basic working knowledge of steel and its design concepts.

The author would like to acknowledge Professor Wallace W. Sanders, Jr., of Iowa State University and Professor Elliot Colchamiro of the New York City Community College of the City University of New York for their aid in writing the design problems at the end of each chapter.

SAMUEL H. MARCUS, P.E.

basics of
structural
steel design

general information: steel materials

1-1 WHAT STRUCTURAL STEEL IS

Steel is a man-made metal containing 95% or more iron. The remaining constituents are small amounts of elements derived from the raw materials used in the making of steel, as well as other elements added to improve certain characteristics or properties of the product. Sometimes ingredients are added to control potential adverse effects on the product.

Although about 5% of the surface of our planet is composed of iron, it is not found in pure or wholesome form. Iron is found mixed with rock, sand, gravel, clay, and so on, and is bound chemically with other elements, such as oxygen, sulfur, and silicon.

However, when iron is found, mined, and refined, it helps to make one of the most versatile of metals—steel. Steel is durable, malleable, economical to use, strong, noncombustible, weldable, tough, available in large quantities, and ductile. The last characteristic is a unique property of the material, which permits permanent deformation prior to tensile fracture. This unique property will be discussed further in Section 1-14.

We will confine our discussions from here on to structural steel—steel in various shapes and forms utilized to support loads and resist the various forces to which a structure is subjected. This structure might be, among other things, a building, bridge, tower, tank, or transmission pole.

A glossary of steel terms pertaining to the product and structural steel engineering and design is found in Sections 1-15 and 1-16, respectively.

1-2   *ITS MANUFACTURE AND FABRICATION*

After iron ore is mined, crushed, washed, and upgraded, it is fed (sometimes with scrap metal) into the blast furnace, a huge, towering, cylindrical steel-shelled, brick-lined, water-cooled vessel, mixed with limestone and coke, and combined with large quantities of heated air blown in at the bottom of the vessel. Temperatures in excess of 3500°F may be attained. *Coke* is a fuel made from a special grade of bituminous coal with fewer impurities than ordinary fuel coal. Because coke is porous, it burns rapidly, producing a very intense heat. *Limestone,* a gray rock which is quarried, acts as a cleaner or *flux* that soaks up impurities from the iron ore. The hot blast of air aids in changing the iron ore into iron. In fact, the aforementioned ingredients emerge at the bottom of the furnace in a molten state called *pig iron* or "a cast of iron." This amounts to 150 to 350 tons, depending on the size of the furnace. For each ton of pig iron produced, about 2½ tons of iron ore, coke, and limestone are required.

The refining process, which changes the pig iron to steel, is accomplished in steelmaking furnaces. A single furnace load is known as a *heat* of steel. The pig iron, together with scrap metal and limestone, charge the furnace. Some iron ore or other sources of iron oxide can be added. Carbon, silicon, phosphorus, and sulfur are reduced to levels permissible by the steel specification. Other elements are added, as required, to the mix in the furnace or to the molten steel.

There are many processes used to change the pig iron to steel: the *open-hearth, basic oxygen, electric furnace,* and *Bessemer processes.* In the United States structural steel is usually produced by the basic open-hearth or basic oxygen process. The words "acid" or "basic," frequently included in the name of the process, describe the type of refractory lining utilized in the furnace and the slag that covers the molten metal at the conclusion of the refining phase.

Oxygen is necessary in all steelmaking processes to oxidize the excess of various elements such as carbon, which if not removed would have an adverse effect on the final steel product. The melted limestone helps to eliminate impurities such as silicon.

Some open-hearth furnaces have a capacity of 600 tons, but 100- to 300-ton capacities are more common. Temperatures within the furnace climb to 3000°F or more. This entire refining cycle takes about 8 to 10 hours, although in some furnaces the cycle time can be reduced by introducing a rich flow of oxygen onto the molten metal.

The basic oxygen furnace is a new development of the industry. This type of furnace resembles in appearance the oldest process of steel-making, the Bessemer process. The furnace is pear-shaped and tilts sideways for charging and pouring. A high-pressure stream of 99.5% pure oxygen is used to combine with carbon and other unwanted elements, causing a very high temperature, which results in the burning off of impurities from the pig iron, converting it to steel. This more modern and efficient refining process takes 50 minutes or less, and about 300 tons of steel can be made in 1 hour. [For a more thorough discussion of the open-hearth and basic oxygen processes and a detailed account of electric furnaces, as well as the vacuum melting process, the reader is referred to *The Picture Story of Steel* (Washington, D.C.: American Iron and Steel Institute, 1965), pp. 18–25.]

The molten steel is cast into ingots and placed in soaking pits to ensure the temperature required for the rolling process. The rolling process is basically the passing of materials through two rolls revolving at the same speed in opposite directions. This process shapes the steel, reducing its cross section and elongating it and shaping it into semifinished products called slabs, blooms, and billets. The chief differences between the three semifinished products are in their cross-sectional area, the ratio of width to thickness, and in their intended uses. All are hot-rolled from ingots rectangular in cross section and with rounded corners.

A *bloom* has a width generally equal to its thickness with a minimum cross-sectional area larger than 36 in.^{2}. A *billet* has a minimum width and thickness of $1\frac{1}{2}$ in. and a cross-sectional area of $2\frac{1}{4}$ to 36 in.^{2}. The *slab* has a minimum thickness of $1\frac{1}{2}$ in. and a minimum width of twice its thickness, with a cross-sectional area generally of 16 in.^{2} minimum.

The semifinished product goes to other rolling mills, producing structural shapes, bars, wire, wire rods, sheet, strip, plate, pipe, and so on. The rolling mills not only shape the steel; with hot rolling they make it tougher, stronger, and more malleable. Figure 1-1 shows a flow chart of the steelmaking process.

The product is shipped from the mill to *fabricating plants* in ordered quantities, dimensions, configurations, and grades. The fabricator "customizes" the product in accordance with the steel detail drawings,

FLOW CHART OF STEELMAKING

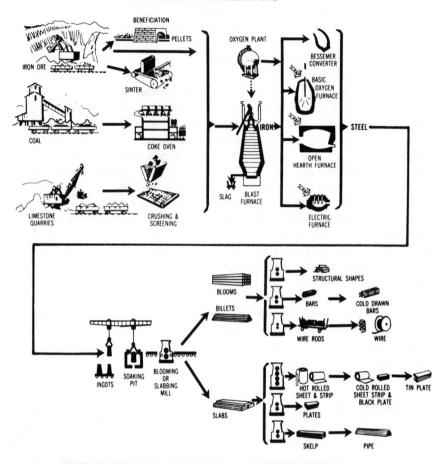

FIG. 1-1 *Courtesy of American Iron and Steel Institute*

which in turn are derived from the engineering drawings. The fabricator shears or cuts to specified lengths, finishes to specified finish, punches, drills, reams as specified, bolts or welds as specified, delivers, and may even erect the steel. In summary, steel is made, rolled, fabricated, and erected according to the specifications of the professional engineer or architect.

1-3   THE "INGREDIENTS" AND THEIR EFFECTS

The making of steel has been likened to the making of a gourmet dish. To achieve the desired characteristics, the adding of certain ingredients in specific quantities or a certain recipe must be carefully followed. A little of this and a little of that but not too much of this or that, or the addition of this would make the resulting dish look very appetizing but would cause it to taste very bitter if added in excessive amounts. And so we add that to minimize the bitter taste caused by adding this. We could go on and on.

The effects of some of the more common alloying agents follow, together with their major purpose. These are based on descriptions by the Canadian Institute of Steel Construction as found in their publication, *General Information on Structural Steel,* and are reproduced by courtesy of the Institute.

Aluminum is often used to promote nitriding, but its major use in steelmaking is as a deoxidizer. It may be used alone, as in low-carbon steels, where exceptional drawability is desired, or more commonly in conjunction with other deoxidizers. It effectively restricts grain growth, and its use as a deoxidizer to control grain size is widely practiced in the steel industry.

Carbon, although not generally considered an alloying element, is by far the most important element in steel. As carbon is added to steel up to about 0.90%, its response to heat treatment and its depth of hardening increase. In the "as-rolled" condition, increasing the carbon content increases the hardness, strength, and abrasion resistance of steel, but ductility, toughness, impact properties, and machinability decrease.

Chromium contributes to the heat treatment of steel by increasing its strength and hardness. Its carbides are very stable, and chromium may be added to high-carbon steels, subject to prolonged anneals to prevent graphitization. Chromium increases resistance to both corrosion and abrasion. Chromium steels maintain strength at elevated temperatures.

Columbium is used in carbon steels to develop higher tensile properties. It also combines with carbon to provide improved corrosion resistance and is often used for this purpose in stainless steels.

Iron is the principal element and makes up the body of steel. In commercial production, iron always contains quantities of other elements. Production of pure iron is accomplished with difficulty and generally in small quantities. Iron does not have great strength, is soft, ductile, and can be appreciably hardened only by cold work.

Manganese by its chemical interaction with sulfur and oxygen makes it possible to roll hot steel. It is next in importance to carbon as an alloying element. It has a strengthening effect upon iron and also a beneficial effect upon steel by increasing its response to heat treatment. It increases the machinability of free machining steels but tends to decrease the ductility of low-carbon drawing steels.

Molybdenum has a pronounced effect in the promotion of hardenability. It raises the coarsening temperature of steel, increases the high-temperature strength, improves the resistance to creep, and enhances the corrosion resistance of stainless steels.

Nickel is soluble in iron and, in combination with other elements, improves the hardenability of steel and toughness after tempering. It is especially effective in strengthening unhardened steels and improving impact strength at low temperatures. It is used in conjunction with chromium and stainless steels.

Phosphorus strengthens steel but reduces its ductility. It improves the machinability of high-sulfur steels and under some conditions may confer some increase in corrosion resistance.

Silicon is one of the principal steel deoxidizers and is commonly added to steel for this purpose. In amounts up to about 2.5% it increases the hardenability of steels. Specified coarse-grain steels are silicon-killed. In lower-carbon electrical steels, silicon is used to promote the crystal structure desired in annealed sheets.

Sulfur added to steel increases machinability. Because of its tendency to segregate, sulfur may decrease the ductility of low-carbon drawing steel. Its detrimental effect in hot rolling is offset by the manganese.

Titanium is an extremely effective carbide former and is used in stainless steels to stabilize the steel by holding carbon in combination. Titanium is used for special single-coat enameling steels. In low-alloy structural steels its use in combination with other alloys promotes fine grain structure and improves the strength of the steel in the as-rolled condition.

Vanadium is a mild deoxidizer, and its addition to steel results in a fine grain structure, which is maintained at high temperature. It has very strong carbide-forming tendencies and very effectively promotes strength at high temperatures. Vanadium steels have improved fatigue values and excellent response to heat treatment. In unhardened steels it is particularly beneficial in strengthening the metal.

1-4   PHYSICAL AND CHEMICAL PROPERTIES

In the United States (we will confine our discussions to steels manufactured in the United States), there are two ways to specify structural steels. The more common way, in particular for the grades of steel specified by the nationally known and respected trade association which is representative of the steel-fabricating industry, the American Institute of Steel Construction (AISC), is to use an ASTM (American Society for Testing and Materials) number. For ferrous metals the designation has the prefix letter A followed by one, two, or three numerical digits (e.g., ASTM A6, ASTM A36, ASTM A514).

The ASTM is a recognized body establishing standards that are nationally accepted and are frequently used as the technical requirements for a purchase agreement between a buyer and a seller, as well as for a specification pertaining to structural steel. Structural steel specified and manufactured by this means is assured of meeting various mechanical as well as chemical stipulations. The ASTM standards for structural steel have a common general format. The scope clause tells what the specification covers (product classification) and what use the steel is intended for. General requirements for delivery state whether ASTM A6 (General Requirements for Delivery of Rolled Steel Plate, Shapes, Sheet Piling and Bars for Structural Use) is applicable. Process states the permissible steelmaking process or processes. Chemical requirements dictate the limits of the various elements of the grade of steel by referral to a table similar to Table 1-2.

Tensile properties give the minimum yield point, minimum percent elongation in a given gage length, and a minimum or range of tensile strength by referral to a table similar to Table 1-1. Bending requirements stipulate the requirements for bend tests for various thicknesses of the material by referral to a table within the specification. Supplementary requirements describe any additional requirements that are to apply only when specified by the purchaser. At times other paragraphs are inserted, which cover other test procedures, such as marking or identification procedures, hardness requirements, rejection requirements, inspection requirements, and packing requirements.

The second manner of specifying steel would be by means of an AISI (American Iron and Steel Institute) number. This assures all involved that the steel meets certain chemical composition as specified. The AISI is to the steel mills what AISC is to the steel-fabricating industry. AISI has readily available numerical designations for various types of steels. These designations were established for the convenience of the potential user. All numerical designations of grades of steel and their meaning may be found in the AISI Steel Products Manual for the particular product. For example, AISI uses a four-numeral series to designate graduation of chemical composition of carbon steel products. The last two digits indicate the approximate average of the percent of carbon in hundredths (e.g., grade designation 1035 represents 0.32 to 0.38% carbon). The first two digits of the four-numeral series have other specific meanings (e.g., 10XX nonresulfurized carbon steel grades, manganese 1.00% maximum). Reference to the aforementioned products manuals should be made for additional information. The AISI number is sometimes specified by an SAE (Society of Automotive Engineers) number. The AISI and SAE number are the same for carbon steels and are used with the letters AISI or SAE preceding the four-digit number.

A Standard Recommended Practice for Numbering Metals and Alloys (UNS), ASTM E527 (SAE J1086), is also in use and covers a

unified numbering system involving a single-letter prefix followed by five
numbers. Reference should be made to the specification for more details.
The UNS system provides a means of correlating many nationally used
numbering systems.

Physical or mechanical properties are those properties of structural
steel that have been determined by such mechanical tests as tension,
bend, and impact tests. The specifications prescribe maximum, minimum,
or a range of values that the grade of steel is required to meet when tested
in accordance with stipulated methods and procedures.

Tension tests, consisting of pulling a standardized specimen apart
until fracture occurs, and the *bend test,* the bending of such a specimen
in a specified manner, are the most common tests required. Some steel
specifications state a maximum as well as a minimum tensile-strength
requirement. The maximum limit is an indirect means of controlling
chemical composition. *Impact tests* are sometimes required and specified
(the Charpy test being the most common). They are a measure of impact
strength or notch toughness (susceptibility of a material to brittleness in
areas containing a groove, scratch, sharp fillet, or notch) and are usually
specified for steels used where low-temperature notch toughness is a
requirement.

From a *tensile test* in which the steel specimen is loaded in a testing
machine at incremental loads during which the corresponding deformations are recorded, a record of load elongation is established from which
yield point or yield strength, yield-stress level, proportional limit, and
tensile strength can be established. Details of the standard tensile test
specimen as well as the procedures of the tensile test are found in ASTM
A6 and A370 (General Requirements for Delivery of Rolled Steel Plates,
Shapes, Sheet Piling and Bars for Structural Use and Mechanical Testing
of Steel Products), respectively.

The load-elongation data provided by the tension test permit the
construction of a *stress-strain diagram* for each particular grade of steel
specimen tested. Stress is usually plotted in kips per square inch (computed as applied load divided by original cross-sectional area of the specimen in square inches) as the ordinate and strain in inches per inch
(recorded deformation divided by the original length) as the abscissa to
give a typical stress-strain diagram, as shown in Fig. 1-2 for various
grades of steel. A small portion of this diagram for an initial range of
strain is shown in Fig. 1-3 drawn to a different scale. This curve, although
for tension tests, would be similar for compression.

The more important properties of steel are as follows:

(1) *Proportional limit or elastic limit.* This is the highest stress at which
     stress is proportional to strain, or at which the stress-strain diagram
     is linear. This proportionality of stress to strain is commonly called

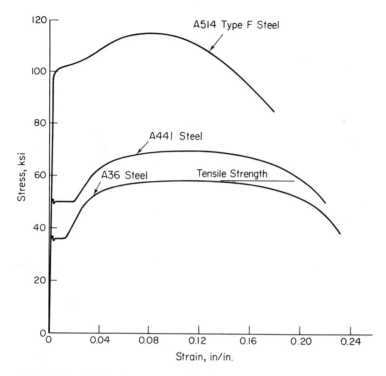

FIG. 1-2 Typical stress-strain curves for structural steels.
Courtesy of U.S. Steel Corporation

Hooke's law, and up to this limit, any deformation is not permanent deformation. When the loading is removed, the specimen returns to its original cross-sectional shape.

(2) *Yield point.* This property is the stress (less than the maximum stress attained) at which the material experiences an increase in deformation (strain) without a further increase in stress. This is shown by the horizontal portion of the curve in Fig. 1-3. Some steels have a sharply defined yield point, producing an upper yield point followed by a lower yield point. This, in turn, is followed by a horizontal (constant increase in strain with no increase in stress) plateau.

(3) *Yield strength.* Other steels do not exhibit a well-defined yield point. Yield strength is then used, which is defined as the stress at which the steel shows a limiting strain of from 0.001 or 0.002 in./in. It is shown in Fig. 1-3 for the 0.2% (0.002) offset method. In this

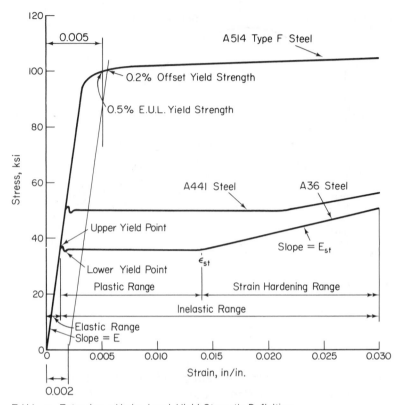

E.U.L. = Extension – Under Load Yield Strength Definition

FIG. 1-3 Typical initial stress-strain curves for structural steels. *Courtesy of U.S. Steel Corporation*

case the value of the yield strength stated should indicate the percent of offset used.

(4) *Yield stress.* A minimum yield stress is the basis for the allowable stress determinations for various grades of steel. This minimum is usually specified in the ASTM Specification for each grade of steel, which ensures that all steel manufactured to a particular grade will have this minimum yield stress. It is usually the stress corresponding to a strain of 0.5% (see Fig. 1-3).

(5) *Tensile strength.* This is the highest stress attained on the stress-strain diagram. It is the maximum axial load in the tension test divided by the original specimen area. For A36 steel, the specified tensile strength ranges from 58 to 80 ksi (kips per square inch, where a kip is 1000 lb) (Fig. 1-2).

Table 1-1
MINIMUM MECHANICAL PROPERTIES[a]

	ASTM No.	Min. Yield Stress (ksi)	Min. Tensile Stress (ksi)
Carbon steels	A36	36	58
		32	58
	A529	42	60
High-strength low-alloy steels	A242 and A441	50	70
		46	67
		42	63
		40[b]	60[b]
	A572 Grade 65	65	80
	Grade 60	60	75
	Grade 50	50	65
	Grade 42	42	60
	A588	50	70
		46	67
		42	63
High-strength quenched and tempered alloy steel	A514	100	110
		90	100
Steel pipe	A53 Grade B[c]	35	60
	Grade C (shaped)	50	62
	Grade C (round)	46	62
Steel tubing	A500 Grade C (shaped)	50	62
	Grade C (round)	46	62
	Grade B (shaped)	46	58
	Grade B (round)	42	58
	Grade A (shaped)	39	45
	Grade A (round)	33	45
	A501	36	58
	A618 Grades I and II	50	70
	Grade III	50	65
Steel sheet and strip	A570 Grade 50	50	65
	Grade 45	45	60
	A606	50-(45)[d]	70-(65)[d]
	A607	45 to 70	60 to 85

[a] The actual ASTM Specifications should be consulted for additional and more complete details.

[b] Only grade A441.

[c] Only grade approved by AISC.

[d] Cut lengths (coils) for hot-rolled or the lower values for cold-rolled material.

The physical or mechanical properties for several grades of steel are shown in Table 1-1. ASTM also specifies (AISI also does) chemical requirements for each of the grades of steel used. Percent maxima, minima, or a percentage range is specified as shown in Table 1-2 for several of the grades of steel.

We discussed in Section 1-3 the effects of each alloying or basic element on the finished steel. We see that the chemical composition of steel plays an important role in the ultimate physical properties of the steel. Chemical composition can control corrosion, strength, and the welding capabilities of steel. These basic requirements assure the potential user of getting a safe and categorized product and require the producer to manufacture a steel with these standards as a minimum. In general, a maximum limit is placed on elements that the steel producer has to reduce in the refining process, such as carbon, sulfur, and phosphorus, and a minimum limit is placed on the elements that must be added, such as the various alloying elements discussed in Section 1-3. Chemical makeup of the steel beyond that specified is at the producer's option. Steels meeting the specified physical and chemical requirements are deemed to be of acceptable standard grades.

Particular producers of steel at times develop their own grades of steel. These grades are called *proprietary* since they were produced by one company and pertain only to that company's material. Many producers manufacture proprietary steels that also meet a standard specification. Some of the more common proprietary names are Mayari-R, Corten, Man-Ten, Tri-Ten, Ex-Ten, T-1, Kaisaloy, Jalloy, Tri-Steel, Hi-Steel, V-Steel, INX Steels, SSS100, RQ100, and N-A-Xtra.

1-5   QUALITY CONTROL AND IDENTIFICATION

Throughout the manufacturing process of steel at the steel mill, various checks and rechecks are performed. A *ladle analysis* is a check of the chemisty of the steel made from samples obtained during the original casting of ingots. A *check analysis* or a further check on the chemistry of the steel is made of drillings taken from the semifinished or finished product. The check analysis usually allows certain specified variations from the ladle analysis. Limitations on both analyses are stated in the ASTM specification for any particular grade of steel. After rolling the steel through various stages to the required dimensions, a hot saw cuts a section to be used for the mechanical or physical test. Coupons or specimens are tested to determine yield point, tensile strength, percent of elongation, bending properties, or other specification requirements.

As many as 170 different tests may be made between the raw-material stage and the finished mill stage to assure that the final product meets the standard performances expected of it. Steel as shipped from the mill is identified by heat number, manufacturer's name, brand or symbol, and size. When a specified yield point exceeds 36 ksi, the specification number is also shown as specified in ASTM A6 and Section 1.26.5 of the AISC Specification. In fact, as part of a very serious quality-assurance program, a written tabular report, a mill test report or certificate, is furnished by the manufacturer. This report shows the results of the mechanical and chemical tests required by the specification and identifies the grade of each piece of steel. This report, by means of good bookkeeping and record keeping, is made available any time from mill to fabricator to erector to structure site. It is a part of the quality-control process. A typical mill test report is shown in Fig. 1-4. Section 1.4.1.1 of the AISC Specification in essence states that certified mill test reports or certified reports of tests made by the fabricator or a testing laboratory in accordance with ASTM A6 constitute sufficient evidence with conformity with one of the acceptable ASTM grades of steel.

Quality control is further assured by formal documents that specify various acceptable tolerances or variations from product dimensions. ASTM A6, General Requirements for Delivery of Rolled Steel Plates, Shapes, Sheet Piling, and Bars for Structural Use, covers a group of common requirements relative to permissible variations. This document applies to products from the mill.

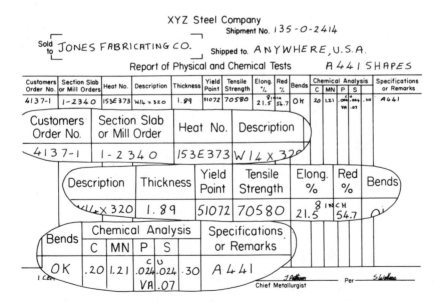

FIG. 1-4 Mill test report. *Courtesy of American Iron and Steel Institute*

Table 1-2
CHEMICAL PROPERTIES[1]

All Requirements are Maximum Percentages Unless Otherwise Indicated (Ladle Analysis)

	ASTM. No.	C	Mn	P	S	Si	Cu	V	Ni	Cr	Other
Carbon Steels	A36										
	Shapes[4]	(3)0.26	—	0.04	0.05	—	0.20 min[6]	—	—	—	—
	Plates	0.25 to 0.29	(2)0.80 to 1.20	0.04	0.05	(2)0.015 to 0.40	0.20 min[6]	—	—	—	—
	Bars	0.26 to 0.29	(2)0.60 to 0.90	0.04	0.05	—	0.20 min[6]	—	—	—	—
	A529	0.27	1.20	0.04	0.05	—	0.20 min[6]	—	—	—	—
High Strength Low Alloy Steels	A242 Type 1	0.15	1.00	0.15	0.05	—	0.20 min	—	—	—	—
	Type 2	0.20	1.35	0.04	0.05	—	0.20 min[5]	—	—	—	—
	A441	0.22	0.85 to 1.25	0.04	0.05	0.40	0.20 min	0.02 min	—	—	—
	A572										
	Grade 42	0.21	(3)1.65 to 0.50	0.04	0.05	(3)0.15 to 0.40	0.20 min[6]	—	—	—	—
	Grade 50	(3)0.21 to 0.23	(3)1.65 to 0.50	0.04	0.05	(3)0.15 to 0.40	0.20 min[6]	—	—	—	—
	Grade 60	(3)0.21 to 0.26	(3)1.65 to 0.50	0.04	0.05	0.40	0.20 min[6]	—	—	—	—
	Grade 65	(3)0.21 to 0.26	(3)1.35 to 1.65	0.04	0.05	0.40	0.20 min[6]	—	—	—	—

Table 1-2 (Continued)

High Strength Low Alloy Steels (Cont.)	A588 Grade A[8]	0.19	0.80 to 1.25	0.04	0.05	0.30 to 0.65	0.25 to 0.40	0.02 to 0.10	0.40 to 0.65 [7]
High Strength Quenched and Tempered Low Alloy Steel	A514 Type A[9]	0.15 to 0.21	0.80 to 1.10	0.035	0.04	0.40 to 0.80	—	—	0.50 to 0.80 [7]
Steel Pipe	A53 Grade B	—	—	[3]0.05 to 0.13	—	—	—	—	—
Steel Tubing	A500 Grade A & B	0.26	—	0.04	0.05	—	0.20 min[6]	—	—
	Grade C	0.23	1.35	0.04	0.05	—	0.20 min[6]	—	—
	A501	0.26	—	0.04	0.05	—	0.20 min[6]	—	—
	A618 Grade I	0.22	1.25	—	0.05	—	—	—	—
	Grade II	0.22	0.85 to 1.25	0.04	0.05	0.30	0.20 min	0.02 min	—
	Grade III	0.23	1.35	0.04	0.05	0.30	—	0.02 min	[7]
Steel Sheet and Strip	A570 Grade 45 & 50	0.25	1.35	0.04	0.05	—	0.20 min[6]	—	—
	A606	0.22	[3]1.25 to 1.40	—	0.05	—	—	—	—
	A607[10] Grade 45[11]	0.22	1.35	0.04	0.05	—	0.20 min[6]	0.01 min —	Cb[10] 0.005 min

(1) Actual ASTM Specifications should be Consulted for Additional & more Complete Details
(2) Not Required in all Thicknesses
(3) Indicates that Percentage Varies between Numbers Shown (Typical)
(4) Mn Content of 0.85-1.35 % and Silicon Content of 0.15-0.30 % Required for Shapes over 426 lb/ft
(5) If Chromium and Silicon Contents are each 0.50 min., then Cu Requirement does not Apply

(6) When Specified
(7) Additional Alloying Elements
(8) One of Nine Grades
(9) One of Fifteen Types
(10) Either V or Cb
(11) One of Six Grades

The AISC Specification for the Design, Fabrication and Erection of Structural Steel for Buildings, the nationally accepted specification, refers to allowable fabrication variations (see Section 1.23, Fabrication to 1.23.8, specifically Section 1.23.8 on Dimensional Tolerances).

The AISC Code of Standard Practice for Steel Buildings and Bridges is a document that states and discusses certain good practices relating to the design, fabrication, and erection of structural steel. In particular, this document relates suggested fabrication and erection tolerances found acceptable to the building industry (Sections 7.11 and 6.4). From mill to fabricator to erector, quality is assured.

Section 1.26.5, Identification of Steel, of the AISC Specification states by reference that all steel having a yield point of more than 36 ksi shall be marked in the fabricator's plant (it is also marked at the mill) to identify its ASTM Specification. This is done by marking with paint the ASTM Specification designation on the piece over any shop coat of paint prior to shipment from the fabricator's plant. ASTM A6 stipulates the means of showing the ASTM Specification number for yield points greater than 36 ksi. The typical color system for identifying individual specifications is designated in ASTM A6 and is shown for typical steels in Table 1-3. These colors are painted on the cut edge or on the rolled surface of flange or legs of shapes and, when specified by the purchaser, on plates.

Table 1-3
ASTM A6 COLOR IDENTIFICATION SYSTEM

ASTM Designation	Color
A242	Blue
A441	Yellow
A514	Red
A529	Black
A572 Grade 42	Green and white
Grade 45	Green and black
Grade 50	Green and yellow
Grade 55	Green and brown
Grade 60	Green and gray
Grade 65	Green and blue
A588	Blue and yellow

A6 also requires the heat number, size of section, length, and mill identification to be on each piece. The manufacturer's brand, name, or trademark is required to be shown in raised letters at intervals along the length. A recommended procedure, Identification of High Strength Steels During Fabrication, is another document authored by AISC to identify structural steel having a yield point greater than 36 ksi.

Quality control and proper identification are dependent on and are only as good as the quality, integrity, and reputability of the mill, fabricator, and erector with whom you are dealing. It is a matter of moral and professional obligation. In other words, it pays to deal with a reputable company rather than with the "cheapest" one.

1-6 THE AISC AND ITS PUBLICATIONS AND SPECIFICATIONS

The American Institute of Steel Construction (AISC) is a nonprofit trade association representing and serving the fabricated structural steel industry of the United States. The institute was founded in 1921. It is headquartered in Chicago, with numerous regional offices located throughout the country.

AISC is supported by three classes of membership: active members, who are directly engaged in the fabrication and erection of structural steel; associate members, who are allied product manufacturers; and professional members, who are individuals or firms engaged in engineering, architecture, or education.

The institute provides services to architects, engineers, contractors, educators, students, builders, developers, building officials, fabricators, and erectors in the form of specifications, technical publications, support of research and development, technical seminars and conferences, fellowship awards, aesthetic award programs, and quality control and safety programs. By so doing, its objectives are met—to improve and advance the use of structural steel through the most efficient and economical design of structures.

The AISC publishes the *Manual of Steel Construction,* the *Specification for the Design, Fabrication and Erection of Structural Steel for Buildings,* the *Code of Standard Practice for Steel Buildings and Bridges,* the *Structural Steel Detailing Manual,* various design manuals, and two quarterly periodicals, *Engineering Journal* and *Modern Steel Construction.* Several other technical publications and reprints, specifications, general publications and general reprints, manuals, and textbooks are made available relative to the subject of structural steel design. A List of Publications is available from the AISC.

It is worth mentioning some other AISC publications:

(1) *QCIS (Quality Criteria and Inspection Standards, AISC).* Recommended quality standards developed for structural steel material. They cover the preparation of materials, fitting and fastening, di-

mensional tolerances, welding, and surface preparation and painting.

(2) *Guide for the Analysis of Guy and Stiffleg Derricks, AISC.* For use in the design and evaluation of guy and stiffleg derricks for steel erection.

(3) *Commentary on Highly Restrained Welded Connections, AISC.* This is a state-of-the-art presentation on potential problems encountered in highly restrained welded connections.

(4) *Iron and Steel Beams 1873–1952, AISC.* This presents in tabular form the design and detailing dimensions and properties for all sections produced during the period.

1-7   OTHER SPECIFICATIONS AND CODES

Listed next is a partial compilation of other specifications and codes relating to structural steel. Depending upon the type of structure involved, the appropriate governing specifications and codes should be consulted. One should realize that regulations exist on virtually every level of government and that one or more of these regulations may take precedence over the others. There are the "model" codes, federal, state, city, county, township, and village regulations. In addition, there are fire, health, and safety regulations, as well as other governing regulations relating to structure—electrical, heating, ventilating, air conditioning, painting, fireproofing, plumbing, and so on. Each specification, regulation, or code may be related in some way to the other.

AISI (American Iron and Steel Institute)   Specification for the Design of Cold-Formed Steel Structural Members. This specification covers the design of structural members which are cold-formed to shape from carbon and low-alloy steel sheet or strip used for load-carrying purposes in buildings. There is similar specification for cold-formed stainless-steel structural members.

AASHTO (American Association of State Highway and Transportation Officials)   This specification governs the design and construction of highways and bridges.

AREA (American Railway Engineering Association)   This is a specification governing the design, fabrication, and erection of steel railway bridges (Chapter 15 of this specification).

AWS (American Welding Society)   A nationally accepted specification for structural welding.

ANSI (American National Standards Institute)—Minimum Design Loads in Buildings and Other Structures—ANSI A58.1 This is a nationally recognized specification and is usually used in the absence of any other applicable building-code load requirements.

Model Codes These building codes are produced by the following four organizations. These model codes become law only when adopted by the state, city, or location involved. Each code is, however, widely used in the region from which its organizational membership is made up. The documents are typically performance-type codes in which performance tests that are published by national test organizations are referred to. In addition, portions of each code are of the specification type, dictating the material to be used, how they are to be used, under what conditions, and so on.

> **SSBC (Southern Standard Building Code)** This code, in existence since 1945, is written by the Southern Building Code Congress, Birmingham, Alabama. It is widely used in the South and Southeast.

> **BOCA (The Basic Building Code)** This code has been in existence since 1950 and is written by the Building Officials and Code Administrators International, Chicago, Illinois. The document is widely used in Eastern and North-central states.

> **ICBO (The Uniform Building Code)** It was first written in 1927 by the International Conference of Building Officials, Pasadena, California. It is widely used in the West.

> **National (The National Building Code)** It has been written by the American Insurance Association, New York City, since 1905. It has been adopted in various localities throughout the country. This code is written by a private enterprise, the insurance industry, whereas the three others have municipal officials as members of their respective organizations.

Specifications for Structural Joints Using ASTM A325 or A490 Bolts (Research Council on Riveted and Bolted Structural Joints of the Engineering Foundation); Endorsed by AISC and IFI (Industrial Fasteners Institute) This specification covers the design and assembly of structural joints using ASTM A325 high-strength carbon steel bolts and ASTM A490 high-strength alloy steel bolts.

SJI (Steel Joist Institute) Standard Specifications and Load Tables for Steel Open Web Joists (DLH-, LH-, and H-series).

ASTM (American Society for Testing and Materials) A multipart, multi-volume book containing standard and tentative test methods, definitions, recommended practices, classifications, specifications, and related material, such as proposed methods. Part 4 covers material on structural steel.

UL (Underwriters Laboratories) UL produces a directory called the Building Materials Directory. Part I, Building Material List, contains the names of participating companies. Part II, Classified Building Materials Index, contains the same company names in connection with evaluated products with respect to specific hazards or performance. UL also produces a Fire Resistance Index with hourly fire ratings for beams, columns, floors, roofs, walls, and partitions tested in accordance with ASTM Standard E119 (Standard for Fire Tests of Building Construction and Materials).

AISE (Association of Iron and Steel Engineers) AISE produces a document, Specifications for the Design and Construction of Mill Buildings (AISE Standard No. 13). This provides an approach to the design and construction of mill buildings and other buildings or structures having related or similar usage.

SSPC (Steel Structures Painting Council) SSPC produces a two-volume manual. Volume I covers good painting practice and Volume 2 covers painting systems and specifications.

1-8   STEEL PRODUCT CLASSIFICATION

In response to a recognized need to improve and standardize the designation for structural steel shapes, the Committee of Structual Steel Producers of AISI developed standard nomenclature for structural steel shapes. These designations enable all mills to use the same identification in ordering, billing, and specifying. This system also allows the now-standard computer and other automated equipment to print out the nomenclature on a standard keyboard. These designations for various types of shapes are presented in Part I of the AISC Manual of Steel Construction and for convenience are repeated in this section (Fig. 1-5) along with the old designations. In addition, all sheet piling sections begin with the letter P for piling, with the succeeding letter or letters defining the configuration followed by a two-digit number, which indicates the weight of the section in pounds per foot (Fig. 1-6). H-piles are designated by HP followed by the H-section depth, an ×, and then the weight in pounds per foot (Fig. 1-7). Steel plate is designated by all dimensions in inches,

Hot-Rolled Structural Steel Designations		
New Designation	Type of Shape	Old Designation
W 24 x 76 W 14 x 26	W Shape	24 WF 76 14 B 26
S 24 x 100	S Shape	24 I 100
M 8 x 18.5 M 10 x 9 M 8 x 34.3	M Shape	8 M 18.5 10 JR 9.0 8 x 8 M 34.3
C 12 x 20.7	American Standard Channel	12 [20.7
MC 12 x 45 MC 12 x 10.6	Miscellaneous Channel	12 x 4 [45.0 12 JR [10.6
HP 14 x 73	HP Shape	14 BP 73
L 6 x 6 x $\frac{3}{4}$	Equal Leg Angle	∠ 6 x 6 x $\frac{3}{4}$
L 6 x 4 x $\frac{5}{8}$	Unequal Leg Angle	∠ 6 x 4 x $\frac{5}{8}$
WT 12 x 38 WT 7 x 13	Structural Tee Cut from W Shape	ST 12 WF 38 ST 7 B 13
ST 12 x 50	Structural Tee Cut from S Shape	ST 12 I 50
MT 4 x 9.25 MT 5 x 4.5 MT 4 x 17.15	Structural Tee Cut from M Shape	ST 4 M 9.25 ST 5 JR 4.5 ST 4 M 17.15
PL $\frac{1}{2}$ x 18	Plate	PL 18 x $\frac{1}{2}$
Bar 1 ⛶	Square Bar	Bar 1 ⛶
Bar 1$\frac{1}{4}$ φ	Round Bar	Bar 1$\frac{1}{4}$ φ
Bar 2$\frac{1}{2}$ x $\frac{1}{2}$	Flat Bar	Bar 2$\frac{1}{2}$ x $\frac{1}{2}$
Pipe 4 Std. Pipe 4 x - Strong Pipe 4 xx - Strong	Pipe	Pipe 4 Std. Pipe 4 x - Strong Pipe 4 xx - Strong
TS 4 x 4 x 0.375 TS 5 x 3 x 0.375 TS 3 OD x 0.250	Structural Tubing: Square Structural Tubing: Rectangular Structural Tubing: Circular	Tube 4 x 4 x 0.375 Tube 5 x 3 x 0.375 Tube 3 OD x 0.250

FIG. 1-5 *Courtesy of American Institute of Steel Construction*

fractions of an inch, or decimals of an inch. As an alternative, thickness may be specified in pounds per square foot.

In 1977–1978, ASTM A6 published revisions to the standard profile shapes for W and HP shapes. This new series of shapes includes 187 W shapes, 111 of which were included in the previous series, and 15 HP shapes, 9 of which were included in the previous series. Eighty-one W shapes of the previous series have been discontinued, while 76 have been added to the new series. Six HP shapes have also been added to the new series. The AISC, to assist designers and fabricators, has published an

Sheet Piling Designations

New Designations	Type of Sheet Piling	Old Designations
PZ38	Piling — Z Shape	ZP38
PZ32	Piling — Z Shape	ZP32
PZ27	Piling — Z Shape	ZP27
PDA27	Piling — Deep Arch	DP2
PMA22	Piling — Medium Arch	AP3
PSA23	Piling — Shallow Arch	SP4
PSA28	Piling — Shallow Arch	SP5
PS28	Piling — Straight	SP6a
PS32	Piling — Straight	SP7a
PSX35	Piling — Straight, Extra Strength Interlock	SP7b

FIG. 1-6

H-Pile Designations

New Designations	Old Designations
HP14 x 117	BP14 x 117
HP14 x 102	BP14 x 102
HP14 x 89	BP14 x 89
HP14 x 73	BP14 x 73
HP12 x 74	BP12 x 74
HP12 x 53	BP12 x 53
HP10 x 57	BP10 x 57
HP10 x 42	BP10 x 42
HP8 x 36	BP8 x 36

FIG. 1-7

interim document that completely lists the properties and dimensions for designing and detailing, respectively, for both the new and old series. The eighth edition of the AISC Manual also includes six new S shapes, which replace six old S shapes.

Structural members are categorized relative to the type of load that they serve to carry: tension members that carry axial tension load only, compression members that carry axial compression load only, bending members or beams that carry loads normal or perpendicular to the longitudinal axis of the member. Members can be subject to both bending loads and axial tension or compressive forces simultaneously.

As discussed, the semifinished product of the mill goes through rolling mills to produce structural steel shapes, plates, bars, pipes, tubes, sheet, and so on, to carry loads. The cross sections of typical rolled shapes are shown in Fig. 1-8 and are classified, designated, and described as shown.

Fig. 1-8a: W-Shape Section This is the most commonly used shape; it has two horizontal elements, called *flanges,* and a vertical member, referred to as a *web*. Until recently this shape has been called a *wide flange* shape, and has been designated by the term WF. Essentially, W shapes

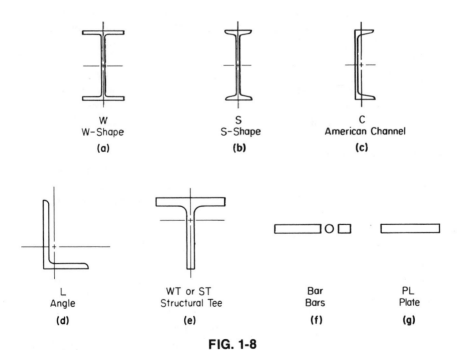

W	S	C
W–Shape	S–Shape	American Channel
(a)	(b)	(c)

L	WT or ST	Bar	PL
Angle	Structural Tee	Bars	Plate
(d)	(e)	(f)	(g)

FIG. 1-8

have the inner and outer edges of the top and bottom flange parallel. The same inside to inside of flange dimension is maintained (with a slight variation by groups) for a given depth category of shape. The designation W 24 × 76 means a W shape, nominally 24 in. deep (outside to outside of flange) and weighing 76 lb per lineal foot of span.

Fig. 1-8b: S-Shape Section   This rolled shape also has two parallel flanges and a web. However, the inner surfaces of the flange have a slope of approximately 16⅔% (2 in. in 12 in.). These shapes have long been called *American Standard* beams. The designation S24 × 100 means an S shape 24 in. deep (outside to outside of flange) and weighing 100 lb per lineal foot of span. Some of the S24 and S20 groupings have depths in excess of 24 in. and 20 in., respectively.

Fig. 1-8c: American Standard Channel Section   This cross section used to be designated by the symbol [or ⌐, depending on whether the section's web was vertical or horizontal. However, the symbol C is now used. As for the S shape, this section has two parallel flanges and a web, with the inner surface of both flanges having a slope of approximately 16⅔% (2 in. in 12 in.). The designation C 12 × 20.7 indicates an American Standard Channel with a depth (outside to outside of flange) of 12 in. and a weight of 20.7 lb per lineal foot of span. To indicate the position of the channel (web horizontal or vertical), the engineer usually indicates the appropriate position by the old symbol [or ⌐ in addition to the usual designation.

The letter HP indicates bearing pile shapes having two parallel flanges with parallel flange surfaces and a web element. The web and flange thicknesses are essentially equal, and the width of flange and depth of section are nominally equal. HP 14 × 73 designates an HP shape (bearing pile) nominally 14 in. in depth (outside to outside of flange) and 73 lb per lineal foot of span.

The letter M refers to shapes that cannot be classified as W, HP, or S shapes. Similarly, MC designates channels that cannot be classified as American Standard Channels. These shapes are not as readily available as the others, and availability should be checked prior to specifying their use.

Fig. 1-8d: Angle Shapes   Shapes with two legs of rectangular cross section that are normal one to another. The inner and outer surface of each leg is parallel. Equal leg or unequal leg angles are available. The thickness of each leg is the same. L is the symbol now used to designate an angle shape, whereas the symbol ∟ was used before. L 6 × 4 × ⅝ designates an unequal leg angle whose large leg is 6 in. long and ⅝ in. thick and whose short leg is 4 in. long and ⅝ in. thick.

Fig. 1-8e: Structural Tees (WT or ST Section) These sections are obtained by splitting the webs of various beams. They may be split from a W shape (WT) or from the S shape (ST). These shapes have a single horizontal flange and a web or stem. WT 12 × 38 designates a structural tee cut from a W shape whose depth (tip of stem to outside flange surface) is nominally 12 in. and weighs 38 lb per lineal foot of span.

Fig. 1-8f: Bars Bars are generally classified as 6 in. or less in width and 0.203 in. and over in thickness or over 6 to 8 in. in depth and 0.230 in. and over in thickness. These sections can be rectangular (flat), circular, or square. Bar 1 ⯀ and bar 1 ⬤ indicate a 1-in.-square and 1-in.-diameter round bar, respectively, whereas bar $2\frac{1}{2} \times \frac{1}{2}$ indicates a flat bar $2\frac{1}{2}$ in. wide and $\frac{1}{2}$ in. thick. The weights of square, round, and flat (rectangular) bars are given in tables found in the AISC Manual of Steel Construction, Part 1. Specifying widths in $\frac{1}{4}$-in. and thicknesses in $\frac{1}{8}$-in. increments is the preferred practice.

Fig. 1-8g: Plates Plates are rectangular in shape, generally over 8 in. in width and 0.230 in. and over in thickness or over 48 in. in width and 0.180 in. and over in thickness. *Sheared plates* are rolled between horizontal rolls and trimmed (sheared or gas cut) on all edges. *Universal plates* (UM) are rolled between horizontal and vertical rolls and trimmed (shear or gas cut) on the ends only.

The extreme width of UM plates is currently 60 in. and for sheared plates is 200 in. Availability and limiting thicknesses and lengths should be verified before specifying. Varying the widths by even inches (although smaller increments are available) and specifying thicknesses by $\frac{1}{32}$-in. increments up to $\frac{1}{2}$-in. plate; $\frac{1}{16}$-in. increments over $\frac{1}{2}$- to 1-in. plate; $\frac{1}{8}$-in. increments over 1- to 3-in. plate; and $\frac{1}{4}$-in. increments over 3-in. plate is the preferred practice. Plate thicknesses may be specified in inches or by weight per square foot.

Figure 1-9 shows in summary form a complete classification of bar and plate as well as strip and sheet. *Strip* (hot-rolled) is flat-rolled steel defined as up to $3\frac{1}{2}$ in. in width and 0.0255 to 0.203 in. in thickness; or over $3\frac{1}{2}$ to 6 in. in width and 0.0344 to 0.203 in. in thickness; or over 6 to 12 in. in width and 0.0449 to 0.2299 in. in thickness.

Sheet (hot-rolled) is flat-rolled steel defined as over 12 to 48 in. in width and 0.0449 to 0.2299 in. in thickness; and over 48 in. in width and 0.0449 to 0.1799 in. in thickness.

Welded sections similar in cross section to rolled W shapes or tees can be made from rectangular plates welded one to another. In addition, box-shaped sections can be fabricated from rectangular plates and W shapes or rectangular plates alone (Fig. 1-10). At times, the upper or

Thickness (inches)	Width (inches)					
	To 3½ incl.	Over 3½ To 6	Over 6 To 8	Over 8 To 12	Over 12 To 48	Over 48
0.2300 & thicker	Bar	Bar	Bar	Plate	Plate	Plate
0.2299 to 0.2031	Bar	Bar	Strip	Strip	Sheet	Plate
0.2030 to 0.1800	Strip	Strip	Strip	Strip	Sheet	Plate
0.1799 to 0.0449	Strip	Strip	Strip	Strip	Sheet	Sheet
0.0448 to 0.0344	Strip	Strip	Hot rolled sheet and strip not generally produced in these widths and thicknesses			
0.0343 to 0.0255	Strip					
0.0254 & thinner						

FIG. 1-9 *Courtesy of American Institute of Steel Construction*

lower (or both) flanges of a shape are required to be strengthened. This is often done by connecting a plate to the flange element(s) (Fig. 1-10c).

Both dimensions for detailing and properties for designing for typical rolled shapes are given in tabular form in Part 1 of the AISC Manual of Steel Construction. The dimensions and properties for pipe sections (standard extra strong and double extra strong), structural tubing (square and rectangular), and other sections are also given in the AISC Manual, Part 1.

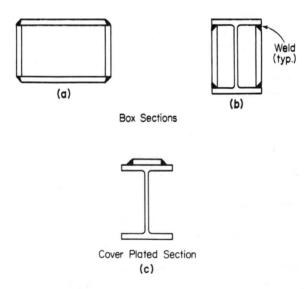

(a)

(b)

Weld (typ.)

Box Sections

Cover Plated Section
(c)

FIG. 1-10 Welded sections.

1-9   TYPES OF STRUCTURAL STEEL

In the pleasant and easy days of a generation ago, there were very few kinds or grades of steel on the market, but research and progress during and since World War II have developed a whole family of steels to facilitate economical design. The AISC Specification for the Design, Fabrication and Erection of Structural Steel for Buildings states that structural steel must conform to any one of a number of ASTM grades of steel.

Carbon and manganese are used to increase the strength of steels but, when used in excessive amounts, can affect adversely the hardness, toughness, weldability, and ductility of steel. *Carbon steel* is a term applied to a broad range of steels, ranging up to 1.70% maximum carbon, 1.65% maximum manganese, and 0.60% maximum silicon. This category includes materials having a carbon level almost as low as ingot iron, which is virtually free of carbon, up to cast iron, which contains greater than 1.7% carbon. Carbon steels are classified in four groups:

Type	% Carbon
Low-carbon steels	0.15 max.
Mild carbon steels	0.15–0.29
Medium-carbon steels	0.30–0.59
High-carbon steels	0.60–1.70

The carbon range for most of the structural steels is 0.15–0.29% (mild carbon steel), with manganese up to 1.60%.

ASTM A36 (structural steel) is a weldable mild carbon steel and has a guaranteed minimum yield of 36 ksi for all shapes and for plates up to 8 in. in thickness. This grade of steel is considered to be the workhorse steel and is the most common steel in use today, having replaced the now-obsolete A7 and A373 steels.

ASTM A529 (structural steel with a 42-ksi minimum yield point) is a higher-strength carbon steel available in plates and bars up to $\frac{1}{2}$ in. in thickness or diameter and shapes in Group 1 (see Section 1-17). Where 0.02% copper is specified, A529 has an atmospheric corrosion resistance equal to twice that of structural carbon steel without copper. This steel is used in the relatively light structural members of standard steel buildings.

High-strength steels all have limited carbon and manganese, plus small quantities of alloying ingredients. They usually also have the characteristic of the yield point varying with thickness (in plates): 50 ksi for plate thicknesses up to $\frac{3}{4}$ in., 46 ksi for plate thicknesses over $\frac{3}{4}$ in. to $1\frac{1}{2}$ in., and 42 ksi for plate thicknesses from over $1\frac{1}{2}$ in. to 4 in. The yield point for shapes varies with groupings (see Section 1-17) from 50 to 42 ksi.

ASTM A242 (high-strength low-alloy structural steel) is a very broad specification stipulating minimum mechanical properties and limits the maxima on carbon and manganese for weldability. The specification is limited to material up to 4-in. plate. Generally, these steels have enhanced atmospheric corrosion resistance of at least two times that of carbon steels with copper, or four times carbon steel without copper. When *self-weathering* (unpainted) steels are specified, one usually specifies A242 with the added requirement that the steel have from four to six times the corrosion resistance of carbon steel. Self-weathering is the term used to describe a steel that has chemical properties allowing it to form a very dense and tight oxide (rust), which in effect seals the base metal from further oxidation and therefore affords a means (other than a coating) of protecting the steel from further corrosion. For this to occur, the steel must be exposed to the elements (alternately dry and wet). The tight oxide, or *patina* as it is called, gives a deep-brown appearance and is frequently used in structures for aesthetic reasons, as well as from the low-maintenance point of view.

ASTM A441 (high-strength low-alloy structural manganese vanadium steel) is a weldable steel with reasonable maxima on carbon and manganese with an added alloy to bring its strength up. It is a substitute for the phased-out A440, which was intended primarily for bolting and riveting. A441 is suitable for welding, riveting, or bolting. The atmospheric corrosion resistance of this steel is about twice that of carbon steel. This specification is limited to material up to 8 in. in plate and bar thicknesses, and for thicknesses over 4 in., the yield point is 40 ksi.

ASTM A572 (high-strength low-alloy columbium-vanadium steels of structural quality) covers six grades or strength levels for shapes, plates, sheet piling, and bars. Yield point varies from 42 to 65 ksi (42, 50, 60, and 65 ksi for grades 42, 50, 60, and 65 respectively). Grades 42 and 50 are intended for riveted, bolted, or welded construction of all structures, while grades 60 and 65 are intended for riveted or bolted construction of bridges and for riveted, welded, or bolted construction of other applications. Available grades vary for groupings of shapes and thicknesses of plates, and reference should be made to the availability table in Section 1-17. When 0.20% minimum copper is specified, the A572 steels provide atmospheric corrosion resistance similar to A242 and A441 steels.

ASTM A588 (high-strength low-alloy structural steel with 50,000-psi minimum yield point to 4 in. thick) is a relatively newer specification specifically created to maintain a higher yield-point level for heavier shapes and thicker plates. The specification covers shapes, plates, and bars for welded, riveted, or bolted construction. It is intended primarily for use in welded bridges and buildings where savings in weight and added durability are important. The atmospheric corrosion resistance is about four times that of carbon steel without copper. The material makes available all shapes at a 50-ksi yield stress level. Plate yield points vary from 42 to 50 ksi, depending upon the thickness of material (50 ksi up to 4 in. in thickness). This grade of steel is also used, as A242 is, for self-weathering applications. A588 also has enhanced toughness characteristics (resistance to sudden fracture in the presence of notches, dynamic loads, and reduced temperatures).

High-Strength Quenched and Tempered Alloy Steel

ASTM A514 (high-yield-strength quenched and tempered alloy steel plate, suitable for welding) is a heat-treated (a combination of heating and cooling operations to produce desired properties) steel in plates in thicknesses up to 6 in. and is primarily intended for use in welded bridges and other structures. Yield-point levels vary from 90 to 100 ksi, depending on plate thicknesses.

Structural Tubing

ASTM A53 (welded and seamless pipe) grade B covers hot-formed seamless and welded black and hot-dipped galvanized round steel pipe in nominal sizes ⅛ in. to 26 in. inclusive with varying wall thicknesses. Grade B furnishes a guaranteed minimum yield of 35 ksi, although 36 ksi is used in the AISC Manual design tables. Type E (electric-resistance welded) and type S (seamless) are both provided. Both are suitable for welding.

ASTM A500 (cold-formed welded and seamless carbon steel structural tubing in rounds and shapes) covers steel round, square, rectangular, or special-shaped structural tubing for welded, riveted, or bolted construction. The tubing is provided in welded sizes with a maximum periphery of 64 in. and a maximum wall thickness of 0.500 in., and in seamless with a maximum periphery of 32 in. and a maximum wall thickness of 0.500 in. It is produced in three grades, A, B, and C, and, depending on whether it is round or shaped (square, rectangular, or special) tubing, the yield point varies from 33 to 50 ksi. Under the specification, the maximum sizes would be about 20 in. diameter round, 16 by 16 in. square, or 20 by 12 in. rectangular (although these maximum sizes may not be produced).

ASTM A501 (hot-formed welded and seamless carbon steel structural tubing) covers square, round, rectangular, or special-shaped structural tubing for welded, riveted, or bolted construction. Square and rectangular (common sizes 3 in. by 2 in. to 10 in. by 6 in.) tubing is furnished in sizes 1 to 10 in. across the flat sides with wall thicknesses 0.095 to 1.000 in., depending on size, and round tubing is furnished in nominal diameters $\frac{1}{2}$ to 24 in. with nominal (average) wall thicknesses 0.109 to 1.000 in., depending upon size. Pipe of other dimensions may be furnished. Yield stresses of 36 ksi are provided.

ASTM A618 (hot-formed welded and seamless high-strength low-alloy structural tubing) covers three grades of square, rectangular, round, and special-shaped tubing for welded, riveted, and bolted application in buildings and bridges. Yield levels of 50 ksi are supplied. For enhanced corrosion resistance, grades I and III should be specified. Dimensions shall be agreed upon at the time of purchase.

Sheet and Strip, Structural Quality

Sheets and strips are flat-rolled products of relatively thin thicknesses. We shall merely mention the common AISC listed grades.

ASTM A570, grades 45 and 50 (hot-rolled carbon steel sheets and strip, structural quality). Minimum yield points are 45 and 50 ksi for grades 45 and 50, respectively.

ASTM A606 (steel sheet and strip, hot-rolled and cold-rolled, high-strength, low-alloy, with improved corrosion resistance). A minimum yield point of 50 ksi is provided.

ASTM A607 (steel sheet and strip, hot-rolled and cold-rolled, high-strength, low-alloy, columbium and/or vanadium). A minimum yield point of 50 ksi is provided.

1-10   DESIGN PROPERTIES

Every structural material has certain physical properties that are important to design engineers. These properties, which are common to structural steel, enable them to predict the behavior of the member. For the most part, these properties cannot be significantly changed by chemical and/or mechanical means.

Figure 1-3 shows an idealized record of elastic deformation (elastic range), followed by plastic deformation (plastic range) and strain hardening. The elastic theory of design is concerned with the elastic range, and the plastic theory of design is concerned with the plastic range. The strain from the elastic limit to the onset of strain hardening (a reserve strength that is not taken into account in classic design) is approximately 15 times the maximum elastic strain. Material in the plastic range accepts large strains without an increase in stress, and yet its ability to sustain stress does not decrease. This characteristic defines a unique property of steel, ductility.

From the actual stress-strain diagram (Fig. 1-2) you can see that, for the basic structural steel, rupture or failure takes place from a strain from 15 to 25 times the strain found in the plastic range. The strains shown in the idealized diagram are very small and represent only the extreme left portion of the full-range curve.

The following sections discuss the important properties.

Modulus of Elasticity (Young's Modulus) The typical range is from 28,000 to 30,000 ksi, but for consistent design values, the *modulus of elasticity, E,* is taken as 29,000 ksi, or 29×10^6 psi (at room temperature). E is the ratio or slope of stress to strain up to the proportional limit. This property defines the stiffness of a material and governs deflections, as well as influences the buckling behavior. The stiffness of structural steel, much higher than other structural materials, creates a very definite advantage for structural steel.

Shear Modulus of Elasticity (Modulus of Rigidity) The theoretical value of the *shear modulus of elasticity, G,* is

$$\frac{E}{2(1 + \mu)}$$

The usual minimum design value is 11,000 ksi. It is the ratio of shearing stress to shearing strain within the elastic range of the stress-strain diagram. *Poisson's ratio* (μ) is the ratio of the transverse to the longitudinal (axial) strain. The design value μ for the elastic range is equal to 0.30, and the design value of μ for the plastic range is equal to 0.50.

Unit Weight of Structural Steel The *unit weight of structural steel* is 0.2833 lb/in.^{3}, 490 lb/ft^3, 40.8 lb/ft^2/in. of thickness, or 3.4 lb/in.^{2}/ft of length.

Specific Gravity of Structural Steel The *specific gravity of structural steel* is 7.85.

Coefficient of Expansion The average *coefficient of expansion, ϵ,* for structural steel between room temperature and 100°F is 0.0000065 for each degree. For temperatures of 100 to 1200°F, the coefficient is given by the following approximate formula:

$$\epsilon = (6.1 + 0.0019t) \times 10^{-6}$$

ϵ is the coefficient of expansion for each degree Fahrenheit and t is the temperature in degrees Fahrenheit.

1-11   SELECTION OF STRUCTURAL STEEL

As discussed in Section 1-9, a variety of grades of steel is available for use. Actual selection may involve the consideration of strength, availability, economy, toughness, corrosion resistance, and so on. In general, higher material costs accompany various improved material properties.

Strength   Higher-strength steels can usually be used to advantage when member size is determined by strength. Tension members are a good illustration. A significant increase in allowable stress can result from a higher-strength steel (higher yield point) in this case, provided, of course, that the resulting increase in elongation can be tolerated.

Higher-strength steels can also be used advantageously in beam design, particularly when steel dead load is a major portion of the design load and when deflection or stability does not control (reduction of deflection through continuous or composite design assumptions).

In column and compression member design, the higher-strength steels can be used to advantage if the slenderness ratio, l/r, is small (below 60). In addition, when steel dead load is a major portion of the design load, the higher-strength steels can also be used to advantage.

Higher-strength steels, although they have a higher initial cost (material cost), can reflect an overall weight savings, which can be translated into a reduced cost of fabrication, erection, and shipping. Furthermore, when a poor foundation condition is encountered, a lighter superstructure can reflect a very significant savings in foundations. Conceivably, using similar-sized columns, by means of the lower level (heavier loaded columns) being of higher-strength steel (the upper columns of standard strength), could increase usable floor space and reflect economy.

However, high-strength steels should not be used indiscriminately, and due consideration and study should be made by the engineer relating to stability, both during and after the construction phase. In summary, at times where stability or deflection controls rather than strength, higher allowable stresses offer no real advantage.

A last point: the quenched and tempered alloy steels, A514, should be selected based on particular needs for strength or weight reduction.

Availability   Some grades of structural steel are more readily available than others. Steel mills have various mill schedules, and the availability or nonavailability for a length of time must be considered. In addition, for other than the more commonly used grades of steel, the tonnage is of importance in considering availability.

Economy In structural steel design, the lightest structure is not necessarily the least expensive structure. Detailing, fabrication, and erection costs should be taken into account. Availability of material comes into the picture and should be checked.

The designer should realize that, in general, high-strength steels may show economy when strength is the governing design factor, whereas the standard carbon steels may show economy where stability, deflection, elongation, or rigidity is the controlling design parameter. Sometimes a choice of steel is made based upon a compromise of strength and stability.

In considering relative economy of the various grades of steel, two parameters are helpful: unit price, or dollars per unit of weight (per ton or per 100 lb), and price-to-strength ratio, expressed as unit price divided by yield strength. A low unit price indicates economy where deflection governs, while a low price-to-yield strength ratio indicates economy where strength governs. A general-application steel, A36 for example, should compare favorably by both parameters. A572, grade 50, may be considered as a good economical choice with a good balance between both parameters also. Interval of maintenance, if any, should also be considered from the economical-choice point of view.

Toughness At room temperatures, all structural steels are extremely tough, but at low temperatures the selection of a particular grade of steel should be made with this characteristic in mind. At low temperatures certain grades of steel have less susceptibility to brittle-fracture types of failures. This will be discussed in more detail in Section 1-13.

Corrosion Resistance This may be an important consideration in the selection process, depending upon the type and severity of exposure of structure to the elements, the cost of periodic maintenance (or elimination thereof), and so on. ASTM A242 and A588 can provide substantial savings by eliminating the need for painting or other protective coatings. Even when painting is required, these grades should be considered because paint life is longer when adequate surface preparations of these grades of steel are performed. Finally, the use of these steels is an advantage where periodic maintenance is costly or impossible. Of course, the use of these steels is impracticable under certain conditions, and the supplier or literature should be consulted where use is contemplated.

Table 1-4 summarizes some of the basic features of several structural steel grades, showing the relative cost and cost-to-strength parameters, and should be of use to the designer as a guide to structural-steel-shape selection.

Table 1-4
GUIDE TO THE SELECTION OF STRUCTURAL STEEL SHAPES[a]

Type	ASTM No.	Relative Cost/cwt[b]	Relative Cost/F_y[b]	Group
Carbon Steel	A36[c,d]	1.00/1.28	1.00/1.28	All
High-strength low-alloy steels	A242[e]	1.33	0.96	1,2
		1.35	1.06	3
		1.35/1.52[c,d]	1.15/1.30[c,d]	4,5
	A441	1.18	0.85	1,2
		1.18	0.85	3
		1.18/1.62	1.01/1.39	4,5
	A572[d] Grade 42[c,d]	1.12/1.51	0.96/1.29	All
	Grade 50[c,d]	1.14/1.57	0.82/1.13	1, 2, 3, 4
	Grade 60	1.21	0.73	1,2
	Grade 65	1.25	0.69	1
	A588[c,d]	1.33/1.87	0.96/1.35	All

[a] The parameters shown should be checked with any of the various mills since they may vary. They have been established from a particular mill, at a particular location, at a particular date with no "extras" figured (i.e., length, test, special mechanical and chemistry, quantity, size and section, etc.).

[b] Relative to A36 (ratio cost of the steel to cost A36).

[c] Average extra for columns W 14 × 455 to W 14 × 730 and specification extra included. For lighter sections in group 4 there is no extra.

[d] Where two figures separated by a slash are shown, the first figure is for sections lighter than W 14 × 455 and the second figure for sections W 14 × 455 and heavier.

[e] Self-weathering characteristics.

34

1-12   PLANS AND DRAWINGS

Complete design drawings, which are prepared in the engineer's office, should show the framing plans with member sizes and relative locations (horizontal and vertical). The plans should indicate the entire results of the design of the structure: the type of framing employed (see Section 2-3); the assumed loading, shears, moments, and other loads to be resisted by the members and their connections; and the type or types of steel to be used as well as where they should be used. The type of connector and connection should be indicated, although the designer should give the fabricator enough flexibility as to choice in this area to obtain maximum economy. Required cambers should be called for on these design drawings. Suitable scales for all plan views, elevation views, and section views should be used.

The design drawings should provide ample design information for the preparation of shop drawings. Shop drawings give all the information necessary for the fabrication of the component parts or elements of the structure—location, type, and size of all connectors; whether the connectors are to be installed in the shop or in the field; and so on. The shop drawings should be made to conform with standard practice to facilitate economical fabrication.

When the complete structural steel plans and specifications are received, the steel fabricator can then proceed to order the steel material from the mill and prepare the shop and erection drawings. In general, erection drawings are usually made to a minimum scale of $\frac{1}{8}$ in./ft. Shop drawings are generally drawn to a vertical scale of 1 in./ft vertically with no particular horizontal scale. However, shop drawings of trusses, braces, or the like are usually drawn to a scale between $\frac{1}{4}$ in./ft to $\frac{1}{2}$ in./ft horizontally as well as vertically to indicate slope and bevel. The individual elements of trusses or bracing are usually drawn to a scale of $\frac{3}{4}$ in./ft to 1 in./ft.

If the shop drawings are made by the fabricator, prints of the drawings are sent to the engineer or owner's representative for study and subsequent approval. They are usually returned to the fabricator within a two-week period marked approved or approved subject to the corrections made. The fabricator makes the corrections (if any), returns the corrected prints to the owner, and the owner then permits the fabrication to commence. The owner is then fully responsible for the design adequacy of any connections designed by the fabricator as a part of the preparation of the shop drawings. The fabricator is responsible for the accuracy of the detail dimensions on the shop drawings as well as the general fit-up in the field.

At times, the owner prepares the shop drawings or has them prepared by an independent steel detailer. The owner delivers the drawings to the fabricator in ample time to permit the fabricator to order material and commence fabrication. The owner in this case is responsible for the accuracy and completeness of the drawings furnished.

Adequate notes on the plans and drawings should be made relative to the sequence of welding and technique of welding (where welding is used) to avoid or minimize distortion and shrinkage. The American Welding Society standardized symbols should be used to designate weld type, size, and location (AWS A2.0).

To facilitate uniformity and conformity, the standard designations for structural steel shapes (Fig. 1-5; see Section 1-8) should be used. The standard nomenclature contained in the joint AISC-SJI (Steel Joist Institute) Standard Specifications for Open Web Steel Joists (all series) should be used for this type of construction.

Erection drawings prepared by the erector (fabricator or an independent) show the erection sequence and location of all the structural elements. Erection designations are assigned to each element and are shown on the drawings. These marks (designations) correspond to those shown on the actual structural components. Erection drawings, as the name indicates, are used to build the structure.

And so it goes, from design to design (engineering) drawings to shop drawings to erection drawings to the building of the structure. Of course, all these documents are accompanied by drawing notes and specifications.

1-13   HIGH- AND LOW-TEMPERATURE EFFECTS

For the most part structural steels under normal use encounter a relatively narrow range of temperatures in which their physical and chemical properties remain consistent. Yield and tensile strength remain constant and the elastic modulus, E, remains constant. However, the behavior of steel under fire conditions, where the steel is exposed to temperatures beyond the normal range for periods of time, should be considered.

Short-term elevated-temperature tests performed on the steels covered by the AISC Specification indicate that steels having similar metallurgical characteristics have similar ratios of elevated- to room-temperature tensile and yield strengths. When steel is heated, there is a change in the metal's structure, which manifests itself in the form of a change in mechanical properties, called *aging* or *strain aging*. Carbon steels exhibit a very definite strain-aging effect in the temperature range 300 to 700°F. At about 500 to 600°F, an increase in tensile strength of about 10% is indicated. The higher-strength steels have a less pronounced strain-aging effect. When temperatures of above 700°F are achieved, the yield and tensile strengths of all steels decrease with increasing temperatures. At about 1000°F, the yield strengths are almost lowered to the working or design stress levels (about 60 to 70% of the yield strength).

During a fire within a structure, the high temperatures achieved are usually short-term temperatures and, more often than not, the steel will sustain fire temperatures with little or no ill effects. Realize that the entire cross section of an element must be subject to high temperatures in order to "damage" its physical capabilities. After a fire, visual inspection of the steel can determine whether it is capable of reuse or whether it should be discarded. For instance, the most basic and obvious is that if the original paint system is intact (not blistered) and merely covered with soot, the element may be reused. If distortion due to accelerated heating and cooling and the inability of the member to expand and contract is apparent, discarding of the member and replacement is probably more economical than trying to straighten it for reuse.

A few facts relative to temperature achieved in steel are in order to give a perspective of the entire subject. For carbon steels, preheat temperatures for welding range from 80 to 200°F; galvanizing temperatures are higher than 800°F; temperatures up to about 2300°F are achieved in the various manufacturing processes of steel. At temperatures in excess of 2300°F, the mechanical properties of steel are altered and the steel is permanently damaged or "burned." Burned steel is characterized by heavy surface scaling and pitting caused by oxidation.

A graphical representation of the effect of elevated temperatures on yield strength and tensile strength is shown in Figs. 1-11 and 1-12, respectively

Quenched and tempered alloy steels (A514) that have achieved temperatures in excess of 1200°F should be heat-treated again prior to reuse.

For steel, under fire conditions, to meet various code stipulations, steel elements must be protected from the potential fire by some type of insulation material. ASTM E119, Standard Methods of Fire Tests of Building Construction and Materials, outlines the procedures and methods for fire testing. This specification is widely used as a standard for requiring structural elements and assemblies to resist certain temperatures for a given amount of time when fully loaded. It requires that an average temperature of steel not exceed 1000° and 1200°F and further that a temperature of not more than 1200° and 1400°F be exceeded at any one point. The ranges described are for unloaded and loaded elements, respectively.

For special steels to be used under high-temperature conditions, reference is made to U.S. Steel, Steels for Elevated Temperature Ser-

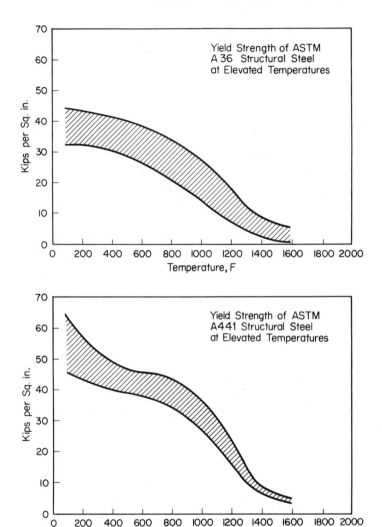

FIG. 1-11   Yield strengths at elevated temperatures.
Courtesy of American Iron and Steel Institute

vice, which is available from the U.S. Steel Corporation. Sometimes structural steel is required to be used under sustained low temperature. The physical characteristics of steel also change under low temperatures and, under a certain combination of conditions, the steel may be subject to brittle fracture. Low temperature, application of a high strain rate and notches or cracks in the steel, in combination, may lead to a brittle failure

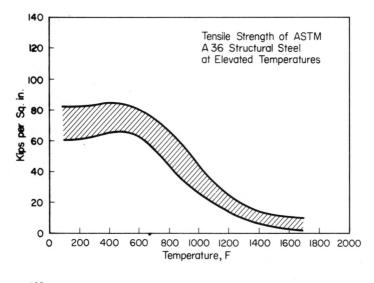

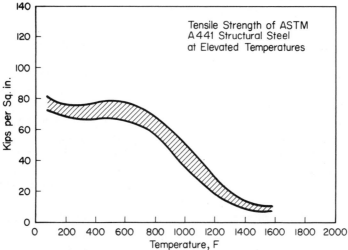

FIG. 1-12   Tensile strengths at elevated temperatures.
Courtesy of American Iron and Steel Institute

of the steel member. Brittle fracture would only be a problem under conditions of tensile stress.

Recent research has established testing zones that relate to the lowest ambient temperature expected for the area in which the structural element will be used. These zones, zones 1, 2, or 3, increase in severity of impact test requirement with the zone number. The impact test requirement is expressed as an average energy absorption in foot-pounds (as determined by Charpy V-notch impact tests) at testing temperatures expressed in degrees Fahrenheit for various grades of steel of certain thickness ranges. For example, for A36 steel at $-30°F$ service temperature (zone 2) with the aforementioned combination of conditions, brittle fracture can be avoided if a standard test specimen absorbs 15 ft-lb of energy at 40°F. These supplemental impact test requirements should only be called for when the combination of conditions occurs with tension stress. A high strain rate is actually of the magnitude of an explosive application of strain, although bridges are considered to fall within this category. The ordinary building usually poses no brittle fracture problem.

ASTM A709, Structural Steel for Bridges, covers five grades of steel available in three strength levels ($F_y = $ 36, 50, and 100 ksi). This specification covers carbon, high-strength, and low-alloy steel for structural shapes, plates, and bars and quenched and tempered alloy steel for structural plates intended for use in bridges.

The American Association of State Highway and Transportation Officials, AASHTO, has similar specifications in their Materials Specifications. Both ASTM A709 and the AASHTO Material Specifications give supplemental impact test requirements in tabular form. The reader is referred to these for further details.

The hull of a ship in cold seawater can be subject to brittle failure. The plates of the hull can conceivably be subject to the combination of loads mentioned and needed to invite brittle failure. Offshore drilling platforms subject to cold sea and wave pounding would be another example of structure possibly vulnerable to brittle failure.

In general, steels having the following characteristics should be used to minimize potential failure: carbon and phosphorus content should be kept low, a high ratio of manganese to carbon should be used, heat treatment, fine grain size, and thin material should be utilized.

Table 1-5 lists several special grades of structural steel for use under low temperatures. Their use is noted for some of the grades listed. Other low-carbon, high-manganese silicon low-temperature steels are available: A572 grade 42 killed fine grain normalized for refrigerated room constant temperature; A131 for ships' plates and shapes: Lukens LT-75 carbon manganese-silicon plate; Armco's Shef-Lo-Temp and Shef-Super-Lo-Temp; USS Char Pac quenched and tempered carbon steel or normalized plate; A203 nickel alloy; and other stainless steels and nickel steels (see Operation Cryogenics, U.S. Steel publication ADUCO 01087A 7/62).

All the aforementioned steels, when contemplated for use, should be checked as to availability for low-tonnage quantities. For further details, reference should be made to the publications cited.

Table 1-5

Grade	Use
Lukens HY-80 plate steel	Thick-walled subsea work enclosures, submarine hulls
3½% Nickel steel (see 3½ Nickel Steel for Low Temperature Service, International Nickel Co., Inc.)	Process equipment and structures operating at −150°F
TRIP (transformation induced plasticity) family of steels; and ESR (electroslag remelting, made by Lukens)	Where low-temperature ductility and toughness are desired
Canadian CSA Standard, G40.8, grade B	Where improved resistance to brittle fracture is desired

1-14   ELASTICITY, DUCTILITY, DURABILITY, REUSABILITY, AND WELDABILITY

Structural steel, within the elastic range, when loaded stretches (strain) as the load (stress) is applied, and when the load is released, the steel returns to its original form. This is *elasticity,* similar to the action of a rubber band when stretched and released.

As the loading is further increased, the steel enters the plastic range where stress is no longer proportional to strain. It is within this range that structural steel exhibits a unique property called *ductility.* Ductility is the ability of a steel element to sustain large strains at no increase of stress, and yet its ability to sustain stress does not decrease. This unique property forms the basis for the plastic behavior of steel and some of the plastic assumptions that are incorporated in the elastic portions of the AISC Specification. It enables a higher-stressed portion of a member to transfer or redistribute a load or loading to a portion of the member with less stress.

Obviously, steel that is bolted may be dismantled and reused. It also is a simpler matter to add to or extend a steel structure rather than one framed with another material. Further, with proper consideration directed toward proper maintenance, steel is *durable*.

The *weldability* of steel relates to the ease and practicability of welding. The scope clause of the ASTM material specification usually states whether the steel is suitable for welding. For economical reasons, a minimal amount of special procedures should be necessary for welding. The weldability of structural steel is often related to the *carbon equivalent* (CE). The CE is a measure of the susceptibility to underbead cracking. A commonly used procedure relating to steels considered to be weldable is shown in Table 1-6. All the grades of structural steel permitted by the AISC Specification are suitable for welding when proper welding procedures are employed.

Table 1-6
CARBON EQUIVALENT OF STRUCTURAL STEELS[a]

Steel Type	Carbon Equivalent (CE)
Carbon steels	$CE = \%C + \%Mn/4 \leq 0.45$
Low-alloy steels	$CE^b = \%C + \%Mn/6 + \%Ni/20 + \%Cr/10$ $- \%Mo/50 - \%V/10 + \%Cu/40 \leq$ 0.40

[a] *Courtesy of Canadian Institute of Steel Construction.*

[b] K. Winterton, "Weldability Prediction from Steel Composition to Avoid Heat-Affected Zone Cracking," *Welding Journal,* June 1961, p. 253–5.

1-15 GLOSSARY OF STEEL TERMS

A glossary of terms related to steel is of value to individuals involved with steel and its use as a construction material. Most of the terms are taken from a publication of the American Society for Metals, *Metals Handbook,* as well as the publication of the Canadian Institute of Steel Construction, *General Information on Structural Steel.*

ACID STEEL. Steel melted in a furnace that has an acid bottom and lining and under a slag that is dominantly siliceous.

AGING. Change in a metal by which its structure recovers from an unstable condition produced by quenching or by cold working such as cold reduction. The change in structure is marked by changes in mechanical properties. Aging that takes place slowly at room temperature may be accelerated by slight increase in temperature. See "Strain Aging."

ALLOY. A substance that has metallic properties and is composed of two or more chemical elements of which at least one is a metal.

ALLOYING ELEMENT. Chemical elements constituting an alloy; in steels, usually limited to the metallic elements added to modify the properties of the steel.

ALLOY STEEL. Steel containing significant quantities of alloying elements (other than carbon and the commonly accepted amounts of manganese, silicon, sulfur, and phosphorus) added to effect changes in the mechanical or physical properties.

ANNEALING. A process involving heating and cooling, usually applied to induce softening. The term also refers to treatments intended to alter mechanical or physical properties, produce a definite microstructure, or remove gases. When applicable, the following more specific terms should be used: black annealing, blue annealing, box annealing, bright annealing, full annealing, graphitizing, isothermal annealing, malleablizing, process annealing, spheroidizing, stabilizing annealing. Definitions of some of these are given in their alphabetical positions in this glossary. When applied to ferrous alloys, the term "annealing," without qualification, implies full annealing. Any process of annealing will usually reduce stresses, but if the treatment is applied for the sole purpose of such relief, it should be designated as "stress relieving."

ARTIFICIAL AGING. An aging treatment above room temperature.

AUSTENITE. A solid solution in which gamma iron is the solvent; characterized by a face-centered cubic crystal structure.

AUSTENITIZING. The process of forming austenite by heating a ferrous alloy into the transformation range (partial austenitizing) or above the transformation range (complete austenitizing).

BASIC OXYGEN PROCESS. The family of named steelmaking processes in which certain oxidizable constituents in the charge serve as fuel for the melting and refining of the charge. High-purity oxygen is injected through a lance against a charge and reacts to physically stir the bath and burn to oxides the carbon, silicon, manganese, and even iron contents to predictable levels, thus creating the heat and refining the steel. Liquid fuels or fluxes may be injected along with the oxygen.

BASIC STEEL. Steel melted in a furnace that has a basic bottom and lining, and under a slag that is dominantly basic.

BEND TESTS. Various tests used to determine the ductility of sheet or plate that is subjected to bending. These tests may include determination of the minimum radius or diameter required to make a satisfactory bend and the number of repeated bends that the material can withstand without failure when it is bent through a given angle and over a definite radius.

BESSEMER PROCESS. A process for making steel by blowing air through molten pig iron contained in a suitable vessel, and thus causing rapid oxidation mainly of silicon and carbon.

BILLET. See "Bloom."

BLAST FURNACE. A shaft furnace in which solid fuel is burned with an air blast to smelt ore in a continuous operation. Where the temperature must be high, as in the production of pig iron, the air is preheated. Where the temperature can be lower, as in smelting copper, lead, and tin ores, a smaller furnace is economical, and preheating of the blast is not required.

BLOOM. (slab, billet) Semifinished products hot-rolled from ingots and rectangular in cross section, with rounded corners. The chief differences are in cross-sectional area, in ratio of width to thickness, and in the intended uses. The American Iron and Steel Institute Steel Products Manual Section 2 (1943) classifies general usage thus:

Type	Width (in.)	Thickness (in.)	Cross-Sectional Area (in.^{2})
Bloom	Width generally equals thickness		36 + (min)
Billet	1$\frac{1}{2}$ (min)	1$\frac{1}{2}$ (min)	2$\frac{1}{4}$ to 36
Slab	2 × thickness (min)	1$\frac{1}{2}$ (min)	16 (min)[a]

[a] Generally.

Blooms, slabs, and billets of rerolling quality are intended for hot rolling into common products such as shapes, plates, strip, bars, wire rod, sheet, and black plate. Blooms, slabs, and billets of forging quality are intended for conversion into forgings or other products to be heat treated.

BLUE BRITTLENESS. Reduced ductility occurring as a result of strain aging, when certain ferrous alloys are worked between 300 and 700°F. This phenomenon may be observed at the working temperature or subsequently at lower temperatures.

BRINELL HARDNESS TEST. A test for determining the hardness of a material by forcing a hard steel or carbide ball of specified diameter into it under a specified load. The result is expressed as the Brinell hardness number, which is the value obtained by dividing the applied load in kilograms by the surface area of the resulting impression in square millimeters.

BRITTLE CRACK PROPAGATION. A very sudden propagation of a crack with the absorption of no energy except that stored elastically in the body. Microscopic examination may reveal some deformation even though it is not noticeable to the unaided eye.

BRITTLE FRACTURE. Fracture with little or no plastic deformation.

BRITTLENESS. A tendency to fracture without appreciable deformation.

BURNT. A term applied to a metal permanently damaged by having been heated to a temperature close to the melting point.

CAMBER. Curvature in the plate of rolled sheet or strip, or in the plane of the web of structural shapes.

CAPPED STEEL. Semikilled steel cast in a bottle-top mold and covered with a cap fitting into the neck of the mold. The cap causes the top metal to solidify. Pressure is built up in the sealed-in molten metal and results in a surface condition much like that of rimmed steel.

CARBIDE. A compound of carbon with one or more metallic elements.

CARBON STEEL. Steel that owes its properties chiefly to the presence of carbon, without substantial amounts of other alloying elements; also termed "ordinary steel," "straight carbon steel," "plain carbon steel."

CARBURIZING. A process that introduces carbon into a solid ferrous alloy by heating the metal in contact with a carbonaceous material—solid, liquid, or gas—to a temperature above the transformation range and holding at that temperature. Carburizing is generally followed by quenching to produce a hardened case.

CAST IRON. An iron containing carbon in excess of the solubility in the austenite that exists in the alloy at the eutectic temperature.

CAST STEEL. Any object made by pouring molten steel into molds.

CEMENTITE. A compound of iron and carbon known as "iron carbide" which has the approximate chemical formula Fe_3C and is characterized by an orthorhombic crystal structure.

CHARGE. (1) The liquid and solid materials fed into a furnace for its operation. (2) Weights of various liquid and solid materials put into a furnace during one feeding cycle.

CHARPY TEST. A pendulum-type single-blow impact test in which the specimen, usually notched, is supported at both ends as a simple beam and broken by a falling pendulum. The energy absorbed, as determined by the subsequent rise of the pendulum, is a measure of impact strength or notch toughness.

CHECK ANALYSIS. Chemical analysis made of drillings taken from semifinished or finished products. The units are subject to certain specified variations from the ladle analysis.

CHIPPING. A method for removing seams and other surface defects with chisel or gouge so that such defects will not be worked into the finished product. Chipping is often employed also to remove metal that is excessive but not defective. Removal of defects by gas cutting is known as "deseaming" or "scarfing."

CLUSTER MILL. A rolling mill where each of the two working rolls of small diameter is supported by two or more backup rolls.

COLD WORK. Plastic deformation at such temperatures and rates that substantial increases occur in the strength and hardness of the metal. Visible structural changes include changes in grain shape and, in some instances, mechanical twinning or banding.

COLD WORKING. Deforming a metal plastically at such a temperature and rate that strain hardening occurs. The upper limit of temperature for this process is the recrystallization temperature.

COMPRESSIVE STRENGTH. Yield: the maximum stress that a metal, subjected to compression, can withstand without a predefined amount of deformation. Ultimate: the maximum stress that a brittle material can withstand without fracture when subjected to compression.

CONTINUOUS CASTING. A casting technique in which an ingot, billet, tube, or other shape is continuously solidified while it is being poured, so that its length is not determined by mold dimensions.

CONTINUOUS MILL. A rolling mill consisting of a number of stands of synchronized rolls (in tandem) in which metal undergoes successive reductions as it passes through the various stands.

CONTROLLED COOLING. A process of cooling from an elevated temperature in a predetermined manner to avoid hardening, cracking, or internal damage, or to produce a desired microstructure. This cooling usually follows the final hot-forming operation.

COOLING STRESSES. Stresses developed by uneven contraction or external constraint of metal during cooling; also those stresses resulting from localized plastic deformation during cooling, and retained.

CORROSION. Gradual chemical or electrochemical attack on a metal by atmosphere, moisture, or other agents.

CREEP. The flow or plastic deformation of metals held for long periods of time at stresses lower than the normal yield strength. The effect is particularly important if the temperature of stressing is in the vicinity of the recrystallization temperature of the metal.

CREEP LIMIT. (1) The maximum stress that will cause less than a specified quantity of creep in a given time. (2) The maximum nominal stress under which the creep strain rate decreases continuously with time under constant load and at constant temperature. Sometimes used synonymously with creep strength.

CREEP STRENGTH. (1) The constant nominal stress that will cause a specified quantity of creep in a given time at constant temperature. (2) The constant nominal stress that will cause a specified creep rate at constant temperature.

CRITICAL COOLING RATE. The minimum rate of continuous cooling just sufficient to prevent undesired transformations. For steel, the slowest rate at which it can be cooled from above the upper critical temperature to prevent the decomposition of austenite at any temperature above the temperature at which the transformation of austenite to martensite starts during cooling.

CROP. The end or ends of an ingot that contain the pipe or other defects to be cut off and discarded; also termed "crop end" and "discard."

CROSS-COUNTRY MILL. A rolling mill in which the mill stands are so arranged that their tables are parallel with a transfer (or crossover) table connecting them. They are used for rolling structural shapes, rails, and any special form of bar stock not rolled in the ordinary bar mill.

CROSS ROLLING. The rolling of sheet so that the direction of rolling is changed about 90° from the direction of the previous rolling.

CROWN. In the center of metal sheet or strip, thickness greater than at the edge.

CRYSTAL. A physically homogeneous solid in which the atoms, ions, or molecules are arranged in a three-dimensional repetitive pattern.

CRYSTALLIZATION. The formation of crystals by the atoms assuming definite positions in a crystal lattice. This is what happens when a liquid metal solidifies. (Fatigue, the failure of metals under repeated stresses, is sometimes falsely attributed to crystallization.)

CUP FRACTURE (CUP AND CONE FRACTURE). Fracture, frequently seen in tensile test pieces of a ductile material, in which the surface of failure on one portion shows a central flat area of failure in tension, with an exterior extended rim of failure in shear.

DEFECT. Internal or external flaw or blemish. Harmful defects can render material unsuitable for specific end use.

DEOXIDATION. Elimination of oxygen in liquid steel, usually by introduction of aluminum or silicon or other suitable element. This term is also used to denote reduction of surface scale (iron oxide).

DIRECTIONAL PROPERTIES. Anisotropic condition where physical and mechanical properties vary, depending on the relation of the test axis to a specific direction of the metal; a result of preferred orientation or of fibering of inclusions during the working.

DUCTILITY. The property that permits permanent deformation before fracture by stress in tension.

ELASTIC LIMIT. The maximum stress that a material will withstand without permanent deformation. (Almost never determined experimentally; yield strength is customarily determined.)

ELECTRIC FURNACE. A melting furnace with a shallow hearth and a low roof in which the charge is melted and refined by an electric arc between one or more electrodes and the charged material. The electrodes normally are suspended through the roof. No liquid or gaseous fuel is usually used; however, gaseous oxygen may be injected into the bath.

ELONGATION. The amount of permanent extension in the vicinity of the fracture in the tension test; usually expressed as a percentage of the original gage length, as 25% in 2 in. Elongation may also refer to the amount of extension at any stage in any process that elongates a body continuously, as in rolling.

EMBOSSING. Raising a design in relief against a surface.

EMBRITTLEMENT. Reduction in the normal ductility of a metal due to a physical or chemical change.

ENDURANCE LIMIT. The maximum stress that a metal will withstand without failure during a specified large number of cycles of stress. If the term is employed without qualification, the cycles of stress are usually such as to produce complete reversal of flexural stress.

EXTRUSION. Conversion of a billet into lengths of uniform cross section by forcing the plastic metal through a die orifice of the desired cross-sectional outline. In *direct extrusion* the die and ram are at opposite ends of the billet, and the product and ram travel in the same direction. In *indirect extrusion* (rare) the die is at the ram end of the billet and the product travels through and in the opposite direction to the hollow ram. A *stepped extrusion* is a single product with one or more abrupt cross-section changes and is obtained by interrupting the extrusion by die changes. *Impact extrusion* (cold extrusion) is the process or resultant product of a punch striking an unheated slug in a confining die. The metal flow may be either between the punch and die or through another opening. *Hot extrusion* is similar to cold extrusion except that a preheated slug is used and the pressure application is slower.

FATIGUE. The tendency for a metal to break under conditions of repeated cyclic stressing considerably below the ultimate tensile strength.

FATIGUE CRACK OR FAILURE. A fracture starting from a nucleus where there is an abnormal concentration of cyclic stress and propagating through the metal. The surface is smooth and frequently shows concentric (seashell) markings with a nucleus as a center.

FATIGUE LIFE. The number of cycles of stress that can be sustained prior to failure for a stated test condition.

FATIGUE LIMIT. The maximum stress that a metal will withstand without failure for a specified large number of cycles of stress. Usually synonymous with endurance limit.

FATIGUE RATIO. The ratio of the fatigue limit for cycles of reversed flexural stress to the tensile strength.

FATIGUE STRENGTH. The maximum stress that can be sustained for a specified number of cycles without failure, the stress being completely reversed within each cycle unless otherwise stated.

FERRITE. A solid solution in which alpha iron is the solvent, and which is characterized by a body-centered cubic crystal structure.

FERRO-ALLOY. An alloy of iron that contains a sufficient amount of one or more chemical elements, such as manganese, chromium, or silicon, to be useful as an agent for introducing these elements into steel by admixture with molten steel.

FILLET. A concave junction of two (usually perpendicular) surfaces.

FINISHED STEEL. Steel that is ready for the market without further work or treatment. Blooms, billets, slabs, sheet bars, and wire rods are termed "semifinished."

FINISHING TEMPERATURE. The temperature at which hot mechanical working of metal is completed.

FLANGE. (1) A projection of metal on formed objects. (2) The parts of a channel at right angles to the central section or web.

FLATNESS. Relative term for the measure of deviation of flat-rolled material from a plane surface: usually determined as the height of ripples or waves above a horizontal level surface.

FRACTURE TEST. Breaking a piece of metal for the purpose of examining the fractured surface to determine the structure or carbon content of the metal or to detect the presence of internal defects.

FULL ANNEALING. A softening process in which a ferrous alloy is heated to a temperature above the transformation range and, after being held for a sufficient time at this temperature, is cooled slowly to a temperature below the transformation range. The alloy is ordinarily allowed to cool slowly in the furnace, although it may be removed and cooled in some medium that ensures a slow rate of cooling.

GRAIN REFINER. Any material added to a liquid metal for the purpose of producing a finer grain size in the subsequent casting, or of retaining fine grains during the heat treatment of wrought structures.

GRAINS. Individual crystals in metals.

HARDENABILITY. In a ferrous alloy, the property that determines the depth and distribution of hardness induced by quenching.

HARDENING. Any process for increasing the hardness of metal by suitable treatment, usually involving heating and cooling.

HARDNESS. Defined in terms of the method of measurement. (1) Usually the resistance to indentation. (2) Stiffness or temper of wrought products. (3) Machinability characteristics.

HEAT TREATMENT. A combination of heating and cooling operations, timed and applied to a metal or alloy in the solid state in a way that will produce desired properties. Heating for the sole purpose of hot working is excluded from the meaning of this definition.

HOMOGENEOUS. Usually defined as having identical characteristics throughout. However, physical homogeneity may require only an identity of lattice type throughout, while chemical homogeneity requires uniform distribution of alloying elements.

HOT FORMING. Working operations, such as bending and drawing sheet and plate, forging, pressing, and heading, performed on metal heated to temperatures above room temperature.

HOT SHORTNESS. Brittleness in hot metal.

HOT QUENCHING. A process of quenching in a medium at a temperature substantially higher than atmospheric temperature.

HOT WORKING. Plastic deformation of metal at such a temperature and rate that strain hardening does not occur. The lower limit of temperature for this process is the recrystallization temperature.

HYDROGEN EMBRITTLEMENT. A condition of low ductility resulting from hydrogen absorption and internal pressure developed subsequently.

IMPACT ENERGY (IMPACT VALUE). The amount of energy required to fracture a material, usually measured by means of an Izod or Charpy test. The type of specimen and testing conditions affect the values and therefore should be specified.

IMPACT TEST. A test to determine the energy absorbed in fracturing a test bar at high velocity. The test may be in tension or in bending, or it may properly be a notch test if a notch is present, creating multiaxial stresses.

INCIDENTAL ELEMENTS. Small quantities of nonspecified elements commonly introduced into product from the use of scrap metal with the raw materials.

INCLUSIONS. Particles of impurities (usually oxides, sulfides, silicates, and such) that are held mechanically, or are formed during solidification or by subsequent reaction within the solid metal.

INGOT. A casting intended for subsequent rolling or forging.

IRON. (1) Element No. 26 of the periodic system, the average atomic weight of the naturally occurring isotopes being 55.85. (2) Iron-base materials not falling into the steel classifications.

KALDO PROCESS. One of the family of basic oxygen steelmaking processes that uses an inclined, rotating cylindrical furnace in which oxygen is injected through a lance in the center line of the furnace. This furnace uses a basic refractory lining and normally no fuels or fluxes are injected with the oxygen.

KILLED STEEL. Steel deoxidized with a strong deoxidizing agent such as silicon or aluminum in order to reduce the oxygen content to a minimum so that no reaction occurs between carbon and oxygen during solidification.

LADLE ANALYSIS. Chemical analysis made from samples obtained during original casting of ingots. This is normally the controlling analysis for satisfying the specification.

LAMINATIONS. Defects resulting from the presence of blisters, seams, or foreign inclusions aligned parallel to the worked surface of a metal.

LAP. A surface defect appearing as a seam, caused by folding over hot metal, fins, or sharp corners and then rolling or forging them into the surface, but not welding them.

L-D PROCESS. One of the basic oxygen steelmaking processes using a vertical cylindrical furnace in which oxygen is injected from above by a lance. The furnace has a basic refractory lining. Some variations of this process include the injection of liquid or gaseous fuels and fluxes along with the gaseous oxygen.

LONGITUDINAL DIRECTION. The direction in a wrought metal product parallel to the direction of working (drawing, extruding, rolling).

LUDER'S LINES OR LÜDER LINES. (stretcher strains, flow figures) Elongated markings that appear on the surface of some materials, particularly iron and low-carbon steel, when deformed just past the yield point. These markings lie approximately parallel to the direction of maximum shear stress and are the result of localized yielding. They consist of depressions when produced in tension and of elevations when produced in compression. They may be made evident by localized roughening of a polished surface or by localized flaking from an oxidized surface.

MACROSCOPIC. Visible either with the naked eye or under low magnification (as great as about 10 diameters).

MACROSTRUCTURE. The structure of metals as revealed by macroscopic examination.

MALLEABILITY. The property that determines the ease of deforming a metal when the metal is subjected to rolling or hammering. The more malleable metals can be hammered or rolled into thin sheet more easily than others.

MANNESMANN PROCESS. A process used for piercing tube billets in making seamless tubing. The billet is rotated between two heavy rolls mounted at an angle and is forced over a fixed mandrel. Billets are called "tube rounds."

MARTENSITE. An unstable constituent in quenched steel, formed without diffusion and only during cooling below a certain temperature known as the M_s (or Ar") temperature. The structure is characterized by its acicular appearance on the surface of a polished and etched specimen. Martensite is the hardest of the transformation products of austenite. Tetragonality of the crystal structure is observed when the carbon content is greater than about 0.5%.

MECHANICAL PROPERTIES. Those properties of a material that reveal the elastic and inelastic reaction when force is applied, or that involve the relationship between stress and strain; for example, the modulus of elasticity, tensile strength, and fatigue limit. These properties have often been designated as "physical properties," but the term "mechanical properties" is preferred.

MECHANICAL WORKING. Subjecting metal to pressure exerted by rolls, dies, presses, or hammers to change its form or to affect the structure and consequently the mechanical and physical properties.

MERCHANT MILL. A mill consisting of a group of stands of three rolls each arranged in a straight line and driven by one power unit, used to roll rounds, squares, or flats of smaller dimensions than would be rolled on the bar mill.

MICROSTRUCTURE. The structure of polished and etched metal and alloy specimens as revealed by the microscope.

MODULUS OF ELASTICITY. The slope of the elastic portion of the stress-strain curve in mechanical testing. The stress is divided by the unit elongation. The tensile or compressive elastic modulus is called "Young's modulus"; the torsional elastic modulus is known as the "shear modulus" or "modulus of rigidity."

NORMALIZING. A process in which a ferrous alloy is heated to a suitable temperature above the transformation range and is subsequently cooled in still air at room temperature.

NORMAL SEGREGATION. Concentration of alloying constituents that have low melting points in those portions of a casting that solidify last.

NOTCH BRITTLENESS. Susceptibility of a material to brittleness in areas containing a groove, scratch, sharp fillet, or notch.

NOTCH SENSITIVITY. The reduction caused in nominal strength, impact or static, by the presence of a stress concentration, usually expressed as the ratio of the notched to the unnotched strength.

OPEN-HEARTH FURNACE. A furnace for melting metal, in which the bath is heated by the convection of hot gases over the surface of the metal and by radiation from the roof.

OUT-OF-ROUND. Deviation of cross section of a round bar from a true circle: normally measured as difference between maximum and minimum diameters at the same cross section of the bar.

OUT-OF-SQUARE. For square bars this is the deviation of cross section from a true square: normally measured as the difference between the two diagonal dimensions at one cross section. For structural shapes, the term out-of-square indicates the deviation from a right angle of the plane of flanges in relation to the plane of webs.

OVERHEATED. A term applied when, after exposure to an excessively high temperature, a metal develops an undesirably coarse grain structure but is not permanently damaged. Unlike a burnt structure, the structure produced by overheating can be corrected by suitable heat treatment, by mechanical work, or by a combination of the two.

PEARLITE. The lamellar aggregate of ferrite and carbide. *Note:* It is recommended that this word be reserved for the microstructures consisting of thin plates or lamellae, that is, those that may have a pearly luster in white light. The lamellae can be very thin and resolvable only with the best microscopic equipment and technique.

PHYSICAL PROPERTIES. Those properties familiarly discussed in physics, exclusive of those described under mechanical properties; for example, density, electrical conductivity, coefficient of thermal expansion. This term has often been used to describe mechanical properties but this usage is not recommended. See "Mechanical Properties."

PICKLE. Chemical or electrochemical removal of surface oxides.

PIG IRON. Iron produced by reduction of iron ore in the blast furnace.

PIPE. A cavity formed by contraction in metal (especially ingots) during solidification of the last portion of liquid metal.

PIT. A sharp depression in the surface of metal.

PLASTIC DEFORMATION. Permanent distortion of a material under the action of applied stresses.

PLASTICITY. The ability of a metal to be deformed extensively without rupture.

POISSON'S RATIO. The absolute value of the ratio of the transverse strain to the corresponding axial strain in a body subjected to uniaxial stress; usually applied to elastic conditions.

POROSITY. Unsoundness caused in cast metals by the presence of blowholes and shrinkage cavities.

PRIMARY MILL. A mill for rolling ingots or the rolled products of ingots to blooms, billets, or slabs. This type of mill is often called a blooming mill and sometimes a cogging mill.

PROOF STRESS. In a test, stress that will cause a specified permanent deformation in a material, usually 0.01% or less.

PROPORTIONAL LIMIT. The greatest stress that the material is capable of sustaining without a deviation from the law of proportionality of stress to strain (Hooke's law).

QUENCHING. A process of rapid cooling from an elevated temperature by contact with liquids, gases, or solids.

QUENCHING CRACK. A fracture resulting from thermal stresses induced during rapid cooling or quenching: frequently encountered in alloys that have been overheated and liquated and are thus "hot short."

RECRYSTALLIZATION. A process whereby the distorted grain structure of cold-worked metals is replaced by a new, strain-free grain structure during annealing above a specific minimum temperature.

RED SHORTNESS. Brittleness in steel when it is red hot.

REDUCTION IN AREA. The difference between the original cross-sectional area and that of the smallest area at the point of rupture; usually stated as a percentage of the original area; also called "contraction of area."

REFINING TEMPERATURE. A temperature, usually just higher than the transformation range, employed in the heat treatment of steel to refine the structure, in particular, the grain size.

RESIDUAL STRESS. Macroscopic stresses that are set up within a metal as the result of nonuniform plastic deformation. This deformation may be caused by cold working or by drastic gradients of temperature from quenching or welding.

RESQUARED. Flat-rolled material (plate, sheet, or strip) first cut to approximate size and finally resheared to very close tolerances; also any material having been cut to equally close tolerances as to dimensions and squareness, by whatever method.

REVERBERATORY FURNACE. A furnace with a shallow hearth, usually nonregenerative, having a roof that deflects the flame and radiates heat toward the hearth or the surface of the charge.

RIMMED STEEL. An incompletely deoxidized steel normally containing less than 0.25% C and having the following characteristics: (a) During solidification an evolution of gas occurs sufficient to maintain a liquid ingot top ("open" steel) until a side and bottom rim of substantial thickness has formed. If the rimming action is intentionally stopped shortly after the mold is filled, the product is termed capped steel. (b) After complete solidification, the ingot consists of two distinct zones: a rim somewhat purer than when poured and a core containing scattered blowholes, a minimum amount of pipe, and an average percentage of metalloids somewhat higher than when poured and markedly higher in the upper portion of the ingot.

ROCKWELL HARDNESS TEST. A test for determining the hardness of a material based upon the depth of penetration of a specified penetrator into the specimen under certain arbitrarily fixed conditions of test.

ROLLER FLATTENING OR ROLLER LEVELING. The process in which a series of staggered rolls of small diameter is used to remove bow and waves from sheet. While passing through the rolls, the sheet is bent back and forth slightly and is delivered approximately flat.

ROLLER STRAIGHTENING. A process involving a series of staggered rolls of small diameter, between which rod, tubing, and shapes are passed for the purpose of straightening. The process consists of a series of bending operations.

ROLL FORMING. (1) An operation used in forming sheet. Strips of sheet are passed between rolls of definite settings that bend the sheet progressively into structural members of various contours, sometimes called "molded sections." (2) A process of coiling sheet into open cylinders.

ROLLING. Reducing the cross-sectional area of metal stock, or otherwise shaping metal products, through the use of rotating rolls.

ROLLING MILLS. Machines used to decrease the cross-sectional area of metal stock and produce certain desired shapes as the metal passes between rotating rolls mounted in a framework comprising a basic unit called a stand. Cylindrical rolls produce flat shapes; grooved rolls produce rounds, squares, and structural shapes. Among rolling mills may be listed the billet mill, blooming mill, breakdown mill, plate mill, sheet mill, slabbing mill, strip mill, and temper mill.

SCAB. (scabby) A blemish caused on a casting by eruption of gas from the mold face or by uneven mold surfaces; or occurring where the skin from a blowhole has partly burned away and is not welded.

SCALING. Surface oxidation caused on metals by heating in air or in other oxidizing atmospheres.

SCARFING. Cutting surface areas of metal objects, ordinarily by using a gas torch. The operation permits surface defects to be cut from ingots, billets, or the edges of plate that is to be beveled for butt welding. See "Chipping."

SEAM. On the surface of metal, a crack that has been closed but not welded; usually produced by some defect either in casting or in working, such as blowholes that have become oxidized or folds and laps that have been formed during working. Seam also refers to lap joints, as in seam welding.

SEGREGATION. In an alloy object, concentration of alloying elements at specific regions, usually as a result of the primary crystallization of one phase with the subsequent concentration of other elements in the remaining liquid. Microsegregation refers to normal segregation on a microscopic scale whereby material richer in alloying element freezes in successive layers on the dendrites (coring) and in the constituent network. Macrosegregation refers to gross differences in concentration (for example, from one area of an ingot to another), which may be normal, inverse, or gravity segregation.

SEMIKILLED STEEL. Steel incompletely deoxidized to permit evolution of sufficient carbon monoxide to offset solidification shrinkage.

SHEARED EDGES. Sheared edge is obtained when rolled edge is removed by rotary slitter or mechanical shear.

SHORTNESS. A form of brittleness in metal. It is designated as "cold," "hot," and "red" to indicate the temperature range in which the brittleness occurs.

SINGLE-STAND MILL. A rolling mill of such design that the product contacts only two rolls at a given moment. Contrast with "tandem mill."

SINKHEAD OR HOT TOP. A reservoir insulated to retain heat and to hold excess molten metal on top of an ingot mold in order to feed the shrinkage of the ingot. Also called "shrink head" or "feeder head."

SKELP. A plate of steel or wrought iron from which pipe or tubing is made by rolling the skelp into shape longitudinally and welding the edges together.

SKIN. A thin surface layer that is different from the main mass of a metal object in composition, structure, or other characteristics.

SLAB. See "Bloom."

SLAG. A nonmetallic product resulting from the mutual dissolution of flux and nonmetallic impurities in smelting and refining operations.

SOAKING. Prolonged heating of a metal at a selected temperature.

SPHEROIDIZING. Any process of heating and cooling that produces a rounded or globular form of carbide in steel. Spheroidizing methods frequently used are:

1. Prolonged holding at a temperature just below Ae_1.
2. Heating and cooling alternately between temperatures that are just above and just below Ae_1.
3. Heating to a temperature above Ae_1 or Ae_3 and then cooling very slowly in the furnace, or holding at a temperature just below Ae_1.
4. Cooling at a suitable rate from the minimum temperature at which all carbide is dissolved to prevent the re-formation of a carbide network, and then reheating in accordance with method 1 or 2 above (applicable to hypereutectoid steel containing a carbide network).

STEEL. An iron-base alloy, malleable in some temperature range as initially cast, containing manganese, usually carbon, and often other alloying elements. In carbon steel and low-alloy steel, the maximum carbon is about 2.0%; in high-alloy steel, about 2.5%. The dividing line between low-alloy and high-alloy steels is generally regarded as being about 5% metallic alloying elements. Steel is to be differentiated from two general classes of "irons": the cast irons, on the high-carbon side, and the relatively pure irons such as ingot iron, carbonyl iron, and electrolytic iron, on the low-carbon side. In some steels containing extremely low carbon, the manganese content is the principal differentiating factor, steel usually containing at least 0.25%; ingot iron contains considerably less.

STRAIGHTNESS. Measure of adherence to or deviation from a straight line, normally expressed as sweep or camber, according to the plane.

STRAIN AGING. Aging induced by cold working. See "Aging."

STRAIN ENERGY. (1) The work done in deforming a body. (2) The work done in deforming a body within the elastic limit of the material. It is more properly elastic strain energy and can be recovered as work rather than heat.

STRAIN HARDENING. An increase in hardness and strength caused by plastic deformation at temperatures lower than the recrystallization range.

STRESS. The load per unit of area. Ordinarily, stress-strain curves do not show the true stress (load divided by area at that moment) but a fictitious value obtained by using always the original area.

STRESS-CORROSION CRACKING. Failure by cracking under combined action of corrosion and stress, either external (applied) or internal (residual). Cracking may be either intergranular or transgranular, depending on metal and corrosive medium.

STRESS RAISERS. Factors such as sharp changes in contour or surface defects that concentrate stresses locally.

STRESS RELIEVING. A process of reducing residual stresses in a metal object by heating the object to a suitable temperature and holding for a sufficient time. This treatment may be applied to relieve stresses induced by casting, quenching, normalizing, machining, cold working, or welding.

STRETCHER FLATTENING OR STRETCHER LEVELING. A process for removing bow and warpage from sheet by applying a uniform tension at the ends so that the piece is elongated to a definite amount of permanent set.

STRETCHER LEVELED FLATNESS. Steel sheets or strip subjected to stretcher leveling thereby acquire a high degree of flatness (together with some increase of stiffness). When the same degree of flatness is procured by other methods like roller leveling, it is then described as "stretcher leveled standard of flatness."

STRETCHER STRAINS. See "Lüder's Lines."

SWEEP. Curvature in structural and other similar shapes normal to the plane of the web.

TANDEM MILL. A rolling mill consisting of two or more stands arranged so that the metal being processed travels in a straight line from stand to stand. In continuous rolling, the various stands are synchronized so that the strip may be rolled in all stands simultaneously. Contrast with "single-stand mill."

TEMPER BRITTLENESS. Brittleness that results when certain steels are held within, or are cooled slowly through, a certain range of temperature below the transformation range. The brittleness is revealed by notched-bar impact tests at room temperature or lower temperatures.

TEMPERING. A process of reheating quench-hardened or normalized steel to a temperature below the transformation range, and then cooling at any rate desired.

TENSILE STRENGTH. The value obtained by dividing the maximum load observed during tensile straining by the specimen cross-sectional area before straining. Also called "ultimate strength."

THERMAL FATIGUE. Fracture resulting from the presence of temperature gradients that vary with time in such a manner as to produce cyclic stresses in a structure.

TOLERANCES. Allowable variations from specified dimensions.

TOUGHNESS. Property of absorbing considerable energy before fracture; usually represented by the area under a stress-strain curve, and therefore involving both ductility and strength.

TRACE. Extremely small quantity of an element, usually too small to determine quantitatively.

TRANSFORMATION RANGE OR TRANSFORMATION TEMPERATURE RANGE. The temperature interval within which austenite forms while ferrous alloys are being heated. Also the temperature interval within which austenite disappears while ferrous alloys are being cooled. The two ranges are distinct, sometimes overlapping but never coinciding. The limiting temperatures of the ranges depend on the composition of the alloy and on the rate of change of temperature, particularly during cooling. See "Transformation Temperature."

TRANSFORMATION TEMPERATURE. The temperature at which a change in phase occurs. The term is sometimes used to denote the limiting temperature of a transformation range. The following symbols are used for iron and steel:

AC_1 The temperature at which austenite begins to form during heating.

AC_3 The temperature at which transformation of ferrite to austenite is completed during heating.

AC_{CM} In hypereutectoid steel, the temperature at which the solution of cementite in austenite is completed during heating.

Ar_1 The temperature at which transformation of austenite to ferrite or to ferrite plus cementite is completed during cooling.

Ar_3 The temperature at which austenite begins to transform to ferrite during cooling.

Ar_{CM} In hypereutectoid steel, the temperature at which precipitation of cementite starts during cooling.

A_4 The temperature at which austenite transforms to delta ferrite during heating; the reverse process occurs during cooling.

M_S (or Ar″) The temperature at which transformation of austenite to martensite starts during cooling.

M_f The temperature at which transformation of austenite to martensite is completed during cooling.

Note: All these changes (except the formation of martensite) occur at lower temperatures during cooling than during heating and depend on the rate of change of temperature. The temperatures of phase changes at equilibrium are denoted by the symbols Ae_1, Ae_3, Ae_{CM}, and Ae_4.

TRANSVERSE. Literally, "across," signifying a direction or plane perpendicular to the direction of working.

ULTIMATE STRENGTH. See "Tensile Strength."

UNIVERSAL MILL. A rolling mill in which rolls with a vertical axis roll the edges of the metal stock between some of the passes through the horizontal rolls.

UNIVERSAL MILL PLATE. Plate rolled on a universal mill having vertical (edge) rolls as well as horizontal rolls; also any plate having characteristics identical to plate produced on a universal mill.

UPSETTING. (1) A metal-working operation similar to forging. (2) The process of axial flow under axial compression of metal, as in forming heads on rivets by flattening the end of wire.

WELDING. A process used to join metals by the application of heat. Fusion welding, which includes gas, arc, and resistance welding, requires that the parent metals be melted. This distinguishes fusion welding from brazing. In pressure welding, joining is accomplished by the use of heat and pressure without melting. The parts that are being welded are pressed together and heated simultaneously so that recrystallization occurs across the interface.

WOODY FRACTURE. Fractures having a fibrous appearance.

YIELD POINT. In mild or medium-carbon steel, the stress at which a marked increase in deformation occurs without increase in load. In other steels and in nonferrous metals this phenomenon is not observed. See "Yield Strength."

YIELD STRENGTH. The stress at which a material exhibits a specified limiting deviation from proportionality of stress to strain. An offset of 0.2% is used for many metals such as aluminum- and magnesium-based alloys, while a 0.5% total elongation under load is frequently used for copper alloys.

YOUNG'S MODULUS. The modulus of elasticity in tension or compression.

1-16   DESIGN NOMENCLATURE

A "Glossary of Terms Pertaining to Structural Steel and Engineering Design" by the Ad Hoc Committee on Nomenclature, Administrative Committee on Metals of the Structural Division, American Society of Civil Engineers (*Journal of the Structural Division, Proceedings of ASCE,* 8322, Aug. 1971, ST8, pp. 2137–2142) is reproduced next for convenience. The AISC Specification and this text may use some of the terms with a slightly different meaning.

ALLOWABLE LOAD. The load that induces the computed maximum allowable or permitted stress at the critical section.

ALLOWABLE STRESS. As used in allowable stress design, it is the maximum stress permitted under working loads. (Values may be obtained from codes and specifications.)

ALLOWABLE STRESS DESIGN. A method of proportioning structures that is based on working loads such that computed stresses do not exceed prescribed values. (The limit of structural usefulness is not specified but is often taken as the yield stress.)

BEAM. A structural member in which the internal stresses on a transverse cross section may be resolved into a resultant shear and bending moment.

BEAM COLUMN. A beam that also functions to transmit axial load.

BIFURCATION. The phenomenon whereby a perfectly straight member under compression may either assume a deflected position or else may remain undeflected; buckling.

BRACED FRAME. A frame in which the resistance to lateral load or frame instability is provided by diagonal, *K,* or other auxiliary system of bracing.

BUCKLING LOAD. The load at which a perfectly straight member under compression assumes a deflected position.

COLD-FORMED MEMBERS. Structural members formed from steel without the application of heat.

COMPACT SAPE. As used in allowable stress design, a cross-sectional shape that will not experience premature local buckling in the inelastic region (to be distinguished from plastic design shapes, which have somewhat more restrictive cross-sectional properties).

COMPOSITE BEAM. A homogeneous metal beam structurally connected to a concrete slab so that the beam and slab respond to loads as a unit.

COMPOSITE HYBRID BEAM. A hybrid beam structurally connected to a concrete slab so that the beam and slab respond to loads as a unit.

DESIGN ULTIMATE LOAD (OR FACTORED LOAD). The working load times the load factor (not design load, which is ambiguous).

DRIFT. Lateral deflection of a building, due to wind or other loads.

DRIFT INDEX. Maximum permitted ratio of lateral sway to story height.

EFFECTIVE LENGTH. The equivalent length (KL_c) used in compression formulas for computing the strength of a framed column.

EFFECTIVE WIDTH. The reduced width of the plate or slab that, with an assumed uniform stress distribution, produces the same effect on the behavior of a structural member as the actual plate width with its nonuniform stress distribution.

FACTOR OF SAFETY. The ratio of a limit of structural usefulness (be it load, stress, or deformation) to the working or service condition.

FATIGUE. A fracture phenomenon resulting from a cyclic stress condition.

FORCE. Resultant of distribution of stress over a prescribed area. An action that develops in a member as a result of loading (formerly called total stress or stress), in kips.

HYBRID BEAM. A fabricated metal beam composed of flanges with a material of a specified minimum yield strength different from that of the web plate.

LATERAL (OR LATERAL-TORSIONAL) BUCKLING. Buckling of a member involving lateral deflection and twist.

LIMIT DESIGN. A design based on any chosen limit of usefulness.

LOAD. An action on a structure or on a member, expressed in kips (where distributed, in kips per foot or kips per square foot).

LOAD FACTOR. A factor by which a working load is multiplied to determine a design ultimate load. (This choice of terms serves to emphasize the reliance upon load-carrying capacity of the structure rather than upon stress.)

LOAD FACTOR DESIGN. A method of proportioning structures for multiples of working load at a chosen limit of structural usefulness. (The limit of structural usefulness can be either a plastic limit, stability limit, elastic limit, fatigue limit, or deformation limit.)

LOCAL BUCKLING. The buckling of a compression element, which may precipitate the failure of a whole member.

MAXIMUM (OR ULTIMATE) LOAD. Plastic limit load or stability limit load, as defined (also the maximum load-carrying capacity of a structure under test).

MECHANISM. An articulated system able to deform without a finite increase in load. It is used in the special sense that the linkage may include real hinges or plastic hinges, or both.

PLASTIC DESIGN. A design method for continuous steel beams and frames that defines the limit of structural usefulness as the maximum load. (The term plastic comes from the fact that the maximum load is computed from a knowledge of the strength of steel in the plastic range.)

PLASTIC HINGE. A yielded zone that forms in a structural member when the plastic moment is attained. The beam rotates as if hinged, except that it is restrained by the plastic moment, M_p.

PLASTIC LIMIT LOAD. The maximum load that is attained when a sufficient number of yield zones have formed to permit the structure to deform plastically without further increase in load. It is the largest load a structure will support, when perfect plasticity is assumed and when such factors as instability, strain hardening, and fracture are neglected.

PLASTIC MODULUS. The modulus of resistance to bending of a completely yielded cross section. It is the combined statical moment about the neutral axis of the cross-sectional areas above and below that axis.

PLASTIC MOMENT. The resisting moment of a fully yielded cross section.

POST-BUCKLING STRENGTH. The load that can be carried by a plate element or structural member after buckling.

PROPORTIONAL LOADING. All loads increase in a constant ratio, one to the other.

REDISTRIBUTION OF MOMENT. A process that results in the successive formation of plastic hinges until the maximum load is reached. As a result of the formation of plastic hinges, less highly stressed portions of a structure may carry increased moments.

RESIDUAL STRESS. The stresses that remain in an unloaded member after it has been formed into a finished product. (Examples of such stresses include, but are not limited to, those induced by cold bending, cooling after rolling, or welding.)

SERVICE LOAD. See "Working Load."

SHAPE FACTOR. The ratio of the plastic moment to the yield moment or the ratio of the plastic modulus to the section modulus for a cross section.

STABILITY LIMIT LOAD. Maximum (theoretical) load a structure can support when second-order instability effects are included.

STRESS. Force per unit area. (Unit as in unit stress is redundant and shall not be used, just as total stress shall not be used when force is meant.)

STUB COLUMN. A short compression-test specimen, sufficiently long for use in measuring the stress-strain relationship for the complete cross section, but short enough to avoid buckling as a column in the elastic and plastic ranges.

SUPPORTED FRAME. A frame that depends upon adjacent braced or unbraced frames for resistance to lateral or frame instability. (This transfer of load is frequently provided by the floor system through diaphragm action.)

TENSILE STRENGTH. The maximum tensile stress that a material is capable of sustaining.

TENSION-FIELD ACTION. The behavior of a plate girder panel under shear force in which diagonal tensile stresses develop in the web and compressive forces develop in the transverse stiffeners in a manner analogous to a Pratt truss.

ULTIMATE STRENGTH. The maximum strength of a cross section. (For example, in a steel beam in bending it is the plastic moment. It should not be used as an alternative for the maximum load-carrying capacity of a structure.)

ULTIMATE STRENGTH DESIGN. As used in concrete design, a method of proportioning cross sections of individual members according to the ultimate bending strength required of the cross section at each critical point. An elastic analysis is necessary first in order to determine the elastic distribution of moments throughout the frame. The nonlinear stress-strain characteristics of concrete and the yield strength of steel are taken into account, but not the redistribution of moment.

UNBRACED FRAME. A frame in which the resistance to lateral load is provided by the bending resistance of frame members and their connections.

UNBRACED LENGTH. The distance between braced points of a member.

WEB CRIPPLING. The local elastic-plastic failure of a web plate in the immediate vicinity of a concentrated load or reaction.

WORKING LOAD. The load expected on a structure and for which it is designed; service load (must be less than or equal to the allowable load).

YIELD MOMENT. In a member subjected to bending, the moment at which an outer fiber first attains the yield stress.

YIELD POINT. The first stress in a material, less than the maximum attainable stress, at which an increase in strain occurs without an increase in stress.

YIELD STRENGTH. The stress at which a material exhibits a specified limiting deviation from the proportionality of stress to strain. The deviation is expressed in terms of strain. [*Note:* It is customary to determine yield strength by (1) offset method (usually a strain of 0.2% is specified), and (2) total-extension-under-load method (usually a strain of 0.5% is specified although other values of strain may be used). Whenever yield strength is specified, the method of test must be stated along with the percent of offset or the total strain under load. The values obtained by the two methods may differ.]

YIELD STRESS. Yield point, yield strength (or yield stress level) as defined.

YIELD STRESS LEVEL. A particular yield strength: the average stress during yielding in the plastic range. It is the stress determined in a tension test when the strain reaches 0.005 in./in.

The AISC Specifications designate allowable stresses as capital letters while the computed or calculated stresses are expressed as lowercase letters. With the ever-increasing family of steels in everyday use, the AISC uses the technique of specifying allowable stresses in terms of the yield point, F_y. For example, the allowable working stress in tension, F_t, is $0.60F_y$. F_y is the specified minimum yield stress of the type of steel being used in kips per square inch (ksi). As used in the AISC Specifications, "yield stress" denotes either the specified minimum yield point for those steels that have a defined yield point, or a specified minimum yield strength for those steel grades that do not have a well-defined yield point.

Nomenclature for the AISC Specifications, as well as general nomenclature for the AISC Manual, can be found in the manual itself.

1-17 PRODUCT AVAILABILITY, SHAPE-SIZE GROUPINGS, AND PRINCIPAL PRODUCERS

The designer must temper his or her thinking relative to shape and grade of steel by considering product availability and the like. Is the steel of particular grade and shape readily available? Of course, this answer depends upon many factors: geographical location of project, tonnages and timing involved, the readiness of availability from mill or fabricator, and so on. Needless to say, these factors change from time to time and the producer should be consulted for accurate determination.

Table 1-7 lists the availability of shapes, plates, and bars according to ASTM structural specifications. Again, accurate determination of availability should be checked with the manufacturers. The structural shape groups as per ASTM A6 are as given in Table 1-8. Tables 1-9 and 1-10 show the principal producers of structural shapes and tubing. As individual companies change their product production from time to time,

Table 1-7
AVAILABILITY OF SHAPES, PLATES, AND BARS ACCORDING TO ASTM STRUCTURAL STEEL SPECIFICATIONS

Steel Type	ASTM Designation		F_y Minimum Yield Stress (KSi)	Shapes Group per ASTM A6					Plates and Bars											
				1[a]	2	3	4	5	To 1/2 Incl.	Over 1/2 to 3/4 Incl.	Over 3/4 to 1 Incl.	Over 1 to 1¼ Incl.	Over 1¼ to 1½ Incl.	Over 1½ to 2 Incl.	Over 2 to 2½ Incl.	Over 2½ to 4 Incl.	Over 4 to 5 Incl.	Over 5 to 6 Incl.	Over 6 to 8 Incl.	Over 8″
Carbon	A36		32					■												■
	A36		36	■	■	■	■		■	■	■	■	■	■	■	■	■	■	■	
	A529		42	■	■				■	■										
High-Strength Low-Alloy	A441		40					■								■	■	■	■	
	A441		42				■									■	■			
	A441		46		■	■									■	■				
	A441		50	■	■				■	■	■	■								
	A572 Gr. 42		42	■	■	■	■	■	■	■	■	■	■	■	■					
	A572 Gr. 50		50	■	■	■	■		■	■	■	■	■	■						
	A572 Gr. 60		60	■	■				■	■	■									
	A572 Gr. 65		65	■					■	■										
Corrosion-Resistant High-Strength Low-Alloy	A242		42				■									■	■			
	A242		46		■	■									■	■				
	A242		50	■	■				■	■	■	■								
	A588		42														■	■	■	■	
	A588		46														■	■			
	A588		50	■	■	■	■	■	■	■	■	■	■	■	■	■				
Quenched and Tempered Alloy	A514		90																■		
	A514		100														■				

■ Available
☐ Not Available

[a] Includes Bar Size Shapes

Courtesy of American Institute of Steel Construction.

64

Table 1-8
STRUCTURAL SHAPE SIZE GROUPINGS FOR TENSILE PROPERTY CLASSIFICATION

Structural Shape	Group 1	Group 2	Group 3	Group 4	Group 5
W-shapes	W 24 × 55, 62	W 36 × 135 to 210 incl	W 36 × 230 to 300 incl	W 14 × 233 to 550 incl	W 14 × 605 to 730 incl
	W 21 × 44 to 57 incl	W 33 × 118 to 152 incl	W 33 × 201 to 241 incl	W 12 × 210 to 336 incl	
	W 18 × 35 to 71 incl	W 30 × 99 to 211 incl	W 14 × 145 to 211 incl		
	W 16 × 26 to 57 incl	W 27 × 84 to 178 incl	W 12 × 120 to 190 incl		
	W 14 × 22 to 53 incl	W 24 × 68 to 162 incl			
	W 12 × 14 to 58 incl	W 21 × 62 to 147 incl			
	W 10 × 12 to 45 incl	W 18 × 76 to 119 incl			
	W 8 × 10 to 48 incl	W 16 × 67 to 100 incl			
	W 6 × 9 to 25 incl	W 14 × 61 to 132 incl			
	W 5 × 16, 19	W 12 × 65 to 106 incl			
	W 4 × 13	W 10 × 49 to 112 incl			
		W 8 × 58, 67			
M-shapes	to 20 lb/ft incl				
S-shapes	to 35 lb/ft incl	over 35 lb/ft			
HP-shapes		to 102 lb/ft incl	over 102 lb/ft		
American Standard Channels (C)	to 20.7 lb/ft incl	over 20.7 lb/ft			
Miscellaneous Channels (MC)	to 28.5 lb/ft incl	over 28.5 lb/ft			
Angles (Structural and Bar size) (L-shapes)	to ½ in. incl	over ½ to ¾ in. incl	over ¾ in.		

NOTE: Structural tees from W, M, and S shapes fall in the same group as the structural shape from which they are cut.
Courtesy of American Society for Testing and Materials (ASTM Standards A6).

65

Table 1-9
PRINCIPAL PRODUCERS OF STRUCTURAL SHAPES

A. Armco Inc.	J. Jones & Laughlin Steel Corp.
B. Bethlehem Steel Corp.	N. Northwestern Steel & Wire Co.
C. CF&I Steel Corp.	U. United States Steel Corp.
I. Inland Steel Co.	

Section and Weight per Ft.	Producer Code	Section and Weight per Ft.	Producer Code
W 36—all	B, U	W 6 × 15—25	A, B, C, I, N, U
W 33—all	B, U	W 6 × 9—16	A, B, I, J[7], N, U[8]
W 30—all	B, U	W 5 × 16—19	A, B, I
W 27—all	B, U	W 4 × 13	A, B, N
W 24—all	A, B, I, U		
W 21—all	A, B, I, U	S 24—all	B
W 18 × 76—119	A, B, I, U	S 20—all	B
W 18 × 50—71	A, B, I, N, U	S 18—all	B, U
W 18 × 35—46	A, B, I, N, U	S 15—all	B, U
W 16 × 67—100	A, B, I, U	S 12—all	B, I, U
W 16 × 36—57	A, B, C[1], I, N, U	S 10—all	B, C, I, U
W 16 × 26—31	A, B, C, I, N, U	S 8—all	B, C, I, U
W 14 × 145—730	B, U	S 7 × 20	C
W 14 × 90—132	B, U	S 7 × 15.3	B, C, I, U
W 14 × 61—82	A, B, I, U	S 6—all	B, C, I, U
W 14 × 43—53	A, B, C, I, U	S 5 × 14.75	C
W 14 × 30—38	A, B, C, I, N, U	S 5 × 10	B, C, I, U
W 14 × 22—26	A, B, C, I, N, U	S 4 × 9.5	C, I, U
W 12 × 65—336	A, B[2], I, U[2]	S 4 × 7.7	B, C, I, U
W 12 × 53—58	A, B, I, U	S 3 × 7.5	B, C, I
W 12 × 40—50	A, B, C, I, U	S 3 × 5.7	B, C, I
W 12 × 26—35	A, B, C, I, N, U		
W 12 × 14—22	A, B, C, I, N, U	M 14 × 18	N
W 10 × 49—112	A, B, I, N[3], U	M 12 × 11.8	J
W 10 × 33—45	A, B, C, I, N, U	M 10 × 9	J
W 10 × 22—30	A, B, C, I, N, U	M 8 × 6.5	J
W 10 × 12—19	A, B, C, I, N, U[4]	M 6 × 20	U
W 8 × 31—67	A, B, C, I, N[5], U	M 6 × 4.4	J
W 8 × 24—28	A, B, C, I, N, U	M 5 × 18.9	B
W 8 × 18—21	A, B, C, I, N, U	M 4 × 13	I, U
W 8 × 10—15	A, B, C, I, N, U[6]		

[1] W 16 × 57 excluded.	[5] W 8 × 48—67 excluded.
[2] W 12 × 210—336 excluded.	[6] W 8 × 10 excluded.
[3] W 10 × 77—112 excluded.	[7] W 6 × 12 and W 6 × 16 excluded.
[4] W 10 × 12 excluded.	[8] W 6 × 9 excluded.

Notes: The W shapes offered by producer J have a 3° taper of the inner flange surfaces, and have design properties about the Y–Y axis somewhat less than those tabulated. The catalog of this producer should be consulted for exact information.

Maximum lengths of shapes obtainable vary widely with producers, but a conservative range for all mills is from 60 to 75 feet. Some mills will accept orders for lengths up to 120 feet, but only for certain shapes and subject to special arrangement. Consult the producers for unusual length requirements.

Courtesy of American Institute of Steel Construction.

Table 1-9 *(Continued)*

PRINCIPAL PRODUCERS OF STRUCTURAL SHAPES			
A. Armco Inc. B. Bethlehem Steel Corp. C. CF&I Steel Corp. I. Inland Steel Co.		J. Jones & Laughlin Steel Corp. N. Northwestern Steel & Wire Co. U. United States Steel Corp.	
Section and Weight per Ft.	Producer Code	Section and Weight per Ft.	Producer Code
HP 14—all	B, U	MC 18—all	B, U
HP 13—all	A, I	MC 13—all	B, U
HP 12—all	A, B[9], I, U[9]	MC 12 X 35—50	B, U
HP 10—all	A, B, I, N, U	MC 12 X 30.9—37	B, U
HP 8 X 36	A, B, I, N, U	MC 12 X 10.6	J
		MC 10 X 28.5—41.1	B, U
C 15—all	B, C, I, N, U	MC 10 X 21.9—28.3	B, U
C 12—all	B, C, I, N, U	MC 10 X 8.4	J
C 10 X 30	B, I, N, U	MC 10 X 6.5	J
C 10 X 15.3—25	B, C, I, N, U	MC 9—all	B
C 9 X 20	C	MC 8 X 21.4—22.8	B, U
C 9 X 13.4—15	B, C, I, U	MC 8 X 18.7—20	B, U
C 8—all	B, C, I, J, N, U	MC 8 X 8.5	J
C 7 X 14.75	C	MC 7 X 19.1—22.7	B, U
C 7 X 9.8—12.25	B, C, I, N, U	MC 7 X 17.6	U
C 6—all	B, C, I, J, N, U	MC 6 X 18	B
C 5—all	B, C, I, J, N, U	MC 6 X 15.3	B, U
C 4—all	A, B, C, I, J, N, U	MC 6 X 15.1—16.3	B, U
C 3 X 6	A, C, J, N	MC 6 X 12	B, U
C 3 X 4.1—5	A, B, C, I, J, N, U		

[9] HP 12 X 63 and HP 12 X 84 excluded.

Note: Maximum lengths of shapes obtainable vary widely with producers, but a conservative range for all mills is from 60 to 75 feet. Some mills will accept orders for lengths up to 120 feet, but only for certain shapes and subject to special arrangement. Consult the producers for unusual length requirements.

Table 1-9 *(Continued)*

PRINCIPAL PRODUCERS OF STRUCTURAL SHAPES			
A. Armco Inc. B. Bethlehem Steel Corp. C. CF&I Steel Corp. I. Inland Steel Co.		J. Jones & Laughlin Steel Corp. N. Northwestern Steel & Wire Co. U. United States Steel Corp.	
Section by Leg Length and Thickness	Producer Code	Section by Leg Length and Thickness	Producer Code
L $8 \times 8 \times 1\frac{1}{8}$	B, I, U	L $3 \times 3 \times \frac{1}{2}$	A, B, C, I, J, N, U
$\times 1$	B, I, U	$\times \frac{7}{16}$	J, N
$\times \frac{7}{8}$	B, I, U	$\times \frac{3}{8}$	A, B, C, I, J, N, U
$\times \frac{3}{4}$	B, I, U	$\times \frac{5}{16}$	A, B, C, I, J, N, U
$\times \frac{5}{8}$	B, I, U	$\times \frac{1}{4}$	A, B, C, I, J, N, U
$\times \frac{9}{16}$	I, U	$\times \frac{3}{16}$	A, B, C, J, N, U
$\times \frac{1}{2}$	B, I, U		
		L $2\frac{1}{2} \times 2\frac{1}{2} \times \frac{1}{2}$	B, I, N
L $6 \times 6 \times 1$	B, C, U	$\times \frac{3}{8}$	A, B, C, I, N, U
$\times \frac{7}{8}$	B, C, U	$\times \frac{5}{16}$	A, B, C, I, N, U
$\times \frac{3}{4}$	B, C, I, U	$\times \frac{1}{4}$	A, B, C, I, N, U
$\times \frac{5}{8}$	B, C, I, U	$\times \frac{3}{16}$	A, B, C, N, U
$\times \frac{9}{16}$	C, I, U		
$\times \frac{1}{2}$	B, C, I, U	L $2 \times 2 \times \frac{3}{8}$	A, B, C, I, N, U
$\times \frac{7}{16}$	C, I, U	$\times \frac{5}{16}$	A, B, C, I, N, U
$\times \frac{3}{8}$	B, C, I, U	$\times \frac{1}{4}$	A, B, C, I, N, U
$\times \frac{5}{16}$	C	$\times \frac{3}{16}$	A, B, C, N, U
		$\times \frac{1}{8}$	A, B, C, N, U
L $5 \times 5 \times \frac{7}{8}$	B, C, I, U		
$\times \frac{3}{4}$	B, C, I, N, U	L $9 \times 4 \times \frac{5}{8}$	B
$\times \frac{5}{8}$	C, I, N	$\times \frac{9}{16}$	B
$\times \frac{1}{2}$	B, C, I, N, U	$\times \frac{1}{2}$	B
$\times \frac{7}{16}$	C, I, N, U		
$\times \frac{3}{8}$	B, C, I, N, U	L $8 \times 6 \times 1$	B, C, I, U
$\times \frac{5}{16}$	B, C, I, N, U	$\times \frac{7}{8}$	C
		$\times \frac{3}{4}$	B, C, I, U
L $4 \times 4 \times \frac{3}{4}$	B, C, I, J, N, U	$\times \frac{5}{8}$	C
$\times \frac{5}{8}$	B, C, I, J, N, U	$\times \frac{9}{16}$	B, C, I, U
$\times \frac{1}{2}$	A, B, C, I, J, N, U	$\times \frac{1}{2}$	B, C, I, U
$\times \frac{7}{16}$	A, C, I, J, N	$\times \frac{7}{16}$	C, I
$\times \frac{3}{8}$	A, B, C, I, J, N, U		
$\times \frac{5}{16}$	A, B, C, I, J, N, U	L $8 \times 4 \times 1$	B, U
$\times \frac{1}{4}$	A, B, C, I, J, N, U	$\times \frac{3}{4}$	B, U
		$\times \frac{9}{16}$	B, U
L $3\frac{1}{2} \times 3\frac{1}{2} \times \frac{1}{2}$	A, C, I, J, N	$\times \frac{1}{2}$	B, U
$\times \frac{7}{16}$	C, J, N		
$\times \frac{3}{8}$	A, B, C, I, J, N, U		
$\times \frac{5}{16}$	A, B, C, I, J, N, U		
$\times \frac{1}{4}$	A, B, C, I, J, N, U		

Note: Maximum lengths of shapes obtainable vary widely with producers, but a conservative range for all mills is from 60 to 75 feet. Some mills will accept orders for lengths up to 120 feet, but only for certain shapes and subject to special arrangement. Consult the producers for unusual length requirements.

Table 1-9 *(Continued)*

PRINCIPAL PRODUCERS OF STRUCTURAL SHAPES			
A. Armco Inc. B. Bethlehem Steel Corp. C. CF&I Steel Corp. I. Inland Steel Co.		J. Jones & Laughlin Steel Corp. N. Northwestern Steel & Wire Co. U. United States Steel Corp.	
Section by Leg Length and Thickness	Producer Code	Section by Leg Length and Thickness	Producer Code
L 7 X 4 X $\frac{3}{4}$	B, I, U	L 4 X 3 X $\frac{5}{8}$	C, I, J
X $\frac{5}{8}$	I	X $\frac{1}{2}$	A, B, C, I, J, N, U
X $\frac{1}{2}$	B, I, U	X $\frac{7}{16}$	A, C, J, N
X $\frac{3}{8}$	B, I, U	X $\frac{3}{8}$	A, B, C, I, J, N, U
L 6 X 4 X $\frac{7}{8}$	C, N	X $\frac{5}{16}$	A, B, C, I, J, N, U
X $\frac{3}{4}$	B, C, I, N, U	X $\frac{1}{4}$	A, B, C, I, J, N, U
X $\frac{5}{8}$	B, C, I, N		
X $\frac{9}{16}$	C, N	L 3$\frac{1}{2}$ X 3 X $\frac{1}{2}$	A, I, J, N, U
X $\frac{1}{2}$	B, C, I, N, U	X $\frac{7}{16}$	J, N
X $\frac{7}{16}$	C, I, N, U	X $\frac{3}{8}$	A, B, I, J, N, U
X $\frac{3}{8}$	B, C, I, N, U	X $\frac{5}{16}$	A, B, I, J, N, U
X $\frac{5}{16}$	C, I, N, U	X $\frac{1}{4}$	A, B, I, J, N, U
		L 3$\frac{1}{2}$ X 2$\frac{1}{2}$ X $\frac{1}{2}$	I, J
L 6 X 3$\frac{1}{2}$ X $\frac{1}{2}$	C, I, N	X $\frac{7}{16}$	J
X $\frac{3}{8}$	B, C, I, N, U	X $\frac{3}{8}$	A, B, C, I, J, N, U
X $\frac{5}{16}$	B, C, I, N, U	X $\frac{5}{16}$	A, B, C, I, J, N, U
		X $\frac{1}{4}$	A, B, C, I, J, N, U
L 5 X 3$\frac{1}{2}$ X $\frac{3}{4}$	B, I, J, U		
X $\frac{5}{8}$	C, I, J	L 3 X 2$\frac{1}{2}$ X $\frac{1}{2}$	I
X $\frac{1}{2}$	B, C, I, J, N, U	X $\frac{7}{16}$	J
X $\frac{7}{16}$	C, J, N, U	X $\frac{3}{8}$	A, B, C, I, J, N, U
X $\frac{3}{8}$	B, C, I, J, N, U	X $\frac{5}{16}$	A, B, C, I, J, N, U
X $\frac{5}{16}$	B, C, I, J, N, U	X $\frac{1}{4}$	A, B, C, I, J, N, U
X $\frac{1}{4}$	C, J, N, U	X $\frac{3}{16}$	A, N
L 5 X 3 X $\frac{5}{8}$	U	L 3 X 2 X $\frac{1}{2}$	I, N
X $\frac{1}{2}$	B, C, I, J, N, U	X $\frac{7}{16}$	J, N
X $\frac{7}{16}$	C, J, N	X $\frac{3}{8}$	A, B, C, I, J, N, U
X $\frac{3}{8}$	B, C, I, J, N, U	X $\frac{5}{16}$	A, B, C, I, J, N, U
X $\frac{5}{16}$	B, C, I, J, N, U	X $\frac{1}{4}$	A, B, C, I, J, N, U
X $\frac{1}{4}$	B, C, J, N, U	X $\frac{3}{16}$	A, C, J, N, U
L 4 X 3$\frac{1}{2}$ X $\frac{5}{8}$	C, I, J	L 2$\frac{1}{2}$ X 2 X $\frac{3}{8}$	A, B, I, N, U
X $\frac{1}{2}$	A, B, C, I, J, N, U	X $\frac{5}{16}$	A, B, N, U
X $\frac{7}{16}$	A, C, J, N	X $\frac{1}{4}$	A, B, I, N, U
X $\frac{3}{8}$	A, B, C, I, J, N, U	X $\frac{3}{16}$	A, B, N, U
X $\frac{5}{16}$	A, B, C, I, J, N, U		
X $\frac{1}{4}$	A, B, C, I, J, N, U		

Note: Maximum lengths of shapes obtainable vary widely with producers, but a conservative range for all mills is from 60 to 75 feet. Some mills will accept orders for lengths up to 120 feet, but only for certain shapes and subject to special arrangement. Consult the producers for unusual length requirements.

Table 1-10
PRINCIPAL PRODUCERS OF STRUCTURAL TUBING

A. Automation Industries, Inc., Harris Tube Div.	Ds. Donovan Steel Tube Co.	T. Tex-Tube Div., Cyclops Corp.
B. Bock Industries of Elkhart, Ind., Inc.	J. James Steel & Tube Co., Div. of Avis Industrial Corp.	V. Van Huffel Tube Corp., Div. of Youngstown Sheet and Tube Company
C. Copperweld Tubing Group	K. Kaiser Steel Tubing, Inc.	W. Welded Tube Co. of America
Dm. Delta Metalforming Co.		

Nominal Size and Thickness	Producer Code	Nominal Size and Thickness	Producer Code
16×16—all	W	$10 \times 4 \times 1$	C, W
14×14—all	W	$10 \times 4 \times \frac{3}{8}, \frac{5}{16}, \frac{1}{4}, \frac{3}{16}$	A, B, C, Dm, W
12×12—all	W	$10 \times 2 \times \frac{3}{8}$	W
$10 \times 10 \times \frac{5}{8}$	C	$10 \times 2 \times \frac{5}{16}$	A, W
$10 \times 10 \times \frac{1}{2}$	C, W	$10 \times 2 \times \frac{1}{4}, \frac{3}{16}$	A, B, W
$10 \times 10 \times \frac{3}{8}, \frac{5}{16}, \frac{1}{4}$	A, C, W	$8 \times 6 \times \frac{1}{2}$	C, W
$8 \times 8 \times \frac{5}{8}$	C	$8 \times 6 \times \frac{3}{8}, \frac{5}{16}, \frac{1}{4}, \frac{3}{16}$	A, B, C, Dm, W
$8 \times 8 \times \frac{1}{2}$	C, W	$8 \times 4 \times \frac{1}{2}$	C, W
$8 \times 8 \times \frac{3}{8}, \frac{5}{16}, \frac{1}{4}, \frac{3}{16}$	A, B, C, Dm, W	$8 \times 4 \times \frac{3}{8}$	B, C, Dm, T, W
$7 \times 7 \times \frac{1}{2}$	C, W	$8 \times 4 \times \frac{5}{16}$	A, B, C, Dm, T, W
$7 \times 7 \times \frac{3}{8}, \frac{5}{16}, \frac{1}{4}, \frac{3}{16}$	A, B, C, Dm, T, W	$8 \times 4 \times \frac{1}{4}, \frac{3}{16}$	A, B, C, Dm, T, V, W
$6 \times 6 \times \frac{1}{2}$	C, W	$8 \times 3 \times \frac{3}{8}$	C, Dm, W
$6 \times 6 \times \frac{3}{8}$	B, C, Dm, T, W	$8 \times 3 \times \frac{5}{16}$	B, C, Dm, W
$6 \times 6 \times \frac{5}{16}$	A, B, C, Dm, T, W	$8 \times 3 \times \frac{1}{4}, \frac{3}{16}$	A, B, C, Dm, W
$6 \times 6 \times \frac{1}{4}, \frac{3}{16}$	A, B, C, Dm, T, V, W	$8 \times 2 \times \frac{3}{8}, \frac{5}{16}$	Dm, W
$5 \times 5 \times 1$	C, W	$8 \times 2 \times \frac{1}{4}, \frac{3}{16}$	A, B, C, Dm, W
$5 \times 5 \times \frac{3}{8}, \frac{5}{16}$	B, C, Dm, T, W	$7 \times 5 \times \frac{1}{2}$	C, W
$5 \times 5 \times \frac{1}{4}, \frac{3}{16}$	A, B, C, Dm, T, V, W	$7 \times 5 \times \frac{3}{8}$	B, C, Dm, T, W
$4 \times 4 \times 1$	C, W	$7 \times 5 \times \frac{5}{16}, \frac{1}{4}, \frac{3}{16}$	A, B, C, Dm, T, W
$4 \times 4 \times \frac{3}{8}, \frac{5}{16}$	B, C, Dm, W	$7 \times 4 \times \frac{3}{8}, \frac{5}{16}$	B, C, Dm, T, W

Section	Codes
$4 \times 4 \times \frac{1}{4}, \frac{3}{16}$	A, B, C, Dm, Ds, J, K, T, V, W
$3\frac{1}{2} \times 3\frac{1}{2} \times \frac{5}{16}$	W
$3\frac{1}{2} \times 3\frac{1}{2} \times \frac{1}{4}, \frac{3}{16}$	A, B, C, Ds, J, K, T, V, W
$3 \times 3 \times \frac{5}{16}$	Dm, W
$3 \times 3 \times \frac{1}{4}, \frac{3}{16}$	A, B, C, Dm, Ds, J, K, T, V, W
$2\frac{1}{2} \times 2\frac{1}{2} \times \frac{1}{4}$	A, C, Ds, J, K, T, V, W
$2\frac{1}{2} \times 2\frac{1}{2} \times \frac{3}{16}$	A, B, C, Ds, J, K, T, V, W
$2 \times 2 \times \frac{1}{4}$	A, C, Dm, Ds, J, K, T, W
$2 \times 2 \times \frac{3}{16}$	A, B, C, Dm, Ds, J, K, T, V, W

Section	Codes
$20 \times 12 \times \frac{1}{2}, \frac{3}{8}, \frac{5}{16}$	W
$20 \times 8 \times \frac{1}{2}, \frac{3}{8}, \frac{5}{16}$	W
$20 \times 4 \times \frac{1}{2}, \frac{3}{8}, \frac{5}{16}$	W
$18 \times 6 \times \frac{1}{2}, \frac{3}{8}, \frac{5}{16}$	W
$16 \times 12 \times \frac{1}{2}, \frac{3}{8}, \frac{5}{16}$	W
$16 \times 8 \times \frac{1}{2}, \frac{3}{8}, \frac{5}{16}$	W
$16 \times 4 \times \frac{1}{2}, \frac{3}{8}, \frac{5}{16}$	W
$14 \times 10 \times \frac{1}{2}, \frac{3}{8}, \frac{5}{16}, \frac{1}{4}$	W
$14 \times 6 \times \frac{1}{2}, \frac{3}{8}, \frac{5}{16}, \frac{1}{4}$	W
$14 \times 4 \times \frac{1}{2}, \frac{3}{8}, \frac{5}{16}, \frac{1}{4}$	C
$12 \times 8 \times \frac{5}{8}$	C, W
$12 \times 8 \times \frac{1}{2}$	A, C, W
$12 \times 8 \times \frac{3}{8}, \frac{5}{16}, \frac{1}{4}$	C, W
$12 \times 6 \times \frac{1}{2}, \frac{3}{8}, \frac{5}{16}, \frac{1}{4}$	C, W
$12 \times 4 \times \frac{1}{2}$	A, B, C, W
$12 \times 4 \times \frac{3}{8}, \frac{5}{16}, \frac{1}{4}$	A, B
$12 \times 2 \times \frac{1}{4}$	C, W
$10 \times 6 \times \frac{5}{8}$	C, W
$10 \times 6 \times \frac{3}{8}, \frac{5}{16}, \frac{1}{4}$	A, B, C, W

Section	Codes
$7 \times 4 \times \frac{1}{4}, \frac{3}{16}$	A, B, C, Dm, T, W
$7 \times 3 \times \frac{3}{8}, \frac{5}{16}$	C, Dm, W
$7 \times 3 \times \frac{1}{4}, \frac{3}{16}$	A, C, Dm, W
$6 \times 4 \times 1$	C, W
$6 \times 4 \times \frac{3}{8}$	B, C, Dm, T, W
$6 \times 4 \times \frac{5}{16}$	B, C, Dm, W
$6 \times 4 \times \frac{1}{4}, \frac{3}{16}$	A, B, C, Dm, T, V, W
$6 \times 3 \times \frac{3}{8}$	Dm, W
$6 \times 3 \times \frac{5}{16}$	B, Dm, W
$6 \times 3 \times \frac{1}{4}, \frac{3}{16}$	A, B, C, Dm, T, V, W
$6 \times 2 \times \frac{3}{8}, \frac{5}{16}$	Dm, W
$6 \times 2 \times \frac{1}{4}, \frac{3}{16}$	A, B, C, Dm, K, T, V, W
$5 \times 4 \times \frac{5}{16}$	B, Dm, W
$5 \times 4 \times \frac{1}{4}, \frac{3}{16}$	A, B, C, Dm, T, V, W
$5 \times 3 \times \frac{1}{2}$	C
$5 \times 3 \times \frac{3}{8}$	C, Dm, W
$5 \times 3 \times \frac{5}{16}$	B, C, Dm, W
$5 \times 3 \times \frac{1}{4}, \frac{3}{16}$	A, B, C, Dm, Ds, J, K, T, V, W
$5 \times 2 \times \frac{5}{16}$	Dm, W
$5 \times 2 \times \frac{1}{4}$	A, B, C, Dm, Ds, J, K, T, W
$5 \times 2 \times \frac{3}{16}$	A, B, C, Dm, Ds, J, K, T, V, W
$4 \times 3 \times \frac{5}{16}$	Dm, W
$4 \times 3 \times \frac{1}{4}, \frac{3}{16}$	A, B, C, Dm, Ds, J, K, T, V, W
$4 \times 2 \times \frac{5}{16}$	Dm, W
$4 \times 2 \times \frac{1}{4}$	A, B, C, Dm, Ds, J, K, T, W
$4 \times 2 \times \frac{3}{16}$	A, B, C, Dm, Ds, J, K, T, V, W
$3 \times 2 \times \frac{1}{4}$	A, C, Dm, Ds, J, K, T, W
$3 \times 2 \times \frac{3}{16}$	A, B, C, Dm, Ds, J, K, T, V, W

71

Tables 1-9 and 1-10 should be used only as a general guide; the manufacturers should be consulted directly when detailed information is
needed on the extent of their most current product production.

PROBLEMS

(1-1) A standard tension test is performed in the laboratory on a cylindrical steel specimen. Appropriate readings are taken and summa-
rized as follows:

	Initial	Final
(a) Specimen diameter	0.882 in.	0.473 in.
(b) Gage length	8.000 in.	9.921 in.

(c) Summary of load readings versus gage length (elastic range
only):

Load (kips)	L (in.)
2	8.000
4	8.001
6	8.002
8	8.003
10	8.004
12	8.004
14	8.005
16	8.006
18	8.007
20	8.008
22	8.012
24	8.015
26	8.021
28	8.032

(d) Critical loads: ultimate strength = 75.5 kips; rupture strength
= 68.3 kips.
1. Plot the stress versus strain curve for the range of data provided.
2. Compute the exact modulus of elasticity for the specimen, using
the curve plotted in step 1.
3. Determine the specimen's elastic limit.
4. Determine the 0.2% yield strength for the specimen.
5. Determine the ultimate tensile strength of the specimen.
6. Determine the "actual" rupture stress (based on final rupture
area).
7. Determine the steel's ductility by
(a) Percent of area reduction.
(b) Percent of increase in length.

buildings

2-1   INTRODUCTION

To function, buildings must be capable of sustaining all imposed loads and forces. These loads and forces are transferred to the vital "spine" of the building, its structural framework. The types of loads and forces and how they are distributed to the frame, along with the elements and types of framing of the structure, will be discussed in this chapter. The means of fastening one structural element to another, the stresses and strains exerted upon these elements, and the general design thereof will be touched upon. All design considerations will parallel the AISC Manual of Steel Construction and the AISC Specification.

2-2   LOADS AND FORCES

The loads a building must be designed for are usually stipulated by an applicable building code. However, where an applicable building code is nonexistent, the American Standard Building Code Requirements for

Minimum Design Loads in Buildings and Other Structures, ANSI (American National Standards Institute) A58.1, is recommended. Excerpts from this code are found in the AISC Manual of Steel Construction, Seventh Edition, Part 5, and in part are incorporated in the discussions of building loads to follow.

Dead load is the term used to describe the weight of the structure itself and all the materials fastened to the structure and permanently supported by it. Examples of dead loads in a building are the weights of floor slabs, floor metal deck, partitions, beams, columns, plumbing, heating equipment, and ceiling material. They are usually expressed as a uniform load in pounds per square foot or per linear foot (psf or plf). At times, an estimate of some of these dead loads has to be made since the actual sizes and thicknesses are not known until after the design process is complete. Unit weights in pounds per cubic foot (pcf) of various construction materials can be found in ANSI A58.1, local building codes, or in the AISC Manual, Seventh Edition, Part 5.

Live load is a superimposed load on the structure usually caused by use and occupancy of the building. Live loads are all loads not connected permanently to the building. ANSI provides minimum live loads related to the occupancy or use of the building and/or specified areas of the building and should be used when no other governing code or specification governs. Live loads are expressed as uniform loads (psf) or concentrated loads in pounds. These loads are usually applied vertically. Snow, wind, and earthquake loads, although they are classified as live loads, will be discussed separately. Examples of live loading in a building include human occupants, furniture, machinery, and stored or stacked materials. Once again, an accurate determination of live loads is difficult, owing to the many variables involved, such as mobility of load, variable weight of the item, changing quantity of storage or stacking, and so on. The governing code or specification conservatively states live loading, ensuring public safety and structural integrity of the structure. These values are based upon years of experience and good engineering practice. Live loads for storage warehouses are often of particular interest to the designer. Reference may be made to U.S. Department of Commerce, National Bureau of Standards, Recommended Live Loads for Storage Warehouses (Part 5, AISC Manual of Steel Construction, Seventh Edition).

With the exception of warehouses and similar structures, the maximum stipulated live loading is not likely to occur on each floor at the same time. Most codes therefore allow a reduction in accordance with a formula that usually involves the ratio of dead load to live load per square foot of area supported by a member. Live loads that are suddenly applied, such as elevators, cranes, and reciprocating machinery, often produce a dynamic effect called *impact*. Section 1.3.3 of the AISC Specification states that for these cases the assumed live loads should be sufficiently increased to provide for this action or that the increase shall be a stipulated value relating to the condition of load application.

In roof structures, *snow load* (when it exists) is of primary concern to the designer. Snow loads in psf are usually given by local codes and are applied on the horizontal projection of the roof surface. Nationally accepted codes specify minimum design snow loads by giving the ground snow load in pounds per square foot for 25-year, 50-year, or 100-year mean recurrence intervals in the form of a map on which isolines (lines joining equal points of snow load) are drawn. ANSI A58.1 is one source for these maps in the United States. These ground snow loads are then multiplied by appropriate snow coefficients, which in effect increase or decrease the ground snow load. These coefficients are based upon factors such as roof slope, roof exposure to wind, roof shape, roof susceptibility to nonuniform accumulation (pitched and curved roofs), and lower level of multilevel roof accumulation. These coefficients, as well as suggested snow-load distribution on various roof shapes, are given in these codes. For further details, reference is made to ANSI A58.1. Special consideration to special conditions should be made by the designer with reference to special snow regions and special conditions prone to snow accumulation (drift, lower levels of multilevel roofs, etc.).

Another live load that should be given careful consideration is *wind load*. The AISC Specification permits allowable stresses to be increased by one third when wind acts alone or in combination with dead and live loads, with the proviso that the section computed on this basis is not less than that required for dead and live load (plus impact if any) computed without an increase in stress. The Specification further states an alternative approach to the aforementioned when allowable stresses are computed on the basis of certain factors applied to the grouping of load types. ANSI A58.1, for example, uses probability factors to increase or decrease allowable stresses for various groupings of load.

In a manner similar to that used to determine snow load, ANSI uses a typical basic 50-year mean recurrence interval of wind speed in miles per hour (annual extreme fastest-mile speed 30 ft above the ground) for the usual type of building, which is read from a map on which isolines (lines joining equal points of observed mph wind) are drawn. The wind speed squared, $(mph)^2$, multiplied by 0.00256, yields a basic wind pressure in pounds per square foot. Various corrective multiplication factors for height above ground and type of exposure are applied to this basic wind pressure to give effective velocity pressures in pounds per square foot for buildings and for portions of buildings. For convenience, effective velocity pressures in pounds per square foot for various basic wind speeds (mph) and height above ground (ft) are given in several tables in ANSI for city and suburban areas.

Finally, the design wind pressure for a building is computed as the product of the effective velocity pressure and a coefficient termed a *pressure coefficient*. A negative pressure coefficient indicates a suction as opposed to a positive pressure. The total design wind load on a building can be found by using a net pressure coefficient when available or by computing the vector sum of the loads on the individual sides of a building. Pressure coefficients are determined from wind-tunnel tests performed on scaled models of buildings in the laboratory and are given in ANSI for different conditions, dependent on building shape, orientation of the building relative to wind direction, windward or leeward building surface (roof or wall surface), and so on. For a more detailed study, reference should be made to the ANSI Code or to Wind Forces on Structures, the Final Report of the Task Committee on Wind Forces of the Committee on Loads and Stresses of the Structural Division, American Society of Civil Engineers (Paper No. 3269, *Transactions,* vol. 126, Part II, 1961, p. 1124).

The discussions on wind loading as heretofore presented are general in nature. The purpose is to make the individual reader aware of the many variables involved in arriving at a wind loading. The applicable codes usually dictate the wind loading to be used in the design of a building. An excellent reference for a typical wind analysis on a high-rise steel-frame building is Design of High Rise Steel Framed Buildings, Continuing Education Program, American Iron and Steel Institute.

Where a building is located is the determining factor of whether the building is subject to another type of lateral force or load, *earthquake*. Usually the applicable local codes account for the severity or nonseverity of earthquake in the particular locale. These codes provide criteria for computing the lateral force and the distribution of this force in the height of the building.

Some areas of the country are more prone to earthquakes than others. Earthquake risk zones are assigned and are shown on maps as determined by earthquake history. Examples are maps compiled by the Seismology Division of the Coast and Geodetic Survey, which plot earthquakes where and when they have occurred. Severity of the quake is noted on the map by the grading of round dots, and risk zone numbers are then assigned.

The total lateral seismic forces assumed to act on a building are dependent on several factors: a horizontal force factor, which varies with type or arrangement of the resistant framing elements of a building, a numerical coefficient, which relates to the earthquake risk zone as determined by maps, an additional coefficient, which depends upon the period of vibration of the building, and last, the total dead load of the building (dead load plus a percentage of floor live load).

These lateral forces simulate the actual horizontal inertia force created when the ground beneath a building suddenly moves during an earthquake and the building's mass tends to resist the ground motion. Building codes specify this simulated lateral force. A more detailed discussion of earthquake forces may be found in the ANSI A58.1 Code. An additional excellent reference is Recommended Lateral Force Requirement, and Commentary, Seismology Committee, Structural Engineers Association of California. Seismic Design for Buildings, Army TM 5-809-10, Navy NAV FAC P-355, Air Force AFM 88-3, Chapter 13, and Earthquake Forces on Tall Structures by Henry J. Degenkolb, Bethlehem Steel Corporation Booklet 651, are additional references on the subject.

Because of the occurrence of earthquake loads being infrequent and the period of sustained load being short, the AISC Specification, as for wind loads, permit a one third increase in allowable stress when earthquake forces occur either alone or in combination with dead and live loads, provided that the section so designed is no less than that required for dead and live loads alone computed with no increase in allowable stress. Other loads that are encountered, such as any earth pressures, fluid pressures, construction or erection loads, and crane loads, should also be taken into account by the designer.

The possible load combinations should be evaluated by the designer, and logical as well as feasible combinations should be investigated. Often, the governing code specifies the combinations to be investigated. The AISC Specification does not dictate load groupings. ANSI A58.1 suggests group loadings to be investigated. Design experience and good engineering judgment often dictate the critical combination.

Gravity loads are dead loads and live loads that are applied in a vertical sense. Live loads are usually given in pounds per square foot and when applied are distributed through the floor to the horizontal structural floor support members beneath, as in Fig. 2-1. For a typical bay $L_1 \times L_2$ with a beam spacing of S and a floor live load of w psf, the beam loading is Sw plf along the beam span length L_1. The contributing loaded area for a typical beam is shown crosshatched. Each beam has an end reaction of $(SwL_1) \frac{1}{2}$ delivered to the girder, which frames normal to the beams. The beam has a weight or dead load of w_2 plf, which accounts for the dead load of the beam itself, as well as flooring and/or slab material. An additional end reaction of $(w_2L_1) \frac{1}{2}$ is delivered to the girder by each beam owing to dead load. The total vertical end reaction of each beam loads the girder with a concentrated load P_1, which is equal to $L_1/2 \, (Sw + w_2)$, as shown in Fig. 2-1b.

The girder loaded (by the beams) as shown loads each column with

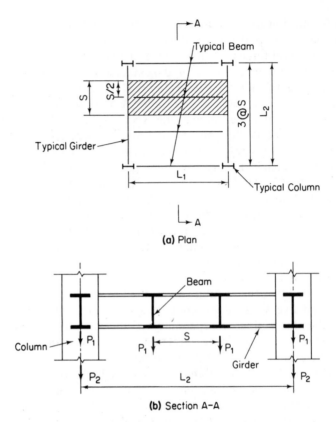

FIG. 2-1 Steel framing plan showing live load distribution.

a load P_2, which is equal to $(P_1 + P_1) \frac{1}{2}$ plus P_1 from the girder spanning from column flange to column flange; or $P_2 = 2P_1$. Of course, an additional vertical axial column load due to the column dead weight, as well as the weight of the skin or facade of the building (the material that encloses the building), exists.

The transfer of load by contributing area from floor to horizontal beam to horizontal girder and then to the vertical column, as illustrated, gives the reader a feel for the basic statics involved. For horizontal forces, wind for example, the load-transfer process is similar: from contributing area of skin to columns or from skin to girts to columns.

Live load and snow load should be applied in any practical or feasible manner such that the chosen pattern of loading creates the most critical results (stresses) in the member. The word "practical" is used with meaning. Checkerboard live loading (alternate bay) is practical,

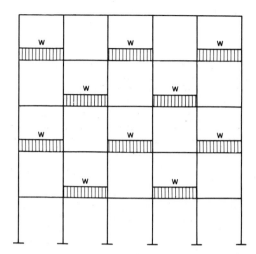

FIG. 2-2 Checkerboard type of loading.

while checkerboard snow load is not practical or realistic. The checkerboard type of loading is shown in Fig. 2-2.

2-3   TYPES OF FRAMING AND MEMBERS

The AISC Specification lists three basic types of steel framing and their related design assumptions. *Type 1,* commonly known as rigid frame or continuous framing, assumes that the beam-to-column connections are sufficiently rigid so as to cause the angle between the intersecting members to remain unchanged throughout the loading cycle. *Type 2,* simple framing, assumes that for gravity loading the ends of the beams are connected to the columns to take no moment (only to take shear). *Type 3,* semirigid framing, is framing that is partially restrained. The moment capacity of this type of connection is somewhere between Types 1 and 2. It assumes that beam and girder connections have a relied upon and known moment capacity. These types of connections, their design and details, will be fully discussed in Chapter 11.

Beams utilized as structural floor and roof members in buildings can be of many configurations (Fig. 2-3), depending upon such factors as economy, aesthetics, and necessary span. *Solid web hot-rolled W shapes* are available in 4-, 5-, 6-, 8-, 10-, 12-, 14-, 16-, 18-, 21-, 24-, 27-, 30-, 33-, and 36-in. nominal depths with a variety of dimensions (flange width and thickness and web thickness) and weights (plf) and are commonly used as beams. The choices are more numerous by far when compared with available shapes in other countries. W shapes are suitable for spans up to about 100 ft.

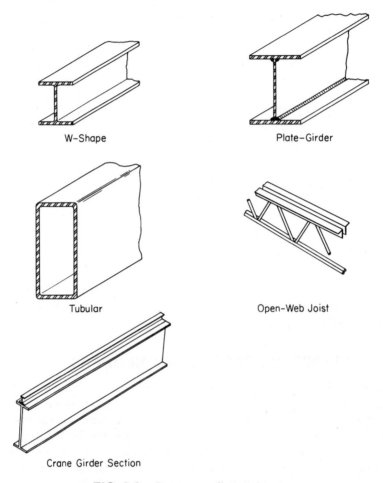

W-Shape

Plate-Girder

Tubular

Open-Web Joist

Crane Girder Section

FIG. 2-3 Beam configurations.

Plate girders, which are beams fabricated from hot-rolled plate material, are used where longer span lengths are encountered. Plate girders, of course, may be very much deeper than the deepest W shape. Although fabrication costs are greater, the savings in weight may be quite substantial. Their webs are thinner and more efficient. The design and details of plate girders will be discussed in more detail in Chapter 7. Plate girders may be of constant depth or of varying depth (tapered web plate). The more common spanning capabilities of the tapered type are in the range 40 to 90 ft, although spans up to 250 ft have commonly been achieved with built-up constant-depth girders. Depths within the range 41 to 86 in. are common for constant-depth web sections.

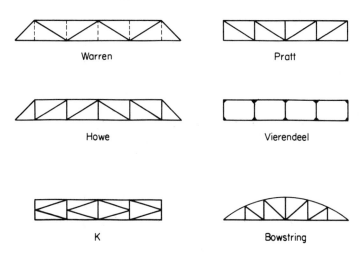

Warren   Pratt

Howe   Vierendeel

K   Bowstring

FIG. 2-4 Truss configurations.

Steel open web joists are actually trusses, usually with a Warren-type web configuration. The top and bottom chords and the webs of the joists vary as to the type of steel material and shape used. The details, design, and specifications jointly adopted by the Steel Joist Institute (SJI) and the AISC will be discussed in Chapter 12. Standardized series offer a range of depth of 8 to 24 in. in the standard series, from 18 to 48 in. in the longspan series, and 58 to 72 in. in the deep longspan series. The Specifications should be consulted to determine the individual depths within these ranges that are available. The standardized series of open web steel joists have the following span capabilities: standard series, 8 to 48 ft; longspan series, 25 to 96 ft; and deep longspan series, 89 to 144 ft. This type of member is relatively inexpensive and lends itself to giving a very light and economical floor system. Depth-to-span ratios for steel open web joists are commonly 1:20.

Trusses are usually of greater depths than the standardized joists and have greater spanning capabilities (Fig. 2-4). Truss elements are the same as those involved in open web joists, two chords and web members. However, the individual members of the chords and webs are usually more substantial—W shapes, channels, or angles. Depth-to-span ratios for a one-way system of trusses (main load-carrying trusses spanning in one direction only) are conservatively 1:10. Feasible and economical spans of from 30 to 300 ft are not unusual for various types of buildings utilizing trusses. When primary trusses span in two directions (Fig. 2-5), a more economical system for carrying loads is provided and a much lighter structural system achieved. In addition, depths of trusses are sub-

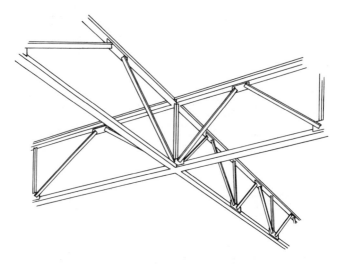

FIG. 2-5 Planar two-way truss system.

stantially less. In a two-way system of framing, the trusses span orthog-
onal to each other. At times the cross section of these trusses is pyramidal, as opposed to the more usual planar cross section, and, in this case,
a three-directional or "space" system is achieved. This system is one of
the lightest framing systems in existence. Roof systems framed in this
manner have been known to weigh from 1 to 4 psf (Fig. 2-6).

The *staggered truss system* of framing (Fig. 2-7) in medium- to
high-rise apartment houses, hotels, and motels is an economical system
of framing. Full-story-height trusses (usually of the Pratt type) span the
transverse or small dimension of a building from perimeter column to
perimeter column (there are no interior columns) and are staggered on
alternate floors. The spans in the typical building are about 60 ft, and the
resulting steel framing weighs approximately 6 psf.

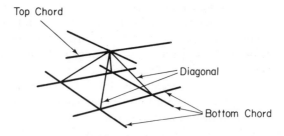

FIG. 2-6 Pyramidal truss.

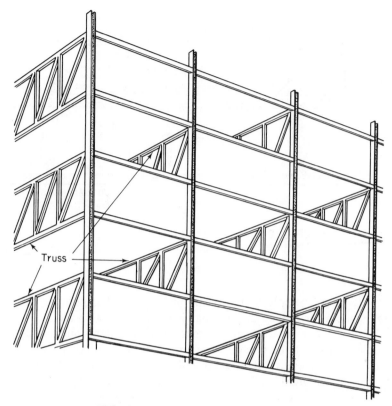

FIG. 2-7 Staggered truss system.

The *interstitial truss system* of framing is popular in modern hospital structures. It provides economically framed and isolated areas for hospital utilities (Fig. 2-8). Obviously, the fabrication and erection of trusses are more costly, but the spanning capabilities and savings of material can be substantial.

Rigid frames, which are fabricated from rolled shapes or built-up shapes, are a popular way of framing a building. In most cases they require more material than truss-type framing. Fabrication and erection costs are usually less than the truss-type framing. The high headroom achieved with this type of construction makes it very desirable and competitive. The system is merely made up from a horizontal or sloped beam supported on two columns, as shown in Fig. 2-9. Rigid frames will be discussed in more detail in Chapter 13. Rigid frames are frequently seen in the more modern industrial buildings, tennis court enclosures, swimming pool enclosures, and so on. Economical span lengths with this type

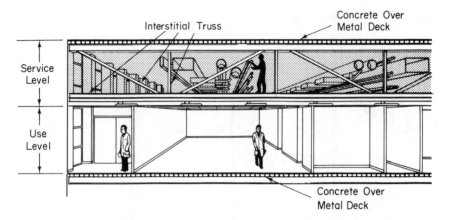

FIG. 2-8 Interstitial truss system.

of framing range from 30 to 100 ft and above. Another popular truss on column system for industrial buildings is shown in Fig. 2-10.

Space frames are also seen framed in steel systems such as domes, folded plates, hyperbolic paraboloids, and cable suspended and supported systems. These systems of framing will not be discussed in any detail. Some typical schematics of these systems are shown in Fig. 2-11.

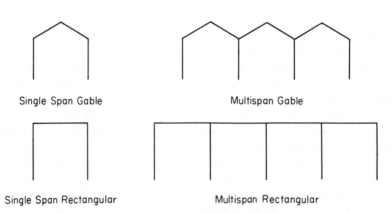

FIG. 2-9 Rigid frames.

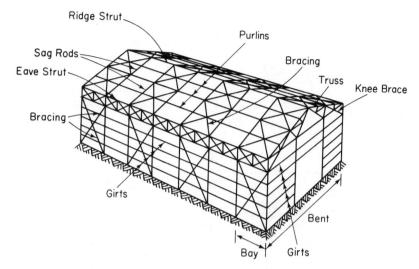

FIG. 2-10 Typical industrial-building framing truss on column system.

The discussion of the structural elements of the framing system and then the systems of framing, together with system capabilities and economies, provides the designer with a basic knowledge of the "guts" of the system and gives an idea of how to investigate which system is best for the structure under study.

To present a clear picture of the typical framing of a multilevel building, Fig. 2-12 is presented for clarification. It shows the horizontal framing system that supports the floors and roof, framed into the support vertical columns. The solid web W shapes can also be open web joists. Details of the typical open web joist-girder-column floor construction are shown in Fig. 2-13.

Up to this point we have only discussed beamed members and have said little about columns. Columns may have many cross-sectional shapes. Some of the more common are shown in Fig. 2-14. Column sections tend to be square in cross section (i.e., nearly equal in depth and width). The design and details of columns will be discussed in Chapters 5 and 9. Bracing members that provide resistance to lateral loads and/or stability to any structure will be discussed in Chapter 8.

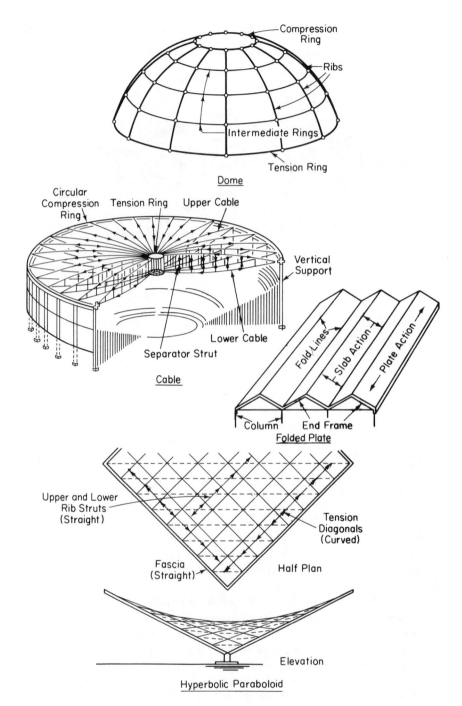

FIG. 2-11 Space structures.

86

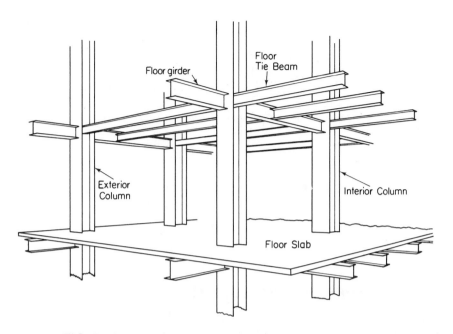

FIG. 2-12 Typical multistory-building framing system.

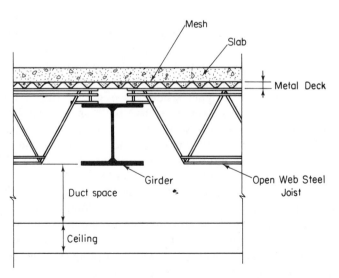

FIG. 2-13 Typical open web steel joist floor system.

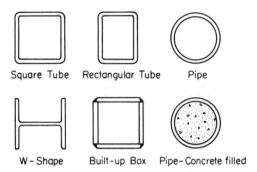

FIG. 2-14 Columns.

2-4   TYPES OF CONNECTORS

Up to this point we have discussed individual elements of the structural frame but have said little of how they are joined together. There are basically three types of connectors in use today: the rivet, bolt, and weld. Field riveting, for the most part, is no longer in use, and riveting in the shop is performed by only a few fabricators throughout the country. Only some bridge elements are shop-riveted today. Bolting and welding are the connectors in most popular use today. Sometimes the two are combined.

Bolting has, for all purposes, replaced riveting as a mechanical connector. There are two types of bolts in use today, the "common" or machine bolt (ASTM A307) and the high-strength bolt (ASTM A325 and A490). Both are headed fasteners with a threaded body or shank, utilizing washers and nuts. The common bolt is by far the cheaper of the two and should be used by the designer whenever possible. Installation procedures are simpler and less expensive than for high-strength bolts. A325 and A490 high-strength bolts have the advantage that their nut will never work its way off the threaded shank of the bolt during the life of the structure. The chemical and physical properties of the A307, A325, and A490 bolts are found in their respective ASTM Specification. The AISC Manual includes Specification for Structural Joints Using ASTM A325 or A490 Bolts as approved by the Research Council on Riveted and Bolted Structural Joints of the Engineering Foundation and as endorsed by the AISC and Industrial Fasteners Institute. These specifications cover the hardware requirements, allowable working stresses, installation procedures and methods, as well as inspection procedures.

The progress that has been made in the process of *welding*, the development of welding equipment, electrodes, and methods, and the science of weld design and detailing have certainly promoted this method of connecting steel to steel to the popular rank. All metals are weldable provided that the proper welding procedures are employed. Welding can be as popular and economical as any mechanical means of connecting. The American Welding Society (AWS) has long been considered the authoritative body in the field of welding. AWS D1.1 is the nationally accepted specification that covers all the facets of welding. It is referred to in the AISC Specification, as well as in most other specifications and codes.

To use welding or bolting, or a combination of both, is a design decision. An excellent discussion of the factors to be weighed may be found in an article by Henry J. Stetina in the November 1963 issue of *Civil Engineering*. Reprints of this article are available through the AISC.

As a general rule, the option to weld or bolt or to use a combination of both should be left to the decision and choice of the fabricator and/or erector of the steel (under the guidance of the designer, of course).

The particular economies that can be derived from each varies with the geographic area in which the project can be built (does the available labor in the area work well with bolting or welding and is the labor available?), with which method the fabricator is used to working with and can perform more economically, and with the compatibility of the connection and frame with the means of connecting.

Exotic methods of connecting are currently under development but are not yet practical nor economical. Epoxies and adhesive plastics may very well play a future role along with bolts and welds. A more thorough discussion relative to connector design and detail will be given in Chapter 11.

2-5 STRESSES, STRAINS, AND FACTORS OF SAFETY

Loads and forces have already been defined and discussed. When these external forces act on a structural element, they are resisted by internal forces called *stresses*. The accompanying dimensional changes of the structural elements are called *strains*. Internal force per unit of area defines unit stress. Strain per unit of length in the direction of dimensional change defines unit strain. In structural design or analysis, the designer is concerned with a point (or points) on the member that would cause damage to the member if the loading was further increased because of excessive stress or deformation (strain).

The design load must be controlled and so it is, by the use of *allowables:* allowable stresses, allowable deformations, and so on. Allowables are used in the design of structural elements through two major procedures that are now in common use: allowable stress design and plastic design. The procedures and limitations of both will be discussed in Chapter 6. The control of stress is a strength concept of design. In addition to this consideration, control of potential instability of the structure, which could prevent attainment of full strength, should be investigated. This concept is discussed throughout many of the succeeding chapters.

A working stress or an allowable stress is the maximum unit stress that is considered to be safe for the material involved when the material is resisting applied loads and forces. This allowable stress is less than the ultimate strength and the yield stress (in the case of allowable stress design) of the member and assures that the member will not rupture or collapse. The ratio of ultimate or yield stress to allowable stress is called the *factor of safety*. In allowable stress design, as well as in plastic design, the factor of safety is usually based upon yield stress. In plastic design, the factor of safety is applied in a different manner. The applied loads are multiplied by a constant called a *load factor* (equivalent to a factor of safety), and the yield stress is used as an allowable stress.

Section 2-7 discusses allowable stresses (as per the AISC Specification), which are expressed as a multiple of the yield stress (the two exceptions to this are for tension members where the allowable stress is also expressed as a multiple of the ultimate stress and for shear at beam end connections when the top flange is coped, where the allowable stress is expressed as a multiple of the ultimate stress). The ratio of yield stress to the allowable stress is the factor of safety for allowable stress design. In plastic design the factor of safety or load factor is the ratio of yield stress to allowable stress multiplied by the shape factor, f. The shape factor is equal to the ratio of the plastic modulus, Z, to the elastic section modulus, S. The shape factor is related to cross-sectional shape and varies with this. For W shapes bent about their strong axis, the shape factor is 1.12, which means that there is a 12% reserve between yield-point stress at the outermost fiber of the cross section and yield stress throughout the cross section (full plasticity). This will be discussed in more detail in Chapter 6.

2-6   GENERAL DESIGN CONSIDERATIONS

In addition to the primary structural design considerations of strength and stability, the structural designer must see the overall picture. He or she must consider the architectural planning and layout; the mechanical engineer's layout of heating, ventilating, and air conditioning (ductwork); the electrical engineer's layout of conduit and raceways (electrification requirements may determine the choice of structural member or type of metal floor deck); and the fireproofing requirements, if any (depends upon occupancy or use classification, height of structure, floor area, amount of combustibles, location of building, type of construction material, etc.). Two final points that the designer should consider are economy and adherence to the controlling building codes and specifications.

2-7 ALLOWABLE STRESSES

The AISC Specification for the Design, Fabrication and Erection of Structural Steel for Buildings either in whole or in whole with minor modifications is usually the governing specification for structural steel and is incorporated into any of the many local building codes. As already mentioned, the AISC Specification (Part 1, Allowable Stress Design) expresses all allowable stresses as a multiple of the yield stress F_y or the ultimate stress, F_u. F_y is the specified minimum yield stress and F_u is the specified minimum tensile strength (ultimate strength) of the type of steel used, expressed in kips per square inch (ksi). Yield stress as used in the Specification denotes either the specified minimum yield point (for those steels having a yield point) or the specified minimum yield strength (for those steels that do not have a yield point). Allowable stresses are expressed by capital letters, usually F, with an appropriate subscript. Calculated or actual stresses are expressed by lowercase letters.

Allowable stresses are given for the various types of loading: F_t for tension, F_v for shear, F_a for compression, F_{as} for compression of axially loaded bracing and secondary members, F_b for tension and compression on extreme fibers of bending members, F_p for bearing, and so on. Interaction formulas covering combined axial compression or axial tension and bending are presented. These formulas involve ratios of computed to allowable stresses, and involve f_a, F_a, f_b, F_b, and F'_e (Euler stress divided by a safety factor). More will be said of allowable stresses in the chapters that discuss beams, columns, bracing, and connections subject to various types of loading.

Part 2 of the AISC Specification on plastic design expresses permissible stresses in the form of allowable maximum moments, shears, and loads. Parts 1 and 2 specify various parameters which ensure that local and lateral buckling do not occur prior to full-member-strength development. These parameters will also be discussed in subsequent chapters.

2-8   AISC MANUAL OF STEEL
         CONSTRUCTION

The AISC Manual totals more than 800 pages. As much as it weighs and as thick as it is, it is worth its weight in gold, for it contains a great wealth of material. The manual is basically a handbook, with tables, charts, and various other design aids. Preceding each section or table in each of the six parts of the manual is a short amount of text material explaining the use of the information that follows.

Part 1 contains tables listing dimensions and properties for designing rolled shapes. These tables are in two parts, the first of use to detailers, and the second of use to engineers and designers. The first gives dimensions for detailing, and the second gives properties for designing: moment of inertia, section moduli, radii of gyration, plastic moduli, and other useful design parameters. Tables of properties of double angles and split tees are presented which give design parameters that assist the designer in determining allowable axial stress when the width-to-thickness ratio (a stability parameter discussed in the specification) exceeds the limiting value in the specification. Steel pipe and structural tubing tables are also present, which give all dimensions and design properties. The areas and weights of rectangular sections and of square and round bars are given in tabular form for the convenience of the designer. A summary of the provisions of ASTM A6, which covers standard mill practice, is presented also.

Most of the tables in *Part 2* cover beam and girder design. Allowable Stress Design and Plastic Design Selection Tables (facilitating the selection of structural beam members), Moment of Inertia Selection Tables, Allowable Loads on Beams Tables (for $F_y = 36$ and 50 ksi) for laterally supported beams, Allowable Moments in Beams Charts (for $F_y = 36$ and 50 ksi) for various laterally unbraced intervals, tables to aid the designer in the design of plate girders and composite design, and beam diagrams and formulas (loading and support conditions, shear, and moment) are given.

Part 3 of the manual presents tables ($F_y = 36$ and 50 ksi) giving allowable axial load in kips for various column shapes of different unsupported lengths.

Information on the use of bolts, rivets, and welds in several types of connections is given in *Part 4*. Tables covering framed, seated, end plate, eccentric connections suitable for steels with an F_y up to 100 ksi are presented. The tables of framed and seated connections cover all fasteners: A325 and A490 high-strength bolts and A307 common bolts, rivets, and welds. Combinations of shop-welded and field-bolted beam-to-column connection tables, a diagrammatic presentation of complete and partial penetration and fillet welded joints (accepted by AISC and AWS) are presented. Suggested details, various tables on fastener clearances, gages, and weights are also given in this part of the manual. Connections in tension and moment connections are discussed. Suggested details of beam framing, column base plates, column splices, and so on, are presented.

Part 5 of the manual is a collection of specifications and codes that relate directly to steel design, fabrication, and erection. The AISC Specification, AISC Code of Standard Practice, AISC Quality Certification Program, and Bolt Specifications are given.

Part 6 presents mathematical tables and frequently used data related to structural steel and structural steel products.

The manual has the basic key information needed to design in steel. It is an invaluable tool and an essential reference work for designers. Valuable design examples are found throughout, each of which illustrates the design of beams, columns, and connections. The reader is urged to study this valuable handbook. Most of the tables, charts, aids, and text will be discussed in detail in the chapters that follow.

PROBLEMS

(2-1) A one-story framed structure has a roof that is to be supported by steel beams, spaced 12 ft on centers, and spanning 28 ft. Roof loading consists of the following: (a) live load and snow load estimated at 30 psf, (b) dead load (including beams) estimated at 20 psf. Compute the following for a typical interior roof beam:

    (a)   Total distributed load on each, w.
    (b)   Total vertical load on each, W.
    (c)   Maximum end reaction.
    (d)   Maximum beam shear.
    (e)   Maximum bending moment.

(2-2) The figure below shows a typical framed steel floor system for a storage warehouse, incorporating beams framing into girders, which in turn frame into columns. Using *Recommended Live Loads for Storage Warehouses,* draw complete shear and bending moment diagrams for each of the beams and girders shown in the sketch and determine the load in column C. Assume that the warehouse will be used exclusively for automobile tires piled at a height of 8 ft. Assume the dead load of the steel beams to be 10 psf. The floor is 6 in. of concrete (assume 150 pcf).

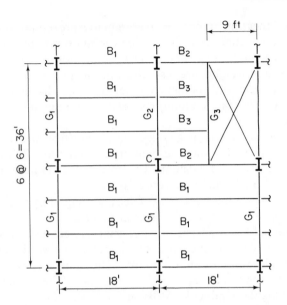

(2-3) Explain briefly why it is sometimes reasonable to assume a check-
        erboard loading on bays carrying live load, whereas snow loads
        may never be analyzed in this fashion.

(2-4) Differentiate between the term *factor of safety* as used in *allow-
        able stress design* and *plastic design*.

3

tension members

3-1 INTRODUCTION

Tensile structures have recently become of more interest to structural designers. Extremely light structures of a variety of aesthetically pleasing forms have been created. Cable structures have enabled the designer, with a minimal amount of material, to create cable-supported or -suspended and tentlike shapes.

Tension members' load-carrying capabilities are independent of the shape of the cross section. The design of this type of member is the simplest of all types of members and as such is suitable to discuss first. Primarily, strength considerations are considered, with stability a secondary consideration.

The discussions of this type of structural member will emphasize the use of hot-rolled shapes made from a single shape or built-up sections made from several shapes, as opposed to cables. The discussions will be further confined to tension in the form of a pull or longitudinal stress. Tension members are sometimes used as the bottom chord of a truss and are subject to a combination of axial tension and bending. This will be discussed in Chapter 9. This chapter will discuss direct tension action only.

3-2   TYPES

The most commonly used tension members were eyebars, pin-connected plates, rods, or cables. More recently, single or double angles, single or double channels, tees, W shapes, and rolled members built up from plates and shapes have been used. Pin-connected plates (plates of constant width), eyebars (plates with widened ends), and round bars with threaded ends (rods) are also in use today as tension members. Figure 3-1 shows several of the more common shapes used as tension members.

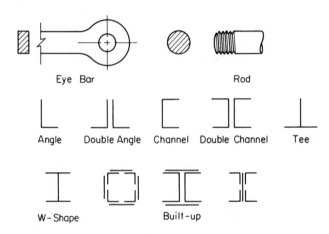

Eye Bar   Rod

Angle   Double Angle   Channel   Double Channel   Tee

W-Shape   Built-up

FIG. 3-1

3-3   NET SECTION / GROSS SECTION / EFFECTIVE NET SECTION

Allowable stresses of tension members will be discussed in the next section. Related to stress will be area, which will require a clear definition of *net area and effective net area of the section* and *gross area of the section*. When the end connections of a tension member are welded, the entire area or gross cross-sectional area of the member can be used. However, when the end connections of the tension member are riveted or bolted, and holes through the section for the fasteners are required, the area lost by the holes must be deducted from the gross area. This remaining area or net area must then be used for strength computations.

In computing the net area of a section, A_{net}, the holes to be deducted from the gross area shall be taken as $\frac{1}{16}$ in. greater than the nominal dimension of the hole normal to the direction of applied stress (AISC Specification, Section 1.14.4). Standard holes for fasteners are usually $\frac{1}{16}$ in. larger than the fastener nominal diameter to facilitate installation (AISC Specification, Section 1.23.4). Holes are punched when the thickness of the material is not greater than the nominal fastener diameter plus $\frac{1}{8}$ in. When the material thickness is greater than this, the holes are either drilled or subpunched and reamed. Subpunched holes are made with dies at least $\frac{1}{16}$ in. smaller than the nominal fastener diameter. For oversized and slotted holes, reference should be made to Section 11-11.1.

The gross section of a member is the summation of the products of the gross width and thickness of each element comprising the cross section. The net section is determined in the same manner, with the exception of substituting the net width for the gross width (AISC Specification, Section 1.14.1). Figure 3-2a illustrates the method of computing the gross area of a single angle and Fig. 3-2b shows the method of computing the

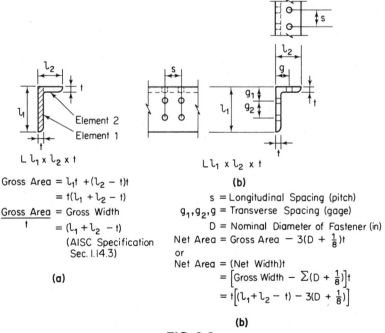

$L\,l_1 \times l_2 \times t$

Gross Area $= l_1 t + (l_2 - t)t$

$\qquad = t(l_1 + l_2 - t)$

$\underline{\text{Gross Area}} = $ Gross Width

$\quad t$

$\qquad = (l_1 + l_2 - t)$

(AISC Specification
Sec. 1.14.3)

(a)

$L\,l_1 \times l_2 \times t$

(b)

s = Longitudinal Spacing (pitch)

g_1, g_2, g = Transverse Spacing (gage)

D = Nominal Diameter of Fastener (in)

Net Area = Gross Area $- 3(D + \frac{1}{8})t$

or

Net Area = (Net Width)t

$\qquad = \left[\text{Gross Width} - \Sigma(D + \frac{1}{8}) \right] t$

$\qquad = t\left[(l_1 + l_2 - t) - 3(D + \frac{1}{8}) \right]$

(b)

FIG. 3-2

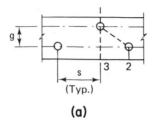

(a)

Net Section of Tension Members

Curves are values of stagger, s, in inches

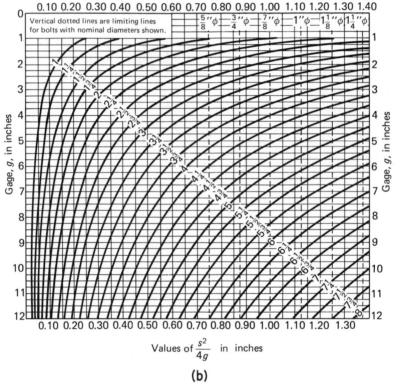

Values of $\dfrac{s^2}{4g}$ in inches

(b)

FIG. 3-3 *Courtesy of the American Institute of Steel Construction*

net area of the same angle when the holes are not staggered in longitudinal spacing (holes are on a line normal to the axis of the angle).

In the case where the holes are staggered, as shown in Fig. 3-3a, the failure plane may occur from 1 to 2 rather than from 1 to 3. The reduction in area along path 1 to 3 is one hole, whereas the reduction

The above chart will simplify the application of the rule for net width, Sections 1.14.2 and 1.14.3 of the AISC Specification. Entering the chart at left or right with the gage g and proceeding horizontally to intersection with the curve for the pitch s, thence vertically to top or bottom, the value of $s^2/4g$ may be read directly.

Step 1 of the example below illustrates the application of the rule and the use of the chart. Step 2 illustrates the application of the 85% of gross area limitation. The restriction that the net area shall be less than 85% of the gross area is limited to relatively short fitting (splice plates, gusset plates, beam-to-column fittings, and so on).

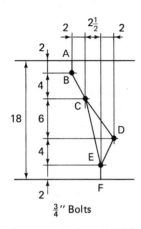

Step 1: Chain A B C E F

Deduct for 3 holes @ $(\frac{3}{4} + \frac{1}{8})$	$= -2.625$
BC, $g = 4$, $s = 2$; add $s^2/4g$	$= +0.25$
CE, $g = 10$, $s = 2\frac{1}{2}$; add $s^2/4g$	$= +0.16$
Total deduction	$= -2.215''$

Chain A B C D E F

Deduct for 4 holes @ $(\frac{3}{4} + \frac{1}{8})$	$= -3.50$
BC, as above, add	$= +0.25$
CD, $g = 6$, $s = 4\frac{1}{2}$; add $s^2/4g$	$= +0.85$
DE, $g = 4$, $s = 2$; add $s^2/4g$	$= +0.25$
Total deduction	$= -2.15''$

Net width = $18.0 - 2.215 = 15.785''$.

Step 2: Net width = 18.0 x 0.85 = 15.3''
(Governs in this example)

In comparing the path CDE with the path CE, it is seen that if the sum of the two values of $s^2/4g$ for CD and DE exceed the single value of $s^2/4g$ for CE, by more than the deduction for one hole, then the path CDE is not critical as compared with CE.

Evidently if the value of $s^2/4g$ for one leg CD of the path CDE is greater than the deduction for one hole, the path CDE cannot be critical as compared with CE. The vertical dotted lines in the chart serve to indicate, for the respective bolt diameters noted at the top thereof, that any value of $s^2/4g$ to the right of such line is derived from a non-critical chain which need not be further considered.

(c)

FIG. 3-3 *(Continued)*

along path 1 to 2 is something more than one hole. It is rather obvious that the area to be deducted relates to the staggered pitch, s, and the gage, g.

The AISC Specification, Section 1.14.2, gives a method by which, for a chain of holes extending across a part in any diagonal or zigzag line, the net width of the part can be computed. The net width is the sum of the diameters of all holes in the chain subtracted from the gross width and for each gage space in the chain, the quantity $s^2/4g$ shall be added. The controlling net area is arrived at by following the path of least net width. The Specification further states that the net section shall not exceed 85% of the corresponding gross section. Section 1.14.2 critera are illustrated in Fig. 3-3b, which is a design aid, making it easier to obtain

$s^2/4g$, and Fig. 3-3c, which illustrates the use of the aid and the application of the 85% rule.

The effective net area, A_e, of an axially loaded tension member where the load is transmitted by bolts or rivets through some but not all of the cross-sectional elements of the member, shall be determined as follows:

For W, and M, shapes with flange widths $b_f \geq 2d/3$, and where the connection is to the flanges with at least three fasteners per line in the direction of stress,

$$A_e = 0.90A_{net}$$

For all the shapes not meeting the preceding conditions and structural tees cut from these shapes and all other shapes (including built-up cross sections), where the connection has at least three fasteners in line in the direction of stress,

$$A_e = 0.85A_{net}$$

For all members whose connections have only two fasteners per line in the direction of stress,

$$A_e = 0.75A_{net}$$

When all the parts of the tension member profile are fully connected, the transfer of stress is uniform. For tension members other than flat plates or bars, when failure by fracture happens within the net area, the ratio of failure load to net area is generally less than the tensile strength of the steel when all the aforementioned parts are not fully connected. This is caused by a concentration of shear stress, referred to as *shear lag*, in the vicinity of the connection. This can be illustrated for an angle that is connected by one leg only. An increase in $\bar{x}$, the distance from the angle profile centroid to the shear plane of the connection, results in shear lag. As the length of connection, l, increases, the amount of shear lag is decreased. This situation can be expressed as $C_t \approx 1 - \bar{x}/l$. The values of C_t given as the decimal multiplier of A_{net} (a reduction of coefficient) in the aforementioned effective net area formulas are reasonably lower bounds for the cases discussed.

Pin-Connected Members, Section 1.14.5 of the AISC Specification, covers the design of eyebars and pin-connected members. This section discusses the required geometry of both types. Net sections through the pinhole are also discussed in this section of the Specification.

The critical area of *threaded parts* of steel tension members is the nominal area based on the major diameter (nominal size).

3-4 ALLOWABLE STRESSES

Section 1.5.1.1 of the AISC Specification states that the allowable tensile stress, F_t, on the gross section is equal to $0.60F_y$ (except for pin-connected members). It goes on to say that this value shall not exceed 0.5 times the minimum tensile strength, F_u, of the steel on the *effective net* area. The reasoning behind two allowable tensile stresses ($0.60F_y$ and $0.5F_u$) concerns the possibility of two modes of failure, yielding of the entire member and fracture at its weakest effective net area. Each allowable assures that this will not occur. When the ratio of effective net area to gross area $\geq F_y/0.833F_u$, yielding of the member as a failure mode controls, whereas when this ratio of areas $< F_y/0.833F_u$, fracture at the weakest net area as failure mode controls.

The ratio of the minimum yield stress, F_y, to the allowable tensile stress, F_t, is 1.667, which may be termed the *safety factor*. A further requirement applicable to the *effective* net section of axially loaded members stipulates that a safety-factor value of 2.0 applies relative to the ratio of minimum tensile strength to allowable tensile stress.

The allowable tensile stress on the *net* section at pin-connected members is $F_t = 0.45F_y$, which is based upon research and experience.

For allowable tension on threaded parts, a value of $0.33F_u$ applied to a tensile area is applicable (Section 3-3). Table 1.5.2.1 of the AISC Specification gives allowable tensile-stress values for rivets and various types of bolts. The latter values are applied to the nominal bolt area. F_u is the specified minimum tensile strength of the threaded part.

The AISC Specification, Section 1.15.3, allows the eccentricity between the gravity axis and the gage line of single and double angles and similar members to be considered small enough to neglect. The actual line of action of the axial force would be on the gage line, whereas the assumed line of action of the axial force would be on the center-of-gravity axis. For cases where the bending caused by eccentricity should be accounted for, the AISC Specification, Section 1.6.2, suggests the use of an interaction formula. The interaction formula, which accounts for axial tensile stress as well as bending tensile stress, is related to the discussion of combined axial compressive stress and bending compressive stress in Chapter 9.

The actual tensile stress is computed using the relationship

$$f_t = \frac{P}{A} \qquad (3\text{-}1)$$

where f_t = computed axial stress, ksi
$\quad\;\; P$ = applied axial tensile load, kips
$\quad\;\; A$ = applicable cross-sectional area, in.²

Of course, f_t shall not exceed F_t and in design one is seeking a member size to resist an axial load and

$$A = \frac{P}{F_t} \qquad\qquad (3\text{-}2)$$

is usually used.

3-5   SLENDERNESS RATIO

The *slenderness ratio, l/r,* is defined as the ratio of the unbraced length (the distance measured along the member between points where it is braced against lateral movement) to the least radius of gyration. The radius of gyration is equal to

$$r = \left(\frac{I}{A}\right)^{1/2}$$

where I = moment of inertia of the cross section, in.^{4} (resistance to rotation)

A = cross-sectional area, in.^{2}

Gross values of the least radius of gyration and of member area are listed in the AISC Manual's Properties for Designing Tables, found in Part 1, and need not be computed for the common rolled shapes. Gross values are typically used to compute l/r values. The higher the value of the slenderness ratio, the more limber (flexible or less stiff) the member.

The AISC Specification, Section 1.8.4, suggests a maximum slenderness ratio of tension members other than rods of l/r. These suggested ratios are not a requirement for structural strength of the tension member but are a recommendation to prevent undesirable vibration or flutter (lateral movement) of the member. They are suggested limitations rather than a mandatory requirement. The limiting ratios are main members, 240; bracing and other secondary members, 300.

The AISC recommendations of maximum slenderness ratios involve some terms not yet explained. They are main members and bracing and other secondary members. *Main members* are structural members whose primary purpose is to support loads imposed upon the structure. They are essential to the structure. *Secondary or bracing members,* as the name indicates, are members that serve to brace against lateral movement due to lateral loads (wind, earthquake, etc.). They may brace an entire structure or a frame or another member. As is apparent, the limiting slenderness ratios are more liberal for secondary members.

3-6   DESIGN

Each of the illustrative examples that follows illustrates the design procedures, the concerned sections of the AISC Specification, and the use
of the various sections of the AISC Manual of Steel Construction.

EXAMPLE 3-1

(a)   Determine the capacity in tension of an A36 1-in.-
diameter threaded rod.

(b)   What should be its maximum length when used as a
main tension member?

(c)   Answer part (b) for a bracing member.

(d)   For the given nominal size (basic major diameter),
how many threads per inch should be specified?

(e)   In part (d), which thread series should be specified?

(f)   In part (d), which thread class symbol should be specified?

solution:

(a)   For a 1-in.-diameter A36 threaded part, the allowable
tension on the tensile stress area is 15.03 kips. An
allowable tensile stress of $F_t = 0.33_u$, or 19.14 ksi, is
used. ($F_u = 58.0$ ksi for A36 material.) Tension stress
area = 0.7854 in.^{2}.

(b)   AISC Specification, Section 1.8.4:

For main tension members: $l/r = 240$
For bracing or secondary tension members: $l/r = 300$

For a 1-in.-diameter rod, $A = 0.7854$ in.^{2} (AISC Manual,
Part 1, Square and Round Bars, Weight and Area Tables). Radius of gyration for a circle: $r = d/4$ in. =
$\frac{1}{4}$ in. (AISC Manual, Part 6, Properties of Geometric
Sections). $l/r = 240$; $l = 240r/12 = 5$ ft.

(c) $l/r = 300$ and $l = 300r/12 = 6.25$ ft.

(d) $n = 8$ threads per inch for a basic major diameter, D
(AISC Manual, Part 4, Screw Threads, Unified Stan-
dard Series, UNC and 4 UN).

(e)   For D values of $\frac{1}{4}$ to 4 in. inclusive, the thread series
is UNC (coarse). For D values of $4\frac{1}{4}$-in. diameter and

larger, the thread series is 4 UN. Therefore, use UNC [same reference as in part (d)].

(f) Thread class symbol: 2A denotes class 2A fit applicable to external threads. 2B denotes corresponding class 2B fit for internal threads. Use class 2A [same reference as in part (d)]° *Use* 1–8UNC 2A, which means: 1 in. nominal size (basic major diameter); 8 threads per inch (*n*); UNC (coarse) thread series symbol: class 2A applicable to external threads, thread class symbol.

EXAMPLE 3-2

Determine the allowable tensile load that a $6 \times 3\frac{1}{2} \times \frac{3}{8}$ single angle can accommodate. Assume A36 steel and $\frac{3}{4}$-in.-diameter bolts. Use the AISC Specification.

solution:

(a) To minimize fabrication costs, use standard gage distances for the size angle given (Fig. 3-4)(AISC Manual, Part 4, Usual Gage for Angles, Inches).

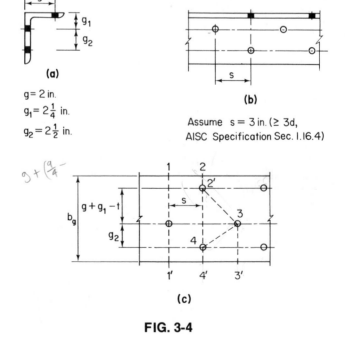

(a)

$g = 2$ in.
$g_1 = 2\frac{1}{4}$ in.
$g_2 = 2\frac{1}{2}$ in.

(b)

Assume s = 3 in. (≥ 3d,
AISC Specification Sec. 1.16.4)

(c)

FIG. 3-4

(b) From Section 1.14.3, AISC Specification: b_g = gross width of angle = $6 + 3\frac{1}{2} - \frac{3}{8} = 9\frac{1}{8}$ in.

(c) b_n = net width (Fig. 3-4c).

Possibilities:

Path	No. Holes Deducted
1–1'	1
2–2'–4–4'	2
2–2'–3–4–4'	3
2–2'–3–3'	2

Applying Section 1.14.2 of the AISC Specification:

Path 1–1': $b_n = b_g -$ (no. holes)(hole diameter) $+ \Sigma s^2/4g$

$b_n = 9.125 - (2)(0.75 + 0.125) = 8.250$ in.

Path 2–2'–4–4': $b_n = 9.125 - (2)(0.75 + 0.125) = 7.375$ in.

Path 2–2'–3–4–4': $b_n = 9.125 - (3)(0.75 + 0.125)$

$$+ \left(\frac{9}{15.5} + \frac{9}{10}\right) = 9.125 - 2.625 + 0.581$$

$$+ 0.9 = 7.981 \text{ in.}$$

Path 2–2'–3–3': not critical, since path 2–2'–4–4' has an equal number of holes to deduct but a smaller path length

Therefore, the critical net width is path 2–2'–4–4' or $b_n = 7.375$ in.

(d) Allowable tensile load, $P = F_t A$. $P = (0.60F_y)(b_g)t = 22.0(9.125)(0.375) = 75.3$ (governs) kips. $P = 0.5F_u b_n t = 29.0(7.375)0.375 = 80.2$ kips. (*Note:* Any of the following design aids from Part 4 of the AISC Manual could have been used in Example 3-2: Tension Members, Net Areas of Two Angles Tables; Reduction of Area for Bolt and Rivet Holes Table; Net Section of Tension Members Chart.)

EXAMPLE 3-3

Design an eyebar as a tension member in accordance with the AISC Specification. Use A441 steel and assume a tensile load of 100 kips.

solution:

Eyebars are now flame-cut from plate material. Thus, our eyebar will be flame-cut. In accordance with Section 1.14.5

of the AISC Specification: The net section of the head, A_h, through the pinhole, transverse to the axis of the eyebar, shall not be less than 1.33 or more than 1.50 times the cross-sectional area of the body of the eyebar, A_b, or

$$1.33A_b \leq A_h \leq 1.50A_b\,;$$
$$D_p = \text{diameter of pin} \geq 7w/8;$$
$$D_{ph} = \text{diameter of pinhole} \leq D_p + \tfrac{1}{32}\text{ in.};$$
$$A_b = P/F_t = 100/0.60\,F_y \text{ (Eq. 3-2).}$$

Additionally, Fig. 3-5 shows other provisions of Section 1.14.5 of the AISC Specification.

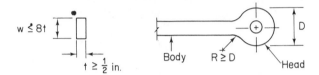

FIG. 3-5

Assume that $F_y = 50$ ksi and $t = \tfrac{3}{4}$ in. $> \tfrac{1}{2}$ in.

$$A_b = 100/0.60(50) = 3.33 \text{ in.}^2 = wt$$
$$w = 3.33/0.75 = 4.44 \text{ in.} < 8t$$

Use an eyebar body: bar $4\tfrac{1}{2} \times \tfrac{3}{4}$ ($A_b = 3.375$ in.²).
   $D_p \geq 7w/8 = 7(4.5)/8 = 3.94$ in. Diameter of pin, $D_p =$ 4 in.
   On the net section at the pinhole: $F_t = 0.45F_y = 0.45(50) = 22.5$ ksi (AISC Specification, Section 1.5.1, Appendix A).
   Required net width: $w_h = P/F_t t = 100/22.5(0.75) = 5.926$ in. *Use* 6.0 in.
   Therefore, $D = D_{ph} + w_h = 4\tfrac{1}{32} + 6 = 10\tfrac{1}{32}$ in. $D_{ph} = D_p + \tfrac{1}{32} = 4\tfrac{1}{32}$ in.
   Check: $w_h = D - D_{ph} = 10\tfrac{1}{32} - 4\tfrac{1}{32} = 6 = 6.0$ in. and actual net area at head $= w_h t = 6.0(0.75) = 4.5$ in.² > 4.44 in.² required. $1.33A_b = 4.489$ in.² $\leq (A_h = 4.5$ in.²$) \leq 1.50A_b = 5.063$ in.².
   Use $R = D = 10\tfrac{1}{32}$ in.

Answer:

   All the provisions in the AISC Specification were arrived at through long experience and extensive testing.

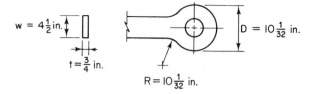

$w = 4\frac{1}{2}$ in. $t = \frac{3}{4}$ in. $D = 10\frac{1}{32}$ in. $R = 10\frac{1}{32}$ in.

EXAMPLE 3-4

The bottom chord of a Pratt roof truss is a tee section. Design the tension diagonal using the AISC Specification. Use a grade of steel such that $F_y = 50$ ksi. Find the size of the tension diagonal subjected to a tensile load of 90 kips: **(a)** for a single angle, and **(b)** for double angles. Use welded connections and assume the stem of the tee chord to be $\frac{3}{8}$ in. thick and the length of a diagonal to be 15 ft.

solution (Fig. 3-6):

(a) Single angle: Required $A = P/F_t = 90/0.60(50) = 3.00$ in.². Since welded connections have been specified, gross areas are used. Maximum slenderness ratio (main member) $= l/r = 240$ and $r \geq 12(15)/240 = 0.75$ in. From the AISC Manual, Part 1, Properties for Designing, Unequal and Equal Angles, $r \geq 0.75$ in. and $A \geq 3.00$ in.²:

Size	A	$r_{min}(r_{zz})$	lb/ft
L5 × 5 × $\frac{5}{16}$	3.03	0.994	10.3
L6 × 4 × $\frac{5}{16}$	3.03	0.882	10.3
L5 × 3½ × $\frac{3}{8}$	3.05	0.762	10.4

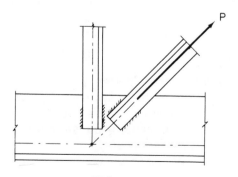

FIG. 3-6

Use L5 × 5 × $\frac{5}{16}$. This choice was made primarily because it is stiffest (largest r_{min}), has the least weight (10.3 lb/ft), and provides ample area (3.03 in.^{2}).

(b) Double angles (back to back, $\frac{3}{8}$ in. = back-to-back dimension): The requirements are the same: $r \geq 0.75$ in. and $A \geq 3.00$ in.^{2}. From the AISC Manual, Part 1, Properties of Sections, Double Angles, Equal and Unequal Angles:

Size	A	r_{min}	lb/ft
(1) 2L $3\frac{1}{2}$ × $3\frac{1}{2}$ × $\frac{1}{4}$	3.38	1.09	11.6
(2) 2L $3\frac{1}{2}$ × 3 × $\frac{1}{4}$ ⊓⌐	3.13	0.914	10.8
(3) 2L $3\frac{1}{2}$ × 3 × $\frac{1}{4}$ ⊓⌐	3.13	1.11	10.8
(4) 2L $2\frac{1}{2}$ × 2 × $\frac{3}{8}$ ⊓⌐	3.09	0.768	10.6

Cases (2) and (3) are identical angles with the same area ($A = 3.13$ in.^{2}), weighing the same amount (10.8 lb/ft). However, case (3) gives a stiffer result ($r_{min} = 1.11$ in.) than case (2) ($r_{min} = 0.914$ in.). Case (2) has the longer ($3\frac{1}{2}$ in.) legs outstanding, while case (3) has the shorter legs (3 in.) outstanding, which accounts for the difference in the r_{min} values. Case (4) meets all the requirements and weighs a little less (10.6 lb/ft). It also is a bit more limber (less stiff) than the other cases ($r_{min} = 0.768$ in. for the 2-in. leg outstanding). *Use* 2L $3\frac{1}{2}$ × 3 × $\frac{1}{4}$ (10.8 lb/ft)(⊓⌐).

Referring to Fig. 3-6, and selecting the most economical section (L 5 × 5 × $\frac{5}{16}$), the slight eccentricity between the gravity axis of the angle member and the center of gravity of its connection welds is ignored. This is in accordance with Section 1.15.3 of the AISC Specification. The use of the equation $A = P/F_t$ assumes that the load is actually applied through the centroid of the cross section. The same assumption is valid for riveted or bolted connections and for double angles as well.

Some engineers prefer to use double angles for main members to create symmetry about the stem of the tee-chord member. In this case, a choice of 2L $3\frac{1}{2}$ × 3 × $\frac{1}{4}$ would be made, because of the stiffness that would be gained. Many designers limit their choice to the use of single angles for bracing or secondary members.

EXAMPLE 3-5

Given 2C 15 × 33.9 in tension with a top and bottom tie plate to ensure that the channels behave as a unit and maintain their spacing as shown in Fig. 3-7, find values for each of the letters found in Fig. 3-7. Use the AISC Specification and Manual.

solution:

$g = 2$ in. (AISC Manual, Part 1, American Standard Channels, Dimensions for Detailing). $d = 6 + 2g = 10$ in. From Section 1.18.3.2, AISC Specification:

$l \geq 2d/3$ in.; *use* 7 in.
$t \geq d/50$ in.; *use* $\frac{1}{4}$ in.
$s \leq 6$ in.; *use* 3 in.
$c \leq 240r$ in. $= 240(0.904) = 216.96$ in.; *use* 18 ft. 0 in.

Since the flange width of a C 15 × 33.9 is 3.400 in. (AISC Manual, Part 1, Channels American Standard, Properties for Designing), $a = 3.400 - g = 1.40$ in., which should be equal to or more than the minimum edge distance, as per Section 1.16.5 of the AISC Specification. Therefore,

$b = d + 2a = 10 + 2.80 = 12.80$ in.

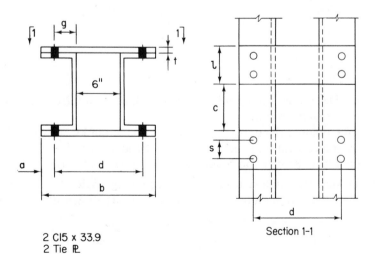

2 C15 x 33.9
2 Tie ℞

Section 1-1

FIG. 3-7

PROBLEMS

(3-1) Compute the critical net width for each of the following sections:
(a) Using the chart in Fig. 3-3 where applicable.
(b) Using the appropriate AISC section directly.

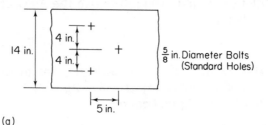

(a)

(b)

(c)

(3-2) (a) Determine the permissible tensile load on a fully threaded $1\frac{1}{8}$ in. diameter rod of A36 steel.
(b) Compute the permissible lengths (maximum) if the rod will be used as a main member, and as a secondary member.
(c) Fully identify the required thread series and class.

(3-3) Design the lightest back-to-back equal leg angles to be used as diagonal web members in a welded truss as shown in the figure.

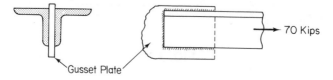

Gusset Plate   70 Kips

Use A36 steel, and assume a total tensile load of 70 kips will be transferred to a $\frac{7}{8}$ in. thick gusset plate. Assume a diagonal web length of 18 ft.

(3-4) Four back-to-back angles having plates alternately bolted on along the *x-x* and *y-y* axes are to carry 270 kips in total tension. Using 4 in. × 4 in. angles and $\frac{1}{4}$ in. tie plates, complete the design by fully specifying the angles, tie plate frequency and length, etc. Assume $F_y = 50$ ksi steel ($F_u = 70$ ksi) and $\frac{5}{8}$ in. diameter bolts.

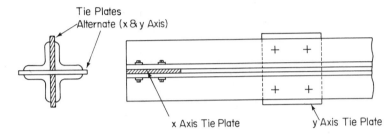

(3-5) The roof truss for a building is fabricated of A36 steel angles ($F_u = 58$ ksi). The connections consist of a single line of $\frac{3}{4}$ in. diameter bolts (standard holes). Design the main tension member (lightest section) with a load of 160 kips (length = 10 ft).
(a) As a single equal-leg angle (one leg of angle connected to gusset with 13 bolts).
(b) As double equal-leg angles (one leg of each angle connected to gusset with 7 bolts).

(3-6) Select the lightest single angle to carry a tensile load of 70 kips. Assume a single line of four $\frac{7}{8}$ in. diameter bolts (standard holes). Length is 12 ft. Only one leg is connected to gusset plate. $F_y = 50$ ksi and $F_u = 70$ ksi.

(3-7) What is the allowable tensile load for a pair of L 6 × 6 × $\frac{5}{8}$ of A36 steel ($F_u = 58$ ksi). The member is connected (using standard gage) as shown with $\frac{7}{8}$ in. diameter bolts (standard holes). If the member is 30 ft long, do the angles need to be connected along their length? If so, where? (Only bracing connected to outstanding legs.)

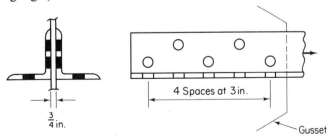

(3-8) Select a W12 tension member to carry a load of 300 kips. It is connected to two adequate $\frac{1}{2}$ in. gusset plates as shown.
(a) With an all-welded connection.
(b) Connected by two lines of $\frac{3}{4}$ in. bolts in each flange (standard holes). (More than two bolts per line.)

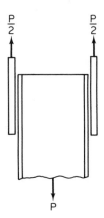

(3-9) What is the allowable tensile load for the connection shown? A36 steel ($F_u = 58$ ksi). Adequate number of 1 in. diameter bolts (standard holes).

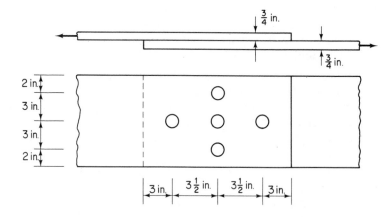

(3-10) What is the allowable tensile load on the connection shown? A36 steel ($F_u = 58$ ksi). Adequate $\frac{7}{8}$ in. diameter bolts (standard holes).

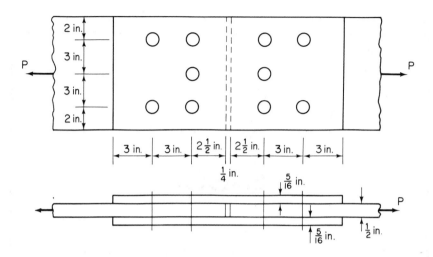

(3-11)   A 20 ft long member is composed of two L 6 × 4 × $\frac{5}{8}$ as shown.
         With 1 in. diameter bolts (standard holes), what is the allowable
         tensile load? Do the angles need to be connected intermittently?
         If so, indicate the spacing of connections. Use standard gages.

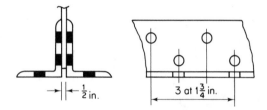

4

shear

4-1   INTRODUCTION

In structural steel design the term *shear* may mean several things. In statics, "shear" means the value of the algebraic sum of vertical loads at any point along a beam span. This is usually expressed in terms of a shear diagram, which is arrived at by knowing the geometry and the loading configuration of the member. Various beam diagrams and their corresponding formulas for the vertical shear, V, at any point along the span of the beam, are given in Part 2 of the AISC Manual. In the laboratory, the word "shear" is used to describe a type of failure in a specimen.

In composite design, which will be discussed in greater detail in Chapter 10, "shear" refers to the horizontal shear, V_h, which is to be resisted by mechanical shear connectors under composite action (e.g., steel beam and concrete slab resisting loads as a single unit).

In the design of bending members, the term "shear center" is used to describe a point through which a load must pass so that no twisting occurs. The point is referred to the cross section of a structural element. The design of bending members is covered in Chapter 6 and shear center will be discussed there.

The internal shear stress present due to transverse loads on bending members is VQ/It, and the shear force per unit length or *shear flow* in a cross section is VQ/I. In the expressions, I is the moment of inertia about the neutral axis of bending calculated for the same loading that causes the vertical shear force, V (in.[4]); Q is the statical moment (first moment) of the cross-sectional area (beyond the point where the shear stress is calculated) taken about the neutral axis (in.[3]); t is the thickness of the element on which shear stress is calculated (in.). The shear flow acts parallel to the sides of the element.

Figure 4-1a illustrates the shear stresses in a typical channel cross section. Figure 4-1b shows the shear forces created by a vertical load through the channel's shear center. From the vertical shear, V, we can compute shear stress, $f_v = VQ/It$, and we can get the horizontal shear forces, S, and the vertical shear force, V, by multiplying f_v by the area. The S forces cause a twisting couple, Sd, that is overcome if the applied force, V, acts at a distance, e, as shown in Fig. 4-1b.

Summarizing, the formula used to compute shear stress introduced by transverse forces is

$$f_v = \frac{VQ}{It} \tag{4-1}$$

Shearing stresses of equal intensity act both vertically and horizontally at any one point in a beam.

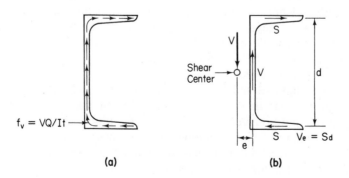

(a)   (b)

FIG. 4-1

4-2   GROSS SECTION

In computing shear stresses, the gross section is used. For the web of a
W shape, the gross section may be taken as the product of the overall
depth of section, d, and the thickness of web, t_w (Section 1.5.1.2, AISC
Specification). No connector hole reduction is taken into account.

4-3   ALLOWABLE STRESS

Our discussions from this point on in this chapter will be confined to
shear and shear stress caused by transverse loads in beams or bending
members. Vertical shears caused by vertical loads, as discussed in Section 4-1, are resisted by internal shear stresses, which are expressed by
Eq. 4-1.

The maximum value of shear stress is at the neutral axis of the
cross section. This will be illustrated in Section 4-4, where we will also
see that for W shapes the shear stress in the flange is small and the
variation of the shear stress throughout the web depth is slight. Therefore, the average shear stress in the web of a W shape can be approximated by

$$f_v = \frac{V}{A_w} = \frac{V}{t_w d} \qquad\qquad (4\text{-}2)$$

where d is the full depth of the section and t_w is the thickness of the web.
The allowable shear stress on the gross section is

$$F_v = 0.40\, F_y \qquad\qquad (4\text{-}3)$$

where the gross section can, as discussed before, be taken as dt_w.

The webs of rolled shapes are of substantial thicknesses such that
shear stress usually is not the controlling design parameter. When two or
more members whose webs lie in a common plane are rigidly connected,
a very high web shear stress usually occurs. Such webs often require
reinforcing and should be investigated by the method suggested in the
Commentary to the AISC Specification (Section 1.5.1.2) or any other
rational design approach.

High-strength bolted beam end connections with coped webs on
which high bearing stresses act may cause a portion of the web to tear
out along the end connection hole perimeter (block shear failure). This
can be prevented by using an allowable shear stress, $F_v = 0.30 F_u$, where
F_u is the specified minimum tensile strength of the steel. F_v is applied to
the minimum net failure surface bounded by the bolt holes.

The same approach may be applied to similar situations, where
failure may occur by shear along a plane through the fasteners or by a

combination of shear along a plane through the fasteners plus tension along a perpendicular plane. Section 1.5.1.2 of the AISC Specification discusses the aforementioned in detail.

4-4   DESIGN EXAMPLES

The design examples selected illustrate the text discussion and also serve to show design procedures more clearly.

EXAMPLE 4-1

For each of the sections shown in Fig. 4-2, find the horizontal shear acting on the bottom horizontal surface of the shaded element. Assume a vertical shear force, V, to act equal to 200 kips and 20 kips for sections **(a)** and **(b)**, respectively. All dimensions are in inches.

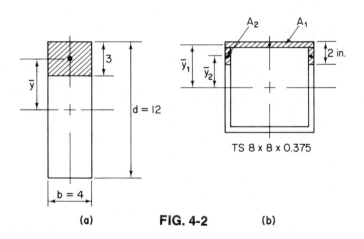

(a)   **FIG. 4-2**   (b)

solution:

(a)

$$V = 200 \text{ kips}$$
$$\bar{y} = 6 - 1.5 = 4.5 \text{ in.}$$
$$A = 3(4) = 12.0 \text{ in.}$$
$$Q = A\bar{y} = 12(4.5) = 54.0 \text{ in.}^3$$
$$I = \frac{bd^3}{12} = \frac{4(12)^3}{12} = 576 \text{ in.}^4$$
$$t = b = 4 \text{ in.}$$
$$f_v = \frac{VQ}{It} = \frac{200(54)}{576(4)} = 4.69 \text{ ksi}$$

(b) TS $8 \times 8 \times .375$:

$I = 102$ in.^{4} (AISC Manual, Part 1, Structural Tubing,
    Square, Dimensions and Properties)
$V = 20$ kips
$\bar{y}_1 = 4.0 - \dfrac{0.375}{2} = 3.812$ in.
$A_1 = 8(0.375) = 3.000$ in.^{2}
$\bar{y}_2 = 4.0 - 0.375 - (2.0 - 0.375(\frac{1}{2}) = 2.813$ in.^{4}
$2A_2 = 2(2.0 - 0.375)(0.375) = 0.609$ in.^{2}
$Q = A_1\bar{y}_1 + 2A_2\bar{y}_2 = 3.0(3.812) + 0.609(2.8125)$
$\quad = 11.436 + 1.713 = 13.15$ in.^{3}
$t = 2(0.375) = 0.750$ in.

At each web, 2 in. down from the top of the section:

$$f_v = \frac{VQ}{It} = \frac{20(13.15)}{102(0.750)} = 3.44 \text{ ksi}$$

EXAMPLE 4-2

For the section in Fig. 4-2a, determine the shearing stress
at horizontal planes located **(a)** 4.5 in. and **(b)** 6 in. from
the top of the section. Plot, to a suitable scale, the shear-
stress distribution from the top to the bottom fiber of the
cross section.

solution:

(a) At 4.5 in.:

$V = 200$ kips
$\bar{y} = 3.75$ in.
$A = 4.5(4) = 18.0$ in.^{2}
$Q = A\bar{y} = 18(3.75) = 67.5$ in.^{3}
$I = 576$ in.^{4}
$t = 4$ in.
$f_v = \dfrac{200(67.5)}{576(4)} = 5.86$ ksi

(b) At 6 in.:

$V = 200$ kips
$\bar{y} = 3.0$ in.
$A = 6.0(4) = 24.0$ in.^{2}

$$Q = A\bar{y} = 72.0 \text{ in.}^3$$
$$I = 576 \text{ in.}^4$$
$$t = 4 \text{ in.}$$
$$f_v = \frac{200(72)}{576(4)} = 6.25 \text{ ksi}$$

The plot of the shear-stress distribution is shown in Fig. 4-3 to vary parabolically across the cross section. Its maximum value is at the neutral axis, varying to zero at the uppermost and lowermost fibers of the cross section. The maximum shear stress is 50% greater than the average shear stress, $v/bd = 200/4(12) = 4.17$ ksi.

EXAMPLE 4-3

Assume a W 12 × 65 beam and a vertical shear force V. Compute (**a**) the maximum shear stress, (**b**) the shear stress just above the bottom of the top flange, (**c**) the shear stress just below the bottom of the top flange, and (**d**) the (average) shear stress in the web. Plot the profile of the shear-stress distribution throughout the cross section.

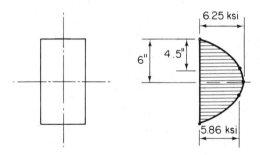

FIG. 4-3

solution:

W 12 × 65:

$$b_f = 12.000 \text{ in.}$$
$$I_{xx} = 533 \text{ in.}^4$$
$$t_f = 0.605 \text{ in.}$$
$$d = 12.12 \text{ in.}$$
$$t_w = 0.390 \text{ in.}$$

(a) At the neutral axis (maximum shear stress):

$V = V$ kips

$A_f = b_f t_f = 12.000(0.605) = 7.260$ in.²

$\bar{y}_f = \dfrac{d}{2} - \dfrac{t_f}{2} = \dfrac{12.12}{2} - \dfrac{0.605}{2} = 5.758$ in.

$\dfrac{A_w}{2} = \left(\dfrac{d}{2} - t_f\right) t_w = \left(\dfrac{12.12}{2} - 0.605\right)(0.390)$

$= 2.127$ in.²

$\dfrac{\bar{y}_w}{2} = \left(\dfrac{d}{2} - t_f\right)\left(\dfrac{1}{2}\right) = \left(\dfrac{12.12}{2} - 0.605\right)(0.5)$

$= 2.727$ in.

$t = t_w = 0.390$ in.

$Q = A_f \bar{y}_f + \dfrac{A_w}{2}\dfrac{\bar{y}_w}{2} = 7.260(5.758) + (2.127)2.727$

$= 47.603$ in.³

$f_v = \dfrac{V(47.603)}{533(0.390)} = 0.229V$ ksi

(b) Above the bottom of the top flange:

$V = V$ kips

$A_f = 7.260$ in.²

$\bar{y}_f = 5.758$ in.

$t = b_f = 12.00$ in.

$Q = A_f y_f = 7.260(5.758) = 41.803$ in.³

$f_v = \dfrac{V(41.803)}{533(12.000)} = 0.00654V$ ksi

(c) Below the bottom of the top flange:

$V = V$ kips

$A_f = 7.260$ in.²

$\bar{y}_f = 5.758$ in.

$t = t_w = 0.390$ in.

$Q = A_f \bar{y}_f = 7.260(5.758) = 41.803$ in.³

$f_v = \dfrac{V(41.803)}{533(0.390)} = 0.201V$ ksi

(d) Average shear stress in web:

$f_v = \dfrac{V}{dt_w} = \dfrac{V}{12.12(0.390)} = 0.212V$ ksi

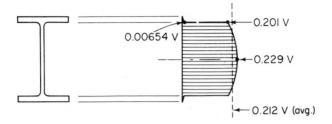

FIG. 4-4

This proves the validity of using the average shear stress (Eq. 4-2), in that the maximum computed shear stress at the neutral axis is only slightly larger than the average shear stress. This holds true for W and S shapes, the most common of the structural shapes.

EXAMPLE 4-4

If V is 60 kips in Example 4-3, is the section adequate for shear stress? Use the AISC Specification's allowable shear stress, F_v and $F_y = 50$ ksi.

solution:

If $V = 60$ kips, then $f_v = 0.212V = 0.212(60) = 12.72$ ksi and $F_v = 0.40F_y = 0.40(50) = 20.0$ ksi. Since $f_v < F_v$, use W 12 × 65.

PROBLEMS

(4-1) Compute and plot graphically the theoretical shear stress variation at 1-in. increments from top to bottom on the steel plate shown based on a total shear of 240 kips. Compare this variation in stress with the average value obtained by using the simple P/A formula.

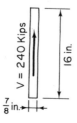

(4-2)   Compute and plot graphically the shear stress variation on a W 16
        × 100 carrying a total shear of 240 kips. Use approximately 1-in.
        increments, and be sure to include the special points at the junction
        between the flange and web and at the centroid. Compare this
        stress variation with the value obtained when the formula $f_v = V/dt_w$ is used.

(4-3)   A simple span beam (W 24 × 84) has a maximum end shear of 100
        kips. Determine the average shearing stress (using the full depth
        of the section) and the maximum theoretical shear stress in the
        beam.

(4-4)   A steel section W 18 × 119 is to be strengthened by the addition
        of two flange cover plates of $\frac{7}{8}$ in. × 10 in., one on the top and the
        other on the bottom flange. After computing the revised moment
        of inertia for this section, compute the maximum shear flow that
        must be resisted between the section and each cover plate due to
        a total external shear of 175 kips.

(4-5)   Explain fully in your own words the following statement: "Short
        beams tend to be *shear* critical."

(4-6)   Compute the vertical shear capacity of a channel C 12 × 30 of 50
        ksi steel. Compare your computed answer with that obtained from
        the steel handbook chart entitled Allowable Loads on Beams.

(4-7)   For the span and loading shown, determine the average shearing
        stress in a W 33 × 221 at the point of maximum shear.

5

axially
loaded
columns

5-1　INTRODUCTION

Any member subjected to compressive load alone, or compressive load in combination with another load, may be defined as a *compression member*. A compression member whose length is substantially larger than its cross-sectional dimensions may be classified as a *column*. To go one step further, a column subjected to a centrally or axially applied load is classified as an *axially loaded column*. The discussions in this chapter will be confined to the latter type of member loaded solely with axial compressive load. Although most columns are used as vertical members, many are used in other positions. Examples of these structural elements include compressive members in trusses (chord or diagonal elements) and bracing members.

5-2   IDEAL AND REAL COLUMNS

Columns that are perfectly straight, loaded exactly through their centroid, free of any residual stress (this will be discussed in Section 5-2.2), and manufactured from a perfectly isotropic (the same properties in all directions) material are termed *ideal columns*. Such columns do not exist. However, ideal-column theory contributes greatly to our knowledge of column behavior, as we will see.

Because a steel column is a manufactured product, it contains certain flaws. Residual stresses due to cooling after rolling, straightening, and welding are always present. Initial crookedness is always present. Perfectly straight columns do not exist and initial imperfections are to be expected. Further, "axial loads" are rarely axial. Accidental eccentricity is inevitable. Finally, no construction material is perfectly isotropic. This type of column is termed a *real column*. They truly exist.

5-2.1   Buckling or Euler Load

Ideal columns buckle when centrally loaded. At a particular load these columns suddenly bend or bow (buckle). In this case, the *buckling load* is the maximum load that the column can carry. This maximum load is given by a formula derived by Leonhard Euler, a Swiss mathematician. Euler's equation for an ideal, pinned-end, centrally loaded column is

$$P_e = \frac{\pi^2 EI}{l^2} \tag{5-1}$$

where

P_e = Euler or buckling load, kips
E = modulus of elasticity, ksi
I = moment of inertia of the cross section about the bending axis, in.[4]
l = length of the column between pinned ends, in.

Euler was the first to realize that the useful strength of a column member is not solely dependent on strength capacity but may be dependent on stability. Euler's assumptions were based on elastic ideal columns.

Since the Euler load expressed by Eq. 5-1 is a stability load, it is not related to any strength of steel. It is directly related to the slenderness ratio, l/r (Section 3-5), or member stiffness, as can be seen by the following mathematical manipulation of the Euler load formula:

$$P_e = \frac{\pi^2 EI}{l^2}$$

Dividing each side by A, the cross-sectional area

$$\frac{P_e}{A} = \frac{\pi^2 EI}{l^2 A}$$

Since $r^2 = I/A$,

$$\frac{P_e}{A} = \frac{\pi^2 E r^2}{l^2}$$

or

$$\frac{P_e}{A} = \frac{\pi^2 E}{(l/r)^2}$$

and

$$F_e = \frac{\pi^2 E}{(l/r)^2} \tag{5-2}$$

where F_e is the Euler stress in kips per square inch and l/r is the slenderness ratio. Equation 5-2 is convenient for design and is valid for all types of cross sections and all grades of steel.

5-2.2   Residual Stresses

As mentioned before, real columns contain residual stresses that are due to cooling after rolling, straightening, and welding. These stresses vary considerably depending on temperature, rate of cooling, and section geometry. The maximum cooling residual stresses in most rolled sections is 10 to 15 ksi (compression) at the flange tips. The portion that is the last to cool after rolling or fabrication has tensile residual stresses (usually near the web-flange junction), which are balanced internally by compressive stresses elsewhere. Residual stresses in W shapes affect weak axis (y-y) strength more than strong axis (x-x) strength. The magnitude and distribution of residual stresses can affect column strength. The affect of residual stresses on column strength is small for steels of higher yield strength. The cold strengthening of columns produces residual stresses of similar magnitude (although with different distribution) and may be treated similarly.

5-2.3   Initial Crookedness and Eccentrically
Applied Load

All columns have some initial crookedness or out-of-straightness. The AISC Code of Standard Practice, Section 6.4.3, limits the initial deviation from straightness between points that are laterally supported to $l/1000$, where l is the distance between the laterally supported points.

Under real load conditions, this initial deviation from straightness is amplified by applied axial load. Initial crookedness does have an effect on column strength. For example, with an $l/r = 100$ for an A36 H shape with a maximum initial crookedness of $l/100$, yielding of the section would start at an average stress 20% lower than the buckling stress. Initial crookedness produces a reduction in strength relative to ideal columns in a manner similar to that caused by residual stresses, although the reduction due to the latter is more significant for W shapes. The effect of initial crookedness is greater for steels with higher yield strengths.

Accidental eccentricity of applied load affects column capacity in a manner similar to initial crookedness.

5-3   DESIGN FORMULAS

Medium column lengths (l/r from 50 to 150) reflect the effect of residual stress and initial crookedness. The AISC formula for allowable axial compressive stress on the gross section of axially loaded compression members takes into account this greater variation in the strength of a medium column slenderness ratio. Residual stresses, as well as initial crookedness, are considered in AISC design formula 1.5-1, which is used when Kl/r, the largest effective slenderness ratio, is less than C_c:

$$F_a = \frac{[1 - (Kl/r)^2/2C_c^2]F_y}{\frac{5}{3} + 3(Kl/r)/8C_c - (Kl/r)^3/8C_c^3} \qquad (5\text{-}3)$$

where $C_c = (2\pi^2 E/F_y)^{1/2}$ and the denominator represents a safety factor. When kl/r exceeds C_c,

$$F_a = \frac{12\,\pi^2 E}{23(Kl/r)^2} \qquad (5\text{-}4)$$

Equation 5-4 corresponds to AISC design formula 1.5-2, which is the Euler stress (Eq. 5-2) divided by a safety factor. Both formulas are based upon column strength estimates suggested by the Column Research Council (CRC). This estimate assumes that the upper limit of elastic buckling failure occurs at an average column stress of $F_y/2$.

Equation 5-3 is a function of F_y; Eq. 5-4 is independent of F_y. The latter formula is the Euler formula (Eq. 5-2) with an applied factor of safety of 1.92, or $\frac{23}{12}$, and is used in the elastic range. The AISC formulas, Eqs. 5-3 and 5-4, can be plotted as shown in Fig. 5-1.

Residual stresses and initial crookedness are considered in design by a parabolic curve between l/r values of zero to C_c, corresponding to

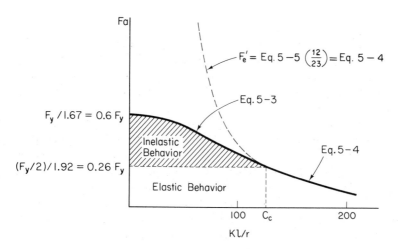

FIG. 5-1

values of F_a of $0.6F_y$ to $0.26F_y$, respectively. C_c defines the dividing point between elastic and inelastic buckling. C_c is derived from the Euler formula as follows:

$$F_e = \frac{\pi^2 E}{(Kl/r)^2} \qquad (5\text{-}5)$$

(Equation 5-5 is Eq. 5-2 with a factor K, which is defined next.) Let $C_c = Kl/r$ and assume that $F_e = F_y/2$. Solving for C_c,

$$C_c = \left(\frac{2\pi^2 E}{F_y} \right)^{1/2}$$

The term K is an effective length factor and is used in the Euler formula, Eq. 5-2, to correct the term l, length, to account for variable end-support conditions of the "Euler column" (the Euler column stress formula was for a pinned-end member that is free to rotate). Effective length factors and effective lengths will be discussed in more detail in Section 5-5.

A variable factor of safety has been applied to Eqs. 5-3 and 5-4. For short columns this factor is 1.67 (Eq. 5-3 at $Kl/r = 0$). Such members are relatively insensitive to accidental eccentricities due to applied load. For longer columns (eq. 5-4) this factor is increased to 1.92 (at C_c and above). Figure 5-1 shows how a smooth transition is provided between the two formulas. From Kl/r valued from zero to C_c, the factor of safety varies according to the denominator of Eq. 5-3.

The allowable compressive stress for axially loaded bracing and secondary members, F_{as}, when l/r exceeds 120 ($K = 1.0$ assumed), is covered by formula 1.5-3 of the AISC Specification:

$$F_{as} = \frac{F_a}{1.6 - (l/200r)} \qquad (5\text{-}6)$$

where F_a is the value from Eq. 5-3 or 5-4.

For a discussion of the AISC Specification width-to-thickness ratio limitations as applied to compression elements subject to axial compression, reference is made to Section 6-8 and Section 1.9 of the AISC Specification.

5-3.1   Allowable-Stress Table/ Allowable Load Table

Equation 5-3, in particular, appears to be difficult to handle arithmetically speaking. The AISC Manual, Appendix A, provides tables that give allowable compressive stresses, F_a, for various Kl/r values for 36 and 50 ksi F_y values. The tables cover main and secondary members for Kl/r values less than or equal to 120, main members for Kl/r values from 121 to 200, and secondary members for l/r values of 121 to 200 (in this case, K is assumed to be unity for all secondary members).

F_{as} is more liberal than F_a and is justified by the lesser importance and the greater effective end restraint of these members. Equation 5-6 does take into account end restraint, and a value of $K = 1.0$ is assumed.

The table for allowable stress (ksi) for compression members of F_y = 36 ksi found in the AISC Manual is reproduced in Fig. 5-2. The AISC column allowable-stress formulas are applicable for all cross-sectional shapes. C_c values for the various F_y values are as listed in Table 5-1. These values are also given in tabular form in the AISC Manual, Appendix A to the AISC Specification.

Table 5-1

F_y (ksi)	C_c
36	126.1
42	116.7
46	111.6
50	107.0
60	97.7
65	93.8
90	79.8
100	75.7

ALLOWABLE STRESS (KSI) FOR COMPRESSION MEMBERS OF 36 KSI SPECIFIED YIELD STRESS

Main and Secondary Members Kl/r not over 120						Main Members Kl/r 121 to 200				Secondary Members* l/r 121 to 200			
$\frac{Kl}{r}$	F_a (ksi)	$\frac{Kl}{r}$	F_a (ksi)	$\frac{Kl}{r}$	F_a (ksi)	$\frac{Kl}{r}$	F_a (ksi)	$\frac{Kl}{r}$	F_a (ksi)	$\frac{l}{r}$	F_{as} (ksi)	$\frac{l}{r}$	F_{as} (ksi)
1	21.56	41	19.11	81	15.24	121	10.14	161	5.76	121	10.19	161	7.25
2	21.52	42	19.03	82	15.13	122	9.99	162	5.69	122	10.09	162	7.20
3	21.48	43	18.95	83	15.02	123	9.85	163	5.62	123	10.00	163	7.16
4	21.44	44	18.86	84	14.90	124	9.70	164	5.55	124	9.90	164	7.12
5	21.39	45	18.78	85	14.79	125	9.55	165	5.49	125	9.80	165	7.08
6	21.35	46	18.70	86	14.67	126	9.41	166	5.42	126	9.70	166	7.04
7	21.30	47	18.61	87	14.56	127	9.26	167	5.35	127	9.59	167	7.00
8	21.25	48	18.53	88	14.44	128	9.11	168	5.29	128	9.49	168	6.96
9	21.21	49	18.44	89	14.32	129	8.97	169	5.23	129	9.40	169	6.93
10	21.16	50	18.35	90	14.20	130	8.84	170	5.17	130	9.30	170	6.89
11	21.10	51	18.26	91	14.09	131	8.70	171	5.11	131	9.21	171	6.85
12	21.05	52	18.17	92	13.97	132	8.57	172	5.05	132	9.12	172	6.82
13	21.00	53	18.08	93	13.84	133	8.44	173	4.99	133	9.03	173	6.79
14	20.95	54	17.99	94	13.72	134	8.32	174	4.93	134	8.94	174	6.76
15	20.89	55	17.90	95	13.60	135	8.19	175	4.88	135	8.86	175	6.73
16	20.83	56	17.81	96	13.48	136	8.07	176	4.82	136	8.78	176	6.70
17	20.78	57	17.71	97	13.35	137	7.96	177	4.77	137	8.70	177	6.67
18	20.72	58	17.62	98	13.23	138	7.84	178	4.71	138	8.62	178	6.64
19	20.66	59	17.53	99	13.10	139	7.73	179	4.66	139	8.54	179	6.61
20	20.60	60	17.43	100	12.98	140	7.62	180	4.61	140	8.47	180	6.58
21	20.54	61	17.33	101	12.85	141	7.51	181	4.56	141	8.39	181	6.56
22	20.48	62	17.24	102	12.72	142	7.41	182	4.51	142	8.32	182	6.53
23	20.41	63	17.14	103	12.59	143	7.30	183	4.46	143	8.25	183	6.51
24	20.35	64	17.04	104	12.47	144	7.20	184	4.41	144	8.18	184	6.49
25	20.28	65	16.94	105	12.33	145	7.10	185	4.36	145	8.12	185	6.46
26	20.22	66	16.84	106	12.20	146	7.01	186	4.32	146	8.05	186	6.44
27	20.15	67	16.74	107	12.07	147	6.91	187	4.27	147	7.99	187	6.42
28	20.08	68	16.64	108	11.94	148	6.82	188	4.23	148	7.93	188	6.40
29	20.01	69	16.53	109	11.81	149	6.73	189	4.18	149	7.87	189	6.38
30	19.94	70	16.43	110	11.67	150	6.64	190	4.14	150	7.81	190	6.36
31	19.87	71	16.33	111	11.54	151	6.55	191	4.09	151	7.75	191	6.35
32	19.80	72	16.22	112	11.40	152	6.46	192	4.05	152	7.69	192	6.33
33	19.73	73	16.12	113	11.26	153	6.38	193	4.01	153	7.64	193	6.31
34	19.65	74	16.01	114	11.13	154	6.30	194	3.97	154	7.59	194	6.30
35	19.58	75	15.90	115	10.99	155	6.22	195	3.93	155	7.53	195	6.28
36	19.50	76	15.79	116	10.85	156	6.14	196	3.89	156	7.48	196	6.27
37	19.42	77	15.69	117	10.71	157	6.06	197	3.85	157	7.43	197	6.26
38	19.35	78	15.58	118	10.57	158	5.98	198	3.81	158	7.39	198	6.24
39	19.27	79	15.47	119	10.43	159	5.91	199	3.77	159	7.34	199	6.23
40	19.19	80	15.36	120	10.28	160	5.83	200	3.73	160	7.29	200	6.22

*K taken as 1.0 for secondary members.

Note: C_c = 126.1

FIG. 5-2 *Courtesy of American Institute of Steel Construction*

From Fig. 5-2, for A36 (F_y = 36 ksi) steel, when Kl/r > 126.1 = C_c, F_a values are determined from Eq. 5-4, which is independent of F_y. Examination of Fig. 5-2 and the other tables for other F_y values found in the AISC Manual discloses the fact that F_a is the same for all F_y values for Kl/r values equal to or larger than C_c = 126.1. To state this in more general terms, for a Kl/r > the C_c value corresponding to a particular F_y value, F_a is the same for all stronger steels for the same Kl/r value. To illustrate. for F_y = 42 ksi and C_c = 116.7:

Kl/r	F_a
116	11.10
117	10.91
118	10.72
119	10.55
120	10.37

and for F_y = 46, 50, 60, 65, 90, or 100 ksi, the values of F_a are identical for the corresponding Kl/r values shown.

Allowable Load Tables for various column W, M, and S shapes, pipe, structural tubing, structural tees, and double angles based on Eqs. 5-3 and 5-4 are found in the AISC Manual, Part 3. These tables enable the designer to find the allowable axial load in kips for a particular section with a particular KL in feet. Tables are presented for both F_y = 36 ksi and 50 ksi for W, M, and S shapes, for F_y = 36 ksi for round pipe and F_y = 36 ksi and 46 ksi for square and rectangular tubing. These tables present solutions to Eqs. 5-3 and 5-4 in terms of load for axially loaded columns.

5-4   END RESTRAINT

The Euler equation (Eq. 5-1 or 5-2) was based upon a member with pinned ends that are free to rotate. In real situations, some columns may have restrained or free ends, which have an important effect on the length term of the Euler formula. This effect is accounted for through the use of a multiplier applied to the length term, l (Eq. 5-5). The term K is the effective length factor. It will be discussed in Section 5-5.

The actual column behavior depends upon the restraining members at the ends, the members that frame into the column ends. Whether or not sidesway occurs is another factor that influences the column behavior. These factors affect column stability because column buckling is really column bending, and end moment, restraints, and rotation can significantly alter the column behavior. Sidesway is illustrated in Fig.

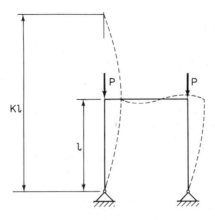

FIG. 5-3

5-3, in which is shown a typical structural frame that is pinned at the base. Frame stability is the ability of the frame to resist sidesway (uninhibited lateral movement of the column tops with respect to their bases) when a *vertical* load or loading is acting. The sidesway is shown dashed in the figure.

5-5 EFFECTIVE LENGTH CONCEPT: K FACTORS

A simple definition of the *effective length factor, K,* is the ratio of the length of an equivalent pinned-end column to the length of the actual column. For the frame shown in Fig. 5-3, this ratio is Kl/l. The equivalent pinned-end column is shown as Kl. Values of K for some ideal end conditions are shown in the Commentary to the AISC Specification (Fig. 5-6). K is not affected by moments applied to the column or by the absence or presence of applied lateral loads to the frame of which the column is a part.

The classical basic case is the concentrically loaded, pinned-end, vertical column of uniform cross section, the Euler column as shown in Fig. 5-4a. It can be seen that Kl is equal to l and therefore $K = 1.0$. The column ends are free to rotate but are restrained from translation.

Figure 5-4b shows a column with ends restrained against both rotation and translation—fixed ends, in other words. There are two points of inflection or contraflexure located at the quarter-points. These points may be thought of as hinges, where $Kl = \frac{l}{2}$ and the ratio Kl/l is $\left(\frac{l}{2}\right)/l$ or $K = 0.5$. Figure 5-4a and b shows the two limiting conditions for columns

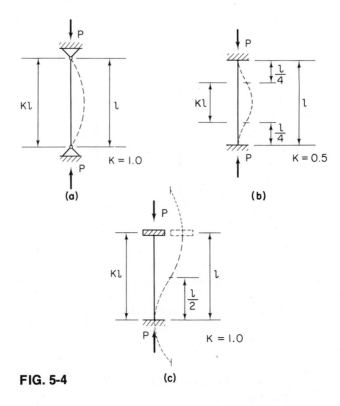

FIG. 5-4

with sidesway inhibited (no translation of column tops relative to their bases). For inhibited sidesway, $0.5 \leq K \leq 1.0$. Conservatively, $K = 1.0$ can be used for all columns where sidesway is inhibited. For tall, slender columns (large l/r ratios) with rigid beam-to-column connections, actual K values (rather than the value of unity) will reflect economy of design, whereas for the usual column proportions, a value of unity for K may be used with little loss of economy. The symbols used in Fig. 5-4 to represent the various column end conditions have the same meaning as those in the Commentary to the AISC Specification (Fig. 5-6). The dashed lines in Fig. 5-4 show the buckled or deflected structure. K values shown in the figure are the theoretical equivalent length factors.

Figure 5-4c represents a case where sidesway is uninhibited. The base of the column is fixed against both rotation and translation, while the top of the column is free to translate but is fixed against rotation. In this case, an inflection point is located at midheight of the column and $Kl = l$, so that $Kl/l = l/l$ or $K = 1.0$. Kl is composed of two halves of an equivalent pinned-end column, shown by the dots in the figure, which represent the extensions of each equivalent pinned-end column. This is the lower bound or envelope for the condition of uninhibited sidesway.

In real columns, full fixity against rotation is nonexistent. The tendency is to approach the hypothetical condition shown in Fig. 5-4c. A more realistic case would be one in which both columns' ends have some partial rotation where the base has fixed translation. For this case, K would always exceed a value of unity. In summary, for the case of uninhibited sidesway, theoretical values of K would be equal to or larger than 1.0.

5-5.1   Braced and Unbraced Frames

It is valuable to extend the theory of effective lengths and effective length factors to frames. A *frame* is defined as a structural member comprising one horizontal member and two vertical members (columns). For frames braced against sidesway, the column is always stronger than the classic pinned-end column of length l. For braced frames (as for columns with inhibited sidesway) it is safe to use the actual length instead of the effective length, and the AISC Specification recommends doing so. The opposite is true for unbraced frames (e.g., columns with uninhibited sidesway). It is not safe to use the actual length; the effective length should always be used. If the girders are very slender and the bases of the columns are pinned, the frame is very unstable. In unbraced frames, Kl may exceed two or more story heights. The stability of unbraced frames is dependent on both the column and the beam.

The range of K values for unbraced and braced frames is shown in Fig. 5-5. Both fixed and pinned column-base-support conditions are

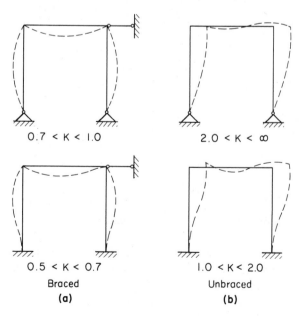

| 0.7 < K < 1.0 | 2.0 < K < ∞ |

| 0.5 < K < 0.7 | 1.0 < K < 2.0 |
| Braced | Unbraced |

FIG. 5-5   (a)   (b)

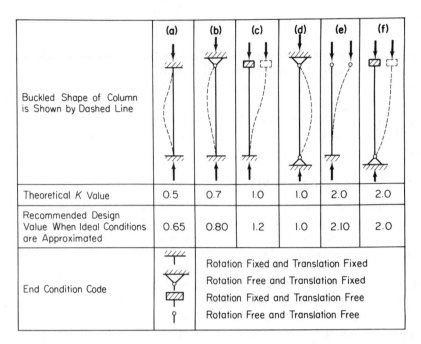

Buckled Shape of Column is Shown by Dashed Line	(a)	(b)	(c)	(d)	(e)	(f)
Theoretical K Value	0.5	0.7	1.0	1.0	2.0	2.0
Recommended Design Value When Ideal Conditions are Approximated	0.65	0.80	1.2	1.0	2.10	2.0
End Condition Code		Rotation Fixed and Translation Fixed				
		Rotation Free and Translation Fixed				
		Rotation Fixed and Translation Free				
		Rotation Free and Translation Free				

FIG. 5-6   *Courtesy of American Institute of Steel Construction*

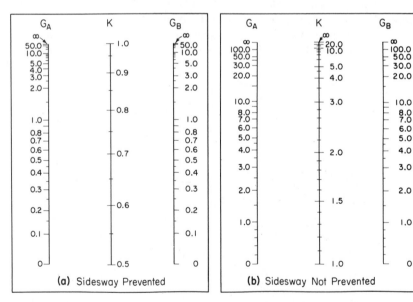

FIG. 5-7   *Courtesy of Jackson and Moreland, Division of United Engineers and Constructors, Inc.,* from *Structural Stability Research Council Guide*

134

shown for each type of frame. Various top-of-column conditions are also shown. Many of the individual columns correspond to the cases shown in Fig. 5-4 or Fig. 5-6.

5-5.2 Design Aids and Methods

Two design methods are suggested within the Commentary to the AISC Specification. Figure 5-6 is a reproduction of a table from the Commentary, which is used for determining the effective length factors for concentrically loaded columns with various idealized end conditions of restraint. Theoretical effective length factors and the recommended design values for these approximated ideal conditions are given. By means of Fig. 5-6 and by simple interpolation between the cases shown, a reasonable K value can be determined.

The other method utilizes an alignment chart in which relative stiffnesses, I/l, of a column and of the beams framing into it are computed. The alignment chart shown in the Commentary to the AISC Specification for the uninhibited sidesway condition for frames with rigidly connected vertical and horizontal members is shown in Fig. 5-7. Also shown in Fig. 5-7 is a similar alignment chart for inhibited sidesway, which can be found in the Structural Stability Research Council Guide to Stability Design Criteria for Metal Structures, Third Edition, as well as in the AISC Commentary.

To enter the charts, evaluation of the relative stiffness of the members of the frame at the ends of the column must be made. The stiffness ratio or relative stiffnesses of the members (beam and column), G, must be evaluated. This ratio is equal to the sum of the column stiffnesses at a joint, divided by the sum of the beam or girder stiffnesses at that joint. This computation is performed at each end of the column. The chart is entered with these two parameters (G at each end of the column), and a K value is then obtained from the chart.

For example, Fig. 5-8a and b illustrates how the G values are computed. All lengths are assumed to be unsupported lengths of column segments and restraining (beam or girder) members. I values are the moment-of-inertia values of the respective members and are taken about axes normal to the plane of buckling under consideration.

From Fig. 5-8a,

$$G_A = \frac{\Sigma(I_c/l_c)}{\Sigma(I_g/l_g)}$$

$$= \frac{(I_{c1}/l_{c1}) + (I_{c2}/l_{c2})}{(I_{g1}/l_{g1}) + (I_{g2}/l_{g2})}$$

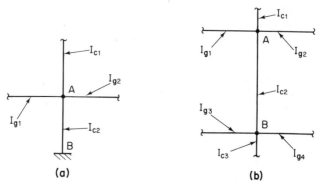

FIG. 5-8

G_B theoretically may be taken as infinity for column ends supported by, but not rigidly connected to, a footing or foundation (pin-connected). However, for practical purposes, $G_B = 10$ is used to account for actual conditions. If the end B were rigidly attached to a properly designed footing, a value of $G_b = 1.0$ is used. (Theoretically, if the base were truly fixed, $G_B = 0$.) In the case shown, a value of $G_B = 1.0$ should be used.

From Fig. 5-8b,

$$G_A = \frac{(I_{c1}/l_{c1}) + (I_{c2}/l_{c2})}{(I_{g1}/l_{g1}) + (I_{g2}/l_{g2})}$$
$$G_B = \frac{(I_{c2}/l_{c2}) + (I_{c3}/l_{c3})}{(I_{g3}/l_{g3}) + (I_{g4}/l_{g4})}$$

Notice that the stiffness ratio, G, is computed for the top and bottom of the column under consideration, the appropriate alignment chart entered with these values, and a K value obtained.

The alignment chart for the case of uninhibited sidesway yields values of K from 1.0 to infinity. For the inhibited sidesway condition, the respective alignment chart gives values of K from 0.5 to 1.0.

The alignment chart is not a direct design method. The member sizes must first be known or estimated. Trial girder sizes may easily be arrived at by assuming that the maximum moment will not be more than the fixed end moment due to uniform loading. This assumption covers a range of 50 to 100% end fixity.

For estimating column sizes, a trial K value must first be made, a column section chosen based upon this K value and various strength requirements, and finally a calculated K value based upon the trial column section arrived at. Figure 5-6 can be used to determine a trial K

value. The determination of the first trial K value for the first column is the most difficult task. After that, experience dictates educated guesses for subsequent values.

5-5.3 Alignment-Chart Assumptions

As discussed in Section 5-5.2, the alignment-chart (Fig. 5-7a) designation of end conditions used in column design to determine the effective length uses a relative stiffness ratio, G, for each end of the column under consideration:

$$G = \frac{\Sigma(I/l)_{\text{column}}}{\Sigma(I/l)_{\text{girder}}} \qquad (5\text{-}7a)$$

This alignment chart is developed on the basis of the following major assumptions:

(1) The columns and girders have the same modulus of elasticity, E. Although E is not in Eq. 5-7a, G should be $\Sigma(EI/l)$ of the column divided by $\Sigma(EI/l)$ of the girder and the E terms cancel. The chart is based upon *elastic* behavior.

(2) It is assumed that for unbraced frames (uninhibited sidesway) in the buckled shape, the girders assume a double (reversed) curvature so that the stiffness is $6EI/l$ (Fig. 5-9a). For braced frames (inhibited sidesway) in the buckled shape, it is assumed that the girder ends rotate as shown in Fig. 5-9b.

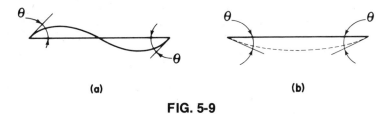

(a)   (b)

FIG. 5-9

(3) The frame's restraining members are in pure bending and are not subject to axial load.

(4) All columns in a story buckle at the same time under the loads applied to those individual columns. It would be incompatible to have only one column moving while all the others do not move.

(5) The total joint restraint (caused by the girders) is equally divided between the columns above and below the joints.

5-5.4   Modifications to the Alignment Chart*

The alignment chart for the uninhibited sidesway condition is a perfectly rational method for the determination of K values provided that each of the five conditions in Section 5-5.3 is adhered to. However, the designer may still use this alignment chart when any of these assumptions are not satisfied. Each of the assumptions will be discussed separately and in the same order as in Section 5-5.3. Each discussion will be followed by an illustrative example. Both discussion and example will show how the chart can effectively be used to determine valid K values under conditions where any or all of the aforementioned assumptions are violated.

(1)   Columns and girders do not have the same stiffness of material: this condition usually happens when the level of axial load on the column superimposed on residual stresses causes local yielding. The column behaves inelastically. Most practical columns fall in the inelastic range.

In the inelastic range, the column stiffness is reduced and an elastic beam framing into it offers more restraint to such a column when compared to an elastic column. This greater restraint can be accounted for in design by altering the G factor to be used in determining the effective length factor, K, from the alignment chart (Fig. 5-7a).

Equation 5-7a is now repeated with the moduli included:

$$G = \frac{\Sigma(EI/l)_{\text{column}}}{\Sigma(EI/l)_{\text{girder}}} \tag{5-7b}$$

Since $(EI)_{\text{column}}$ is reduced for inelastic behavior, a more favorable G value results. This inelastic effect can be approximated by changing G by multiplying by a stiffness reduction factor, F_a/F_e'. F_e' is the Euler stress (Eq. 5-2) divided by a safety factor and is found by use of Eq. 5-4. Notice that this is true for inelastic action and that for elastic action, $F_a = F_e'$, and this factor becomes unity. F_a/F_e' can be replaced by f_a/F_e' to make the solution easier. The term f_a is the computed axial compressive stress and is obtained by dividing the axial load by the area of the column section (P/A).

Disque has compiled a table for $F_y = 36$ and 50 ksi steel from which f_a/F_e' values can be obtained for various values of f_a (Fig. 5-10). G can now be expressed as

$$G_{\text{elastic}} = \frac{\Sigma(I/l)_{\text{column}}}{\Sigma(I/l)_{\text{girder}}} (F_a/F_e') \tag{5-8a}$$

---

*For details, see Joseph A. Yura, "The Effective Length of Columns in Unbraced Frames," *AISC Engineering Journal,* April 1971, and subsequent articles; and Robert O. Disque, "Inelastic K-Factor for Column Design," *AISC Engineering Journal,* April 1973, and Addendum, July 1973.

$$G_{\text{inelastic}} = \frac{\Sigma(I/l)_{\text{column}}}{\Sigma(I/l)_{\text{girder}}} (f_a/F'_e) \qquad (5\text{-}8b)$$

The following illustrative example serves the purpose of showing the procedure of using the alignment chart to determine the K values for both the elastic and inelastic column behavior assumptions, as well as other design aids found in the AISC Manual.

EXAMPLE 5-1

Design the column for the unbraced frame shown in Fig. 5-11 when subjected to an 800-kip axial load. Assume that sidesway is uninhibited about the strong axis of the column and that the column is braced about the minor axis at each floor. Assume A36 steel. Assume (a) elastic and (b)

Stiffness Reduction Factors f_a/F'_e (36 ksi steel)

f_a	$\dfrac{f_a}{F'_e}$	f_a	$\dfrac{f_a}{F'_e}$	f_a	$\dfrac{f_a}{F'_e}$	f_a	$\dfrac{f_a}{F'_e}$
20.5	0.064	17.7	0.387	14.9	0.704	12.1	0.924
20.4	0.073	17.6	0.399	14.8	0.714	12.0	0.929
20.3	0.083	17.5	0.412	14.7	0.724	11.9	0.935
20.2	0.093	17.4	0.424	14.6	0.734	11.8	0.940
20.1	0.103	17.3	0.436	14.5	0.743	11.7	0.944
20.0	0.114	17.2	0.448	14.4	0.752	11.6	0.949
19.9	0.125	17.1	0.460	14.3	0.762	11.5	0.954
19.8	0.136	17.0	0.472	14.2	0.772	11.4	0.958
19.7	0.147	16.9	0.484	14.1	0.781	11.3	0.902
19.6	0.158	16.8	0.496	14.0	0.789	11.2	0.966
19.5	0.169	16.7	0.507	13.9	0.797	11.1	0.970
19.4	0.180	16.6	0.519	13.8	0.805	11.0	0.973
19.3	0.192	16.5	0.530	13.7	0.813	10.9	0.976
19.2	0.204	16.4	0.543	13.6	0.822	10.8	0.979
19.1	0.216	16.3	0.554	13.5	0.830	10.7	0.982
19.0	0.228	16.2	0.565	13.4	0.838	10.6	0.985
18.9	0.240	16.1	0.577	13.3	0.845	10.5	0.987
18.8	0.252	16.0	0.588	13.2	0.853	10.4	0.989
18.7	0.265	15.9	0.599	13.1	0.860	10.3	0.991
18.6	0.277	15.8	0.610	13.0	0.867	10.2	0.993
18.5	0.289	15.7	0.622	12.9	0.874	10.1	0.995
18.4	0.301	15.6	0.633	12.8	0.881	10.0	0.996
18.3	0.313	15.5	0.643	12.7	0.888	9.9	0.997
18.2	0.325	15.4	0.654	12.6	0.894	9.8	0.998
18.1	0.337	15.3	0.664	12.5	0.901	9.7	0.999
18.0	0.350	15.2	0.674	12.4	0.907	9.6	0.999
17.9	0.363	15.1	0.685	12.3	0.912	9.5	0.999
17.8	0.375	15.0	0.695	12.2	0.918	9.4	1.000

FIG. 5-10

Stiffness Reduction Factors f_a/F_e' (50 ksi steel)

f_a	$\dfrac{f_a}{F_e'}$	f_a	$\dfrac{f_a}{F_e'}$	f_a	$\dfrac{f_a}{F_e'}$	f_a	$\dfrac{f_a}{F_e'}$
28.1	0.090	24.4	0.403	20.7	0.704	17.0	0.916
28.0	0.097	24.3	0.412	20.6	0.711	16.9	0.921
27.9	0.104	24.2	0.421	20.5	0.718	16.8	0.925
27.8	0.112	24.1	0.430	20.4	0.726	16.7	0.928
27.7	0.120	24.0	0.439	20.3	0.733	16.6	0.932
27.6	0.128	23.9	0.447	20.2	0.740	16.5	0.935
27.5	0.136	23.8	0.456	20.1	0.746	16.4	0.939
27.4	0.144	23.7	0.465	20.0	0.752	16.3	0.943
27.3	0.152	23.6	0.473	19.9	0.759	16.2	0.946
27.2	0.160	23.5	0.481	19.8	0.766	16.1	0.949
27.1	0.168	23.4	0.490	19.7	0.772	16.0	0.952
27.0	0.176	23.3	0.498	19.6	0.779	15.9	0.955
26.9	0.184	23.2	0.507	19.5	0.785	15.8	0.958
26.8	0.192	23.1	0.516	19.4	0.791	15.7	0.961
26.7	0.201	23.0	0.524	19.3	0.797	15.6	0.964
26.6	0.209	22.9	0.532	19.2	0.803	15.5	0.967
26.5	0.218	22.8	0.541	19.1	0.809	15.4	0.970
26.4	0.227	22.7	0.549	19.0	0.815	15.3	0.972
26.3	0.236	22.6	0.558	18.9	0.821	15.2	0.975
26.2	0.245	22.5	0.566	18.8	0.827	15.1	0.977
26.1	0.253	22.4	0.573	18.7	0.833	15.0	0.979
26.0	0.262	22.3	0.582	18.6	0.838	14.9	0.981
25.9	0.270	22.2	0.590	18.5	0.844	14.8	0.983
25.8	0.279	22.1	0.598	18.4	0.850	14.7	0.985
25.7	0.288	22.0	0.606	18.3	0.855	14.6	0.987
25.6	0.297	21.9	0.614	18.2	0.860	14.5	0.988
25.5	0.306	21.8	0.622	18.1	0.865	14.4	0.990
25.4	0.315	21.7	0.630	18.0	0.870	14.3	0.991
25.3	0.324	21.6	0.637	17.9	0.875	14.2	0.993
25.2	0.333	21.5	0.645	17.8	0.880	14.1	0.994
25.1	0.341	21.4	0.653	17.7	0.885	14.0	0.995
25.0	0.350	21.3	0.661	17.6	0.890	13.9	0.996
24.9	0.359	21.2	0.668	17.5	0.895	13.8	0.997
24.8	0.368	21.1	0.675	17.4	0.899	13.7	0.998
24.7	0.377	21.0	0.682	17.3	0.903	13.6	0.999
24.6	0.386	20.9	0.690	17.2	0.908	13.5	0.999
24.5	0.395	20.8	0.697	17.1	0.912	13.4	1.000

Disque, Robert O. Addendum (July, 1973) to inelastic K-factor for Column Design AISC Engineering Journal, 2nd Quarter 1973 as published in Simplified Steel Design

FIG. 5-10 (Continued)

inelastic column behavior. Assume the same size of column above and below the level considered. Assume all girder sizes to be W 16 × 40 (I_x = 518 in.^{4}).

solution:

(a) Elastic column behavior assumed:
For the y-y axis: bracing provided at column top and bottom; therefore, K = 1.0.

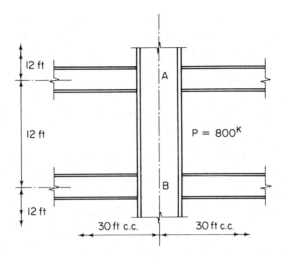

FIG. 5-11

Using the column allowable axial load tables in the AISC Manual, with Kl = 12 ft and assuming a W 14, choose a trial section of W 14 × 145 (P = 832 kips).

For the x-x axis: try W 14 × 145, I_x = 1710 in.⁴, r_x/r_y = 1.59, $G_A = G_B$ = 2(1710/12)/2(518/30) = 8.25.

From the alignment chart (Fig. 5-7a), K = 2.8. The relative major axis Kl = 2.8(12)/1.59 = 21.1 ft,* Kl_y = 12 ft. Therefore, the major axis is critical, and from the column tables (AISC Manual), P_{allow} = 729 < 800 kips, which won't work.

Try W 14 × 176, I_x = 2140 in.⁴, r_x/r_y = 1.60, $G_A = G_B$ = 2(2140/12)/2(518/30) = 10.3 and K = 3.1.

The relative major axis equivalent Kl = 3.1(12)/1.60 = 23.3.

Therefore, the major axis controls. P_{allow} = 853 > 800 kips. *Use* W × 176.

(b) Inelastic column behavior assumed:

For the *y-y axis:* K = 1.0, Kl = 12 ft, and from the column tables, choose a W 14 × 145 (P = 832 kips).

For the *x-x axis: try* W 14 × 145. I_x = 1710 in.⁴, r_x = 6.33 in., A = 42.7 in.², and f_a = P/A = 800/42.7 = 18.74 ksi.

---

* See Example 5-6.

Enter the table of allowable stresses for compression members of 36-ksi specified yield-stress steel found in Appendix A of the AISC Specification (Manual of Steel Construction) (Fig. 5-2) and find the corresponding Kl/r for f_a by interpolation: $Kl/r = 45.50$.

Enter the tables of values of F_e' in Appendix A of the AISC Specification with this Kl/r value and find the corresponding F_e'.

$F_e' = 72.16$ ksi, $f_a/F_e' = 18.74/72.16 = 0.260$, and, from Eq. 5-8b, $G_{\text{inelastic}} = G_{\text{elastic}} (f_a/F_e') = 8.25(0.260) = 2.15$.

From the alignment chart (Fig. 5-7a), $K = 1.62$.

Note that based upon inelastic action, G was reduced from 8.25 to 2.15, resulting in $K = 1.62$ (K_{elastic} was 2.8). $Kl/r = 1.62(12)(12)/6.33 = 36.85$.

Enter the table of allowable stresses for compression members of 36-ksi specified yield-stress steel found in Appendix A of the AISC Specification (Fig. 5-2); $F_a = 19.43$ ksi $> f_a = 18.74$ and $F_a A = P = 19.43(42.7) = 830$ kips > 800 kips. Use W 14 × 145.

Since $F_a < F_e'$, the column behavior can be seen to be inelastic (Fig. 5-1). Another way to determine column behavior would be to determine C_c and compare Kl/r with C_c. If $Kl/r < C_c$, the column action is inelastic. Further, if $f_a > 0.26F_y$, the column behavior is inelastic ($0.26F_y$ corresponds to the value of F_a at $Kl/r = C_c$).

Referring to Fig. 5-1, in the inelastic region the stiffness of the material deviates from the Euler-type curve. The cross section is "deteriorating" and EI does not represent actual stiffness. $E_t I$ is more representative of stiffness in the inelastic region, where E_t is the tangent modulus. Using the AISC Specification, the column stiffness should be based on E_t when $f_a = P/A > 0.26F_y$. This corresponds to the stress level dealing with C_c. In the inelastic region EI/l must be modified by multiplying by the stiffness reduction factor, E_t/E. E_t/E is approximately equal to (a conservative assumption) f_a/F_e' where $f_a = P/A$.

Figure 5-10 could have been utilized to arrive at the reduction factor, f_a/F_e', by entering the figure with $f_a = 18.74$ ksi instead of by the method used.

In summary:

Assumption	G	K	Section	P_{allow} (kips)
Elastic behavior	10.3	3.1	W 14 × 176	853
Inelastic behavior	2.15	1.62	W 14 × 145	830

The design procedure may be itemized:

(1) Choose a column section.

(2) Compute axial stress, $f_a = P/A$.

(3) With f_a, enter the column stress tables and find the Kl/r corresponding to f_a.

(4) With the Kl/r from (3), enter the F_e' tables and find the elastic stress, F_e'.

[Steps (3) and (4) may be eliminated by entering the stiffness-reduction-factor tables found in Addendum to Disque's article (Fig. 5-10) with f_a and reading f_a/F_e'.]

(5) Compute $G_{inelastic} = G_{elastic} (f_a/F_e')$.

(6) Obtain K from the alignment chart using $G_{inelastic}$.

(7) With Kl/r, enter the column stress tables to obtain F_a and compute $F_aA = P_{allow}$ for the chosen column section; if $P_{allow} > P_{actual}$, the choice of section is valid (also if $F_a > f_a$). Another procedure would be to enter the column load tables with Kl to obtain P_{allow} directly.

(2) The girders do not assume a double or reversed curvature with both ends rotating the same amount, so the stiffness represented by $6EI/l$ is not valid. The stiffness of the restraining member must be modified if the far end approaches a pinned or fixed (zero-rotation) condition. The term m, a stiffness modifier, would be, for the following conditions:

Condition	Stiffness	m
Far end pinned	$3EI/l$	3/6
Far end fixed	$4EI/l$	4/6

m is applied as follows in the G computation:

$$G = \frac{\Sigma(I/l)_{column}}{m\,\Sigma(I/l)_{girder}}$$

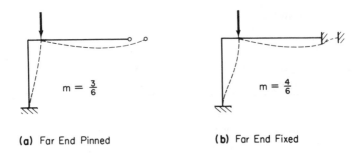

(a) For End Pinned   **(b)** For End Fixed

FIG. 5-12

Figure 5-12a and b shows the cases where the restraining member has the far end pinned and fixed, respectively. m modifies the girder stiffness and is merely the ratio $3EI/l$ or $4EI/l$ to $6EI/l$.

An illustrative example of the use of m will not be presented here, but in the next (third) modification illustrative example, it will be coupled with the necessary modification to G for the case where axial load is in the restraining member. The term "restraining member" is specifically used rather than "girder" because a restraining member need not necessarily be a girder or horizontal member. It could be a vertical member, as in Example 5-2.

(3) The frame's restraining members are subject to axial load. Axial load in a member affects its stiffness; and the larger this axial load, the smaller the stiffness. $6EI/l$, $4EI/l$, or $3EI/l$ cannot be used when there is high compression in the restraining member. If axial tension were introduced in the restraining member, the use of these stiffnesses would be unsafe.

Assume that there is an axial compressive load, P, in the restraining member (the member that restrains the column). By definition, S_0 will be designated as the classical stiffness of the restraining member ($6EI/l$ or $3EI/l$ for one end pinned or $4EI/l$ for one end fixed, etc.). They are the stiffnesses if no axial load is involved. Further, S is defined as the stiffness of the restraining member *with* axial load.

An exact mathematical approach to account for this modification of the restraining members' stiffness is too complex for discussion in this text. Yura* has suggested that the effect of axial load on the bending stiffness of restraining members be represented by

---

* Dr. Joseph A. Yura is an Associate Professor of Civil Engineering, University of Texas, Austin, Texas.

$$S = S_0 \left(1 - \frac{P}{P'_{cr}} \right) \qquad (5\text{-}9)$$

where

$P'_{cr} = \pi^2 E_t I / (Kl)^2 =$ buckling load
$E_t =$ tangent modulus

$K = 1.0$ is assumed for $3EI/l$ stiffness and 0.7 for $4EI/l$ stiffness and is applied to the appropriate stiffness of the restraining member as a multiplier. Equation 5-9 is the lower bound to what amounts to a complex exact solution that accounts for the effect of axial load. When the axial compressive load level is less than $0.26F_y$ (elastic behavior), the stiffness modifier $(1 - P/P'_{cr})$ can safely be approximated by $(1 - f_a/F'_e)$. Similarly, when f_a is larger than $0.26F_y$ (inelastic behavior), the stiffness modifier $(1 - P/P'_{cr})$ can safely be approximated by $(1 - f_a/F_a)$. Equation 5-9 is represented in Fig. 5-13. It is obvious that for an axial load level, P, which is 50% of the buckling load, P'_{cr} (as if the restraining member was a column), the stiffness is one-half. If the axial load level was very low, full stiffness would be used (represented by the typical G). This is the procedure for modifying the stiffness of the restraining member used in the definition for G when axial load is present. The use of this modifer will be illustrated in Example 5-2. This example will also illustrate the use of the stiffness modifer, m, to account for end-support type.

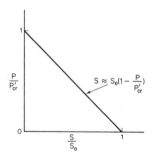

FIG. 5-13

EXAMPLE 5-2

For the structure shown in Fig. 5-14, design the interior column *B*. Assume A36 steel, simple connections in both directions, columns continuous for two stories, and only

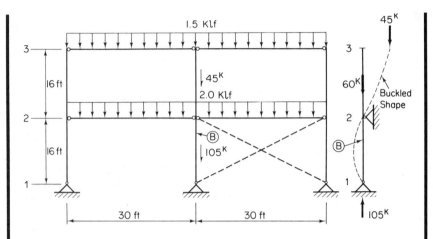

FIG. 5-14

the lower story braced against sidesway. Since the alignment charts can only be used for rigidly connected members, and all connections are simple, girder stiffness is not relied upon.

Note that column section 1–2 of column *B* must restrain section 2–3. Since this is the case, section 1–2 is our restraining member (a vertical girder). There is no relative movement between the ends of this restraining member.

solution:

Try W 10 × 49, A = 14.4 in.², r_y = 2.54 in., P = 105 kips. For restraining member 1-2, $f_a = P/A$ = 7.29 ksi < $0.26F_y$ = 9.36 ksi, which indicates *elastic* behavior and justifies the use of $(1 - f_a/F_e')$ in lieu of $(1 - P/P_{cr}')$. For the braced member 1-2, K = 1.0 and Kl/r = 1.0(12)(16)/2.54 = 75.59 and F_e' = 26.14 ksi. G_3 = 10 and

$$G_2 = \frac{(EI/l)}{(3/6)(EI/l)[1 - (7.29/26.14)]} = 2.77$$

Section 3–2 is considered to be the column (with some restraint at 3 assumed), while section 2–1 is assumed to be the restraining member (girder). The E, I, and l terms for sections 3–2 and 2–1 are the same and cancel in the G_2 computation. $\frac{3}{6}$ is *m*, the stiffness modifier of the restraining section 1–2, to account for the far pinned end at level 1. The ratio 7.29/26.14, or f_a/F_e', is used in lieu of the P/P_{cr}' term (for elastic behavior) to modify the stiffness of the restraining member due to axial load. From the alignment chart (Fig. 5-7b), $K = 2.25$. Therefore, $Kl = 2.25(16)$ = 36 ft, and (from column tables in the AISC Manual) P_{allow} = 74 kips > P_{actual} = 45 kips. *Use* W 10 × 49.

(4)   All columns in a story do not buckle simultaneously. In general, all the columns in a story do not want to buckle simultaneously when each individual column is subjected to a load. Conceivably, when a column in a multibay frame is subjected to a small axial load, while the other columns in the frame are subjected to their buckling loads, the system will not buckle; the lightly loaded column or columns tend to stabilize or brace the heavier loaded columns. Sidesway buckling, as shown in Fig. 5-3, is a story phenomenon with the individual columns interrelated. A single individual column cannot fail by sidesway without all the columns in the same story also buckling in a sway mode. Because of this interaction, the buckling of an individual column in the sway mode in an unbraced frame has no real meaning. Only a *story* can buckle. Frame stability can be checked by $\Sigma P \leq \Sigma P_{cr}$, where ΣP is the total gravity load on the story and ΣP_{cr} is the sum of the individual buckling strengths in a story obtained from the alignment chart. Sidesway will *not* occur until the total frame or load reaches the sum of the potential individual column loads for the unbraced frame. Buckling in a nonsway mode of an individual column is *not* dependent on the buckling load of the other columns. It is an individual phenomenon. The maximum load of an individual column (in an unbraced frame) is that load which corresponds to the nonsway (braced) case ($K = 1.0$). The advantage of this concept can be readily seen by the examples that follow.

EXAMPLE 5-3

Design the columns in the frame shown in Fig. 5-15. Rigid connections are used at the exterior columns, whereas a simple connection is used at the interior column. Assume that all columns are braced top and bottom out of the plane of the frame. Sidesway is permitted in-plane but is not permitted out-of-plane. Assume A36 steel throughout. Assume all girder sizes to be W 16 × 40 ($I_x = 518$ in.^{4}).

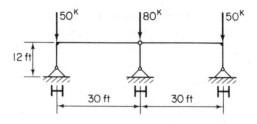

FIG. 5-15

solution:

Interior column: out-of-plane, $K = 1.0$. In-plane: considering the interior column to be truly pinned top and bottom, $G = \infty$ for both ends and $K = \infty$, and the column would be unstable. However, the assumption is made that the exterior columns are designed to stabilize the system and the use of $K = 1.0$ is valid.

From the AISC Manual (Column Allowable Axial Load Tables), $Kl = 1.0(12) = 12$ ft. *Use* W 6 × 20 ($P_{allow} = 79$ kips).

Exterior columns: out-of-plane, $K = 1.0$. In-plane: the exterior columns stabilize the structure for sidesway. In addition to its own load of 50 kips, each exterior column must support an additional hypothetical load of $\frac{80}{2} = 40$ kips for in-plane only. (There is no change for out-of-plane bending, since sidesway is prevented by bracing at individual columns.)

$P = 50 + \frac{80}{2} = 90$ kips for each exterior column. *Try* W 8 × 21, $l = 144$ in. $A = 6.16$ in.^{2}, $r_x = 3.49$ in., $r_y = 1.26$ in., $I_x = 75.3$ in.^{4}.

For the y-y axis: $P = 50$ kips, $Kl/r_y = 1.0(12)(12)/1.26 = 114.3$. From Appendix A of the AISC Specification, $F_a = 11.09$ ksi and $P_{allow} = 11.09(6.16) = 68.3 > 50$ kips.

For the x-x axis: $P = 90$ kips; with sidesway permitted, $G_{top} = \Sigma\, (I/L)_{column}/\Sigma(I/2L)_{girder}$. (*Note:* The girder length was increased by a factor of 2.0 because the far end of beam is pinned.)

$G_{top} = (75.3/12)/[(518/(30)(2)] = 0.73$. $G_{bottom} = 10$ (pinned base).

From the alignment chart (Fig. 5-7b), $K = 1.8$. $Kl/r_x = 1.8(12)(12)/3.49 = 74.3$.

From Appendix A of the AISC Specification, $F_a = 15.97$ ksi. $P_{allow} = 15.97(6.16) = 98.38 > 90$ kips. *Use* W 8 × 21.

Yura discusses the reasoning behind the assumptions in more detail in the reference cited at the beginning of this section.

EXAMPLE 5-4

Design the interior columns in the portion of the large frame shown in Fig. 5-16. Assume rigid connections at all columns. Adjacent columns are turned 90° as shown to equalize sway stiffness in the two main directions of the structure. Assume A36 steel throughout, sidesway permitted in both directions, and extremely stiff roof trusses frame in both directions. Assume **(a)** that all columns buckle simultaneously and then **(b)** that there is no simultaneous buckling of all columns.

Since the roof trusses are extremely stiff, $G_{top} = 0$ (e.g., infinite stiffness of trusses); $G_{bottom} = 10$. From the alignment chart (Fig. 5-7a), $K = 1.65$ and $KL = 1.65(20) = 33.0$ ft in both directions.

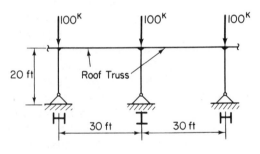

FIG. 5-16

solution:

(a) Simultaneous column buckling:

The weak axis governs; $KL = 33.0$ ft. From the AISC Manual, Column Load Tables, use W 12 × 58 ($P_{allow} = 102$ kips > 100 kips).

(b) Nonsimultaneous column buckling:

Any two adjacent columns do not sway in the same direction simultaneously. The column with the stronger axis in the plane of the frame braces the adjacent column whose weak axis is in the plane of the frame (Fig. 5-17).

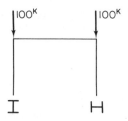

FIG. 5-17

Try W 8 × 40 (r_x/r_y = 1.73).

The strength of two adjacent columns will be checked. They must support a total of 200 kips.

From the AISC Manual Column Load Tables: for KL_y = 33 ft, P_{allow} = 46 kips; KL_x (equivalent) = 33/1.73 = 19.1 ft; P_{allow} = 133 kips. ΣP_{cr} = 179 kips < 200 kips, which won't work.

Try W 8 × 48 (r_x/r_y = 1.74). KL_y = 33 ft, P_{allow} = 58 kips; KL_x (equivalent) = 33/1.74 = 19.0 ft, P_{allow} = 165 kips, $\Sigma\ P_{cr}$ = 223 kips > 200 kips. *Use* W 8 × 48.

Frame stability checks ($\Sigma P \leq \Sigma\ P_{cr}$), where $\Sigma\ P$ is the total gravity load in a story and $\Sigma\ P_{cr}$ is the sum of the individual buckling strengths in a story.

Note that P_{allow} = 58 kips for sidesway buckling about the y-y axis (KL = 33 ft) is less than the applied load of 100 kips, but the 58 kips is the load that can be supported without any bracing. The applied load on the x-x column, 100 kips, is less than P_{allow} = 165 kips, and so the x-x-oriented column can provide some bracing to the y-y-oriented column such that it increases its capacity beyond 58 kips. The x-x column can provide bracing in the order of 165 − 100 = 65 kips for the adjacent column such that the y-y column capacity is 58 + 65 = 123 kips, which is larger than the applied load of 100 kips.

Another requirement is to check that the maximum load permitted on an individual column in an unbraced frame is less than that load capacity which corresponds to the nonsway (braced) case of K = 1.0; from the AISC Manual Column Load Tables, KL_y = 20 ft and the maximum P = 154 kips > 100 kips. *Use* W 8 × 48.

Design (b) permits a lighter column because consideration was given to the bracing effect of less critical columns.

(5) The total joint restraint is not divided equally between the columns above and below the joint. To handle the case where this is true, G will be redefined as

$$G = \frac{(I/l)_{\text{column under consideration}}}{(XI/l)_{\text{restraining members}}}$$

where I/l in the numerator is for one column only (not Σ) and where X is the stiffness distribution factor for that column. Instead of conservatively dividing the stiffness at a joint equally between the column above and below, the stiffness will be divided in the manner that the structure requires it to be. This can best be illustrated by an example.

EXAMPLE 5-5

For the two-story structure shown in Fig. 5-18, design the continuous column 3–2–1, which is subject to the axial load level indicated. Assume girders at levels 2–2 and 3–3 to be W 16 × 40 (I_x = 518 in.⁴) and all steel to be A36. Further assume that the restraint provided by girder 2–2 is divided in proportion to need.

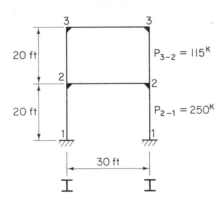

FIG. 5-18

solution:

Column 3–2: P_{3-2} = 115 kips, L = 20 ft. *Try* W 12 × 65 (I_y = 174 in.⁴). From the AISC Manual, Column Allowable Load Tables: for 115 kips and a W 12 × 65 section, KL_y = 39.67

and $K = 39.67/20 = 1.98$. Entering the alignment chart (Fig. 5-7b) with $K = 1.98$ and

$$G_{\text{top } 3-2} = \frac{174/20}{518/30} = 0.50$$

(I/L column segment under consideration).

$$G_{\text{bottom req'd } 3-2} = 30.0 \text{ (needed to stabilize column 3–2)}$$
$$G_{\text{bottom } 3-2 \text{ provided}} = \frac{174/20}{X(518/30)} = 30.0$$

and $X = 0.017$ (this means that to carry 115 kips, you need to use 0.017 of the available beam stiffness at column 2–3, not 50% or 0.50). What is not used for column 2–3 will be used for column 2–1 ($1.0 - 0.017 = 0.983$).
     Column 2–1: $X_{2-1} = 0.983$.

$$G_{\text{top } 2-1} = \frac{174/20}{(0.983)(518/30)} = 0.517$$

[I/L column segment under consideration, not Σ (I/L) columns at joint 2]. $G_{\text{top } 2-1} = 0.517$, $G_{\text{bottom}} = 1.0$, and from the alignment chart (Fig. 5-7b), $K_{2-1} = 1.23$ and $KL_{2-1} = 24.6$ ft and $P_{\text{allow}} = 253.3$ kips > 250 kips. *Use* W 12 × 65.
     Assuming that total joint restraint is divided equally between the columns above and below the joint: *try* W 12 × 72 ($I_y = 195$ in.^{4}). $G_{\text{top } 2-1} = 2(195/20)/(518/30) = 1.13$. $G_{\text{bottom } 2-1} = 1.0$. From the alignment chart (Fig. 5-7b), $K_{2-1} = 1.35$ and $KL_{2-1} = 1.35(20) = 27.0$. $P_{\text{allow}} = 257$ kips > 250 kips. *Use* W 12 × 72.
     Figure 5-19 summarizes the techniques for adjusting the effective length factor, K, when the assumptions of the alignment chart are not satisfied.

5-5.5   *Effect of K on Design*

It is valuable to know the degree of the effect of K on design. In the low-slenderness-ratio range typical of columns in multistory frames, K has a small effect on design. For a value of $l/r = 20$, a $K = 1.7$ causes a reduction (compared to $K = 1.0$) in the allowable axial compressive stress, F_a, of 4.6% for A36 steel. In the medium-slenderness-ratio range, K has a significant effect on design. For a value of $l/r = 70$, a $K = 1.7$ causes a reduction (compared to $K = 1.0$) in the allowable axial compressive stress, F_a, of 36.5 % for A36 steel.

Condition		Note
1. Columns and Girders do not have Same Stiffness	$G_{elastic} = \Sigma(I/L\ Column)/\Sigma(I/L\ Girder)Fa/F_e'$ $G_{inelastic} = \Sigma(I/L\ Column)/\Sigma(I/L\ Girder)fa/F_e'$	fa = P/A When: fa < 0.26Fy, Elastic Behavior fa > 0.26Fy, Inelastic Behavior For Elastic Behavior, Fa = F_e'
2. Girders do not Assume a Double or Reversed Curvature	$G = \Sigma(I/L\ Column)/\Sigma(I/L\ Girder)m$	For Far End of Girder Pinned, m = 1/2 For Far End of Girder Fixed, m = 2/3
3. Restraining Members Subject to Axial Load	$G_{elastic} = \Sigma(I/L\ Column)/\Sigma(I/L\ Girder)(1-fa/Fa)$ $G_{inelastic} = \Sigma(I/L\ Column)/\Sigma(I/L\ Girder)(1-fa/F_e')$	When: fa < 0.26Fy, Elastic Behavior fa > 0.26Fy, Inelastic Behavior
4. All Columns in a Story do not Buckle Simultaneously	For Unbraced Frames: 1. Σ Total Gravity Loads in a Story ≤ Σ Individual Buckling Strengths in a Story. 2. Maximum Load on an Individual Column = That Load Corresponding to the Non-sway (Braced) Case of K=1.0	Less Critical Column Can Brace Other More Critical Columns
5. Total Joint Restraint in not Divided Equally Between Columns Above and Below the Joint	G = I/L Column Under Consideration/ XI/L Restraining Members	X = Stiffness Distribution Factor for Column Under Consideration

FIG. 5-19 Summary: Modifications to alignment chart.

K is a function of the ratio of the moment of inertia of the column to that of the girder. If beam or girder size is increased (by means of composite action for example), K can be reduced. Similarly, if the column size is decreased (by means of high-strength steels, for example), K is reduced. In the low-slenderness-ratio range, the effects of K are small, such that the column design is reasonably independent of the girder size.

In the medium-slenderness-ratio range, improved column strength can be achieved by decreasing the ratio of the moment of inertia of the column to that of the girder. An increase of girder size to improve K (reduce) may be more economical than to increase the column size. An increase of column size without changing the depth would have an unfavorable effect on K (the radius of gyration, r, would remain virtually unchanged and F_a would decrease).

As a general statement, as the column slenderness ratio increases, girder size becomes a significant consideration relative to the stability of the structure. In an unbraced frame, if K exceeds a value of 2 and the column slenderness approaches the medium range, the use of large girders for structural economy should be investigated.

5-5.6   Unity for Certain Unbraced Frames

Following research at Lehigh University on unbraced frames of particular dimensional and loading parameters, a value of $K = 1.0$ can be used. The test frames were 10 to 40 stories high, had $9\frac{1}{2}$- to 14-ft story heights, were two or three bays in depth with bay spans of 20 to 56 ft, and had live loads between 40 and 100 psf, dead loads from 50 to 75 psf, uniform wind loads of 20 psf for the full frame height, and column slenderness ratios from 18 to 42. The story height, bay depth, and bay span parameters are of minor consequence. All beams and columns were compact.

For frames of this type, the AISC Commentary to the Specification suggests a unity value for K for calculating F_a and F'_e when the following conditions are met: f_a/F_a and $f_a/0.60F_y \leq 0.75$; maximum in plane column $l/r \not> 35$; the bare frame working drift index $\not> 0.004$ (roof-level drift divided by total frame height). [Drift is to horizontal lateral load what sidesway is to vertical load (a horizontal translation, Δ, of the frame due to lateral load).] Obviously, when these provisions are met, $K = 1.0$ and the design of a great many high-rise unbraced frames can be simplified. However, pending the results of further research relative to frame stability under gravity loads only, a $K > 1.0$ continues to be recommended for columns in upper stories where gravity load only controls the frame column design.

For frames not encompassed by the Lehigh University research, revised design criteria are required and are currently under study.

5-6 DESIGN AIDS AND TABLES FOR ALLOWABLE STRESSES

Allowable stresses were discussed in Section 5-3 and are expressed by Eqs. 5-3, 5-4, and 5-6. In each of these equations F_a, the allowable stress, is dependent on the slenderness ratio, Kl/r. r, the radius of gyration of the section, is dependent on the section. However, when designing a section, its final size is not immediately known, and therefore the radius of gyration is also unknown. The actual design procedure of an axially loaded column can be illustrated by a design example. The procedure is one that we work with the AISC Manual's Column Allowable Load Tables.

EXAMPLE 5-6

Given the axially loaded pinned-end column as shown in Fig. 5-20, determine the lightest W column size to be used. Assume A36 steel. Also determine F_a and P_{allow}.

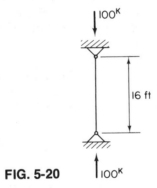

FIG. 5-20

solution:

Since in this case the ends of the column are pinned, $K = 1.0$ and $KL = 1.0(16) = 16$ ft.

If this were not the case, a proper value of K would be selected through the aid of Fig. 5-6.

From the AISC Manual Column Allowable Axial Load Tables, for $KL_y = 16$ ft and $P = 100$ kips, *select* W 8 × 31 ($P_{allow} = 124$ kips).

From the "properties" portion of the Allowable Axial Load Tables: $r_y = 2.02$ in., $r_x/r_y = 1.72$, and $A = 9.13$ in.^{2}. $KL_x = 1.0(16) = 16$ ft. The equivalent (in load-carrying capacity) minor-axis effective length for the major axis is $KL_x/$ $(r_x/r_y) = 16/1.72 = 9.30$ ft, and since 9.30 ft $< KL_y = 16$ ft, the minor-axis effective length controls the design.

The latter procedure is necessary since the Column Allowable Axial Load Tables give the allowable loads with respect to the effective length relative to the least radius of gyration, r_y. Use W 8 × 31. $Kl/r_y = [1.0(16)12]/2.02 =$ 95.0.

From Fig. 5-2 or from Eq. 5-3 ($Kl/r_y < C_c$), the allowable stress is $F_a = 13.60$ ksi and $P_{allow} = F_a A = 13.60(9.13)$ $= 124.2$ kips > 100 kips.

The Column Allowable Axial Load Tables are applicable to primary members with respect to their minor axis. Additional Column Load Tables are found in the AISC Manual for M and S shapes, pipe, structural tubing, and double angles. All loads tabulated are computed in accordance with Eqs. 5-3 and 5-4 for axially loaded members having effective unsupported lengths indicated at the leftmost column of each table. $F_y = 36$ and 50 ksi for all shapes (other than structural tubing, for which $F_y = 36$ and 46 ksi) are tabulated. Values for a maximum Kl/r of 200 (Section 1.8.4, AISC Specification) are tabulated.

As mentioned in Section 5-3, allowable stresses, F_a, may be determined from the tables for allowable stresses for compression members for all values of F_y. F_a values for Eqs. 5-3, 5-4, and 5-6 are tabulated for various Kl/r values.

5-7   DESIGN AIDS FOR TAPERED COLUMNS

Appendix D to the AISC Specification covers the allowable stresses for web-tapered members. For axially loaded tapered compression members, AISC formulas D2-1 and D2-2 govern. These formulas are identical to Eqs. 5-3 and 5-4, respectively, except that S replaces Kl/r. S is defined as the slenderness ratio of the tapered member and

$$S = \frac{Kl}{r_{0y}} \qquad \text{for weak axis bending} \qquad (5\text{-}10)$$

$$S = \frac{K_\gamma l}{r_{0x}} \qquad \text{for strong axis bending} \qquad (5\text{-}11)$$

where S = governing slenderness ratio of a tapered member

K_γ = effective length factor for a tapered member (includes effects of end restraints and tapering)

r_0 = radius of gyration at the smaller end of a tapered member

In addition, $F_{a\gamma}$ replaces F_a as a symbol. In order to use Appendix D, a tapered member must possess symmetry perpendicular to the plane of bending (weak-axis symmetry), have equal and constant flange area, and have a depth that varies linearly as $d_0[1 + \gamma(z/l)]$, where $\gamma = (d_L - d_0)/d_0$ and γ must be less than $0.268l/d_0$ or 6.0, whichever is smaller. d_L and d_0 are the depths at the larger and smaller ends of a tapered member, respectively (or tapered member unbraced segment), and l is the length of the member (or unbraced segment) in inches. z is the distance from the smaller end of a tapered member in inches. A tapered member is shown in Fig. 5-21.

The value of K_γ^* may be determined from Figs. CD1.5.2 to CD1.5.17 of the Commentary on the AISC Specification for cases of sidesway permitted and prevented for tapered column frames for values of $\gamma = 0, 0.5, 1.0, 1.5, 2.0, 3.0, 4.0,$ and 6.0. For cases of $\gamma = 0$, $K_\gamma = K$ and Fig. 5-7 may also be used. Values of γ between the Commentary figures may be arrived at by interpolation. The figures are entered with relative stiffness factors G_{top} and G_{bottom}, and K_γ values are determined.

$$\dot{G}_{\text{top}} = \frac{(I_0/l)_{\text{column}}}{(I/l)_{\text{girder}}}$$

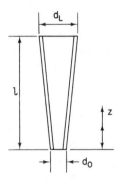

FIG. 5-21

---

*This material is adapted from G. C. Lee, M. L. Morrell, and R. L. Ketter, "Design of Tapered Members," *WRC Bulletin 173*, June 1972.

$$G_{\text{bottom}} = \frac{(I_0/l)_{\text{column}}}{(I/l)_{\text{girder}}}$$

where I_0 is the moment of inertia at the smaller end of the tapered member.

Further information on this subject will be found in AISC Appendix D and the Commentary thereto.

5-8   BASE-PLATE DESIGN

When an axially loaded column is to deliver its load to a concrete support, steel base plates are usually used under the column to distribute the load over an ample area of the concrete foundation. Three major points of consideration are necessary in the design of base plates. Sufficient steel base-plate area must be available such that the calculated bearing stresses on the concrete are less than the allowable bearing stress, F_p (AISC Specification, Section 1.5.5); the bending stresses in the base plate should not exceed the allowable bending stress, F_b; and the means of anchorage should be sufficient to assure proper connection of plate to concrete support. The first two considerations will be discussed in this section.

Part 4 of the AISC Manual gives several suggested details relative to column base plates. For the purpose of simplicity, Fig. 5-22a shows a detail of a base-plate shop-welded directly to the column with the anchor bolts preset in the concrete foundation. Larger and more cumbersome base plates, as shown in Fig. 5-22b, are usually shipped loose to the site, set, and leveled in the concrete foundation prior to steel-column erection.

A complex method of analysis of steel-column base plates is not justified since various assumptions of the distribution of bearing stresses would have to be made. The AISC suggested method of base-plate design is described herein.

The bearing pressure between plate and foundation is assumed to be uniform, and the base plate is designed to distribute the concentrated column force, P, into the concrete foundation uniformly within the rectangle of dimensions $0.95d$ and $0.80b$, as shown dashed in Fig. 5-23.

The allowable bearing pressure, F_p (ksi), on the concrete is dependent upon two parameters: the specified compressive strength of the concrete (ksi), f_c', and the portion of support area occupied by the steel base plate, A_2/A_1, where A_1 is the base-plate bearing area and A_2 is the area of the concrete foundation.

From the AISC Specification, when the entire area of concrete foundation is covered by the base plate,

$$F_p = 0.35 f_c' \tag{5-12}$$

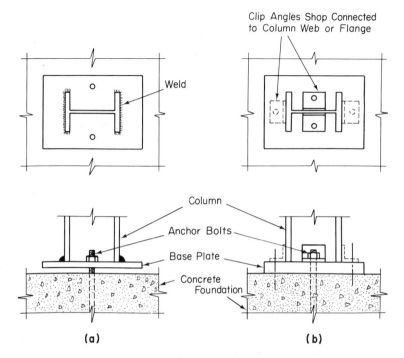

FIG. 5-22

and when less than the entire area of concrete foundation is covered by the base plate,

$$F_p = 0.35 f_c' \left[\frac{A_2}{A_1} \right]^{1/2} \leq 0.7 f_c' \qquad (5\text{-}13)$$

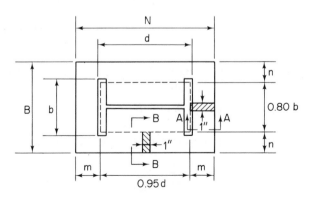

FIG. 5-23

Using $F_p \geq P/A$, where P is the total column load in kips, and substituting in Eq. 5-13, the following equation is obtained:

$$\left(\frac{P}{0.35f_c'}\right)^2 \leq A_1 A_2 \leq 0.7f_c' \qquad (5\text{-}14)$$

Noting that in Eq. 5-13 the second term, $(A_2/A_1)^{\frac{1}{2}}$ is limited to a value of 2 to meet the limitation of $0.7f_c'$. Eq. 5-14 can be rewritten as

$$\left(\frac{P}{0.35f_c'}\right)^2 \leq A_1 A_2 \leq 4A_1^2 \qquad (5\text{-}15)$$

The first two terms of Eq. 5-15 yield

$$A_1 \geq \frac{1}{A_2}\left(\frac{P}{0.35f_c'}\right)^2 \qquad (5\text{-}16)$$

The first and third terms of Eq. 5-14 yield

$$A_1 \geq \frac{P}{0.7f_c'} \qquad (5\text{-}17)$$

From the second and third terms of Eq. 5-15, the lightest base plate can be obtained when $A_2 = 4A_1$ or $A_1 = A_2/4$. Substituting this value of A_1 in terms of A_2 into Eq. 5-15 (first and second terms),

$$A_2 \geq \frac{P}{0.175f_c'} \qquad (5\text{-}18)$$

If the actual conditions permit, the area of the concrete footing or pedestal should be at least $P/(0.175f_c')$ as stated in Eq. 5-18 to obtain the optimum concrete bearing stress.

Referring to Fig. 5-23, the suggested column base-plate design steps are as follows:

(1) Compute $A_1 = 1/A_2\,[P/0.35f_c']^2$, Eq. 5-16, and $A_1 = P(0.7f_c')$, Eq. 5-17, and use the larger value.
(2) Determine $N \approx (A_1)^{\frac{1}{2}} + \Delta$ and $B = A_1/N$ (Fig. 5-23), where $\Delta = 0.5(0.95d - 0.80b)$.
(3) Compute the actual bearing pressure on the concrete f_p.

$$f_p = \frac{P}{BN}$$

(4) Determine m and n (Fig. 5-23) from the following equations:

$$m = \frac{N - 0.95d}{2}$$

$$n = \frac{B - 0.80b}{2}$$

Choose n' from Table 5-2. A definition of n' will be stated in the discussion that follows.

Table 5-2
Values of n'

Column Section Range	n'	Column Section Range	n'
W 14 × 730–145	5.77	W 10 × 45–33	3.42
W 14 × 132–90	5.64	W 8 × 67–31	3.14
W 14 × 82–61	4.43	W 8 × 28–24	2.77
W 14 × 53–43	3.68	W 6 × 25–15	2.38
W 12 × 336–65	4.77	W 6 × 16–9	1.77
W 12 × 58–53	4.27	W 5 × 19–16	1.91
W 12 × 50–40	3.61	W 4 × 13	1.53
W 10 × 112–49	3.92		

Note: The values given are for use in the formula $t = n'\,[f_p/\,(0.25F_y)]^{1/2}$.

(5) Use the larger of the values of m, n, or n' to solve for the base-plate thickness, t, using the applicable formula.

$$t = m \left(\frac{f_p}{0.25F_y} \right)^{1/2} \tag{A}$$

or

$$t = n \left(\frac{f_p}{0.25F_y} \right)^{1/2} \tag{B}$$

or

$$t = n' \left(\frac{f_p}{0.25F_y} \right)^{1/2} \tag{C}$$

The formulas for the thickness of the base plate, t, are derived from the bending moment produced in the cross-hatched segments, n or m, 1 in. wide, as seen in Fig. 5-23. This approach is used when m and n are not small. For cantilever strip n, assume (Fig. 5-24) that

$$F_b = \frac{Mc}{I} = \frac{[f_p(n)(n/2)]t/2}{[(1)(t)^3]/12}$$

where $M = f_p(n)(n/2)$, $c = t/2$, $I = \frac{1}{12}(1)t^3$, $F_b = 0.75F_y$ (AISC Specifications, Sect. 1.5.1.4.3, solid rectangular sections bent about their minor axis), is the allowable bending stress in the base plate (ksi), and Eq. B is obtained when appropriate substitutions are made. Equation A is obtained in a similar manner for a 1-in.-wide m segment.

This method of design can best be illustrated by a design example.

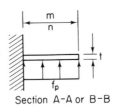

Section A–A or B–B

FIG. 5-24

When the values of m and n are small, the highest stress in the base plate occurs at the face of the column web at half-distance from the inside faces of the flange. For this condition, a flat plate with one edge fixed and one edge free is assumed. The two short edges are assumed to be supported. The following formula* may be used to compute an approximate value for t for this case:

$$t = \left[\frac{3(b_f - t_\omega)^2 f_p}{4(1 + 3.2\alpha^3)F_b} \right]^{1/2}$$

where

$$\alpha = \frac{b_f - t_\omega}{2(d - t_f)}$$

For convenience, this equation for t may be rewritten as Equation C, where

$$n' = \frac{b_f - t_\omega}{2} \left[\frac{1}{1 + 3.2\alpha^3} \right]^{1/2}$$

---

*R. J. Roark, *Formulas for Stress and Strain*, 3rd ed., McGraw-Hill Book Co., New York, 1954, p. 205.

Values of n' are given for a range of column cross sections in Table 5-2.

EXAMPLE 5-7

Assume a controlling $KL_y = 20$ ft for a W 14 × 109 column of A36 steel. The 28-day compressive strength of the concrete foundation, f_c', is 4000 psi. Design a steel A36 base plate for this column. Assume a 28 × 28 in. concrete foundation.

solution:

For W 14 × 109, $d = 14.32$ in., $b = 14.605$ in. and from the Column Allowable Axial Load Tables (Part 3 of the AISC Manual or Fig. 9–5) for $F_y = 36$ ksi and $KL_y = 20$ ft, $P = 544$ kips. $A_2 = 28(28) = 784$ in.²

(1) $A_1 = 1/784 \, [544/0.35(4.0)]^2 = 193$ in.²   (Eq. 5–16)
     $A_1 = 544/0.7(4.0) = 194$ in.² (governs)   (Eq. 5–17)

(2) $N \approx (A_1)^{\frac{1}{2}} + \Delta = (194)^{\frac{1}{2}} + 0.5[0.95(14.32) - 0.80(14.605)] = 14.89$ in.   (use 16.0 in. > d)
     $B = A_1/N = 194/16.0 = 12.13$ in.   (use 16.0 in. > b)

(3) $f_p = P/(BN) = 544/(16)16 = 2.13$ ksi

(4) $m = (N - 0.95d)/2 = [16.0 - 0.95(14.32)]/2 = 1.20$ in.
     $n = (B - 0.80b)/2 = [16.0 - 0.80(14.605)]/2 = 2.16$ in.
     From Table 5-2 for a W 14 × 109, $n' = 5.64$ in. (governs)

(5) $t = n'[f_p/(0.25F_y)]^{\frac{1}{2}} = 5.64[2.13/(9.0)]^{\frac{1}{2}} = 2.74$ in. (Eq. C) (use $t = 2\frac{3}{4}$ in.)
     Use base plate PL $2\frac{3}{4}$ × 16 × 1′–4 16 in.

PROBLEMS

(5-1) Determine the allowable axial load for a W 14 × 120 for the following conditions. Do not use AISC tables or design aids to obtain F_a, but calculate F_a using basic equations (including C_c). Use A36 steel. Use tables only to check calculations.
   (a) $KL = 24$ ft, main member.
   (b) $KL = 24$ ft, secondary member.

    (c) $KL = 48$ ft, main member.

    (d) $KL = 48$ ft, secondary member.

(5-2)   Determine the allowable axial column load for the following. No sidesway.

    (a) $F_y = 36$, W 14 × 145, $L = 20$ ft (pinned ends).

    (b) $F_y = 36$, W 14 × 109, $L = 30$ ft (fixed ends).

    (c) $F_y = 36$, W 14 × 132, $L = 40$ ft (pinned ends).

        (1)   As a main member.

        (2)   As a secondary member.

    (d) $F_y = 50$, W 14 × 311, $L = 30$ ft (one end fixed, one end pinned).

(5-3)   What is the lightest W 12 column shape that can support the axial loads shown under the conditions stated. $F_y = 36$ ksi.

    (a) $K_x = 1.0$, $K_y = 2.1$, $L = 12$ ft, $P = 200$ kips.

    (b) $K_x = 1.0$, $K_y = 0.8$, $L = 18$ ft, $P = 500$ kips.

    (c) $K_x = 1.0$, $K_y = 1.0$, $L = 30$ ft, $P = 500$ kips.

(5-4)   A W 14 × 82 (A36) column 22 ft long is simply supported at each end. Assuming y-y bracing at its midpoint, compute the maximum permissible axial load. Compare your answer with that obtained using the allowable axial column load table provided in the AISC manual.

(5-5)   Choose the most economical column section (W shape, minimum depth 12 in.) to support an axial load of 850 kips. Assume the total length of 26 ft is pinned at one end and fixed at the other (both axes). Use A36 steel.

(5-6)   Select a W 14 shape for a pinned-ended column (both axes) to carry an axial load of 550 kips if the member length is 20 ft. Assume A36 material.

    (a) If no intermediate bracing.

    (b) If braced at mid-height in weak direction.

(5-7)   Design the appropriate W 14 column for an axial load of 490 kips. Length is 30 ft. No sidesway.

    (a) Pinned ends (A36 steel).

    (b) Fixed ends (A36 steel).

    (c) Fixed ends except braced in weak axis at mid-height (A36 steel). Assume brace is pinned.

    (d) Pinned ends ($F_y = 50$).

(5-8)   Design column C in Problem 2-2. The effective length of the member is 12 ft in both axes. Use A36 steel.

(5-9)   Design each column for the rigid braced frame shown. Restrict your design to 14-in. W shapes of A36 steel. Axial design loads are shown in addition to girder moments of inertia. Assume all columns are hinged (rotation free and translation fixed) in the minor axis direction unless otherwise shown.

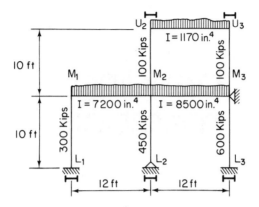

(5-10)   Determine the axial capacity of each of the individual columns in the multistory frame. Assume A36 steel. Girder moments of inertia in in.^{4}, I, are indicated. Assume all columns are hinged (rotation free and translation fixed) in the minor axis direction unless otherwise shown.

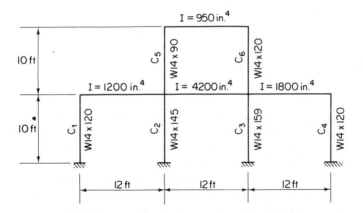

(5-11)   Design a square base plate of A36 steel to fully develop the capacity of a W 14 × 342 column 10 ft long. Assume $f_c' = 3000$ psi and $K_x = K_y = 1$. The base plate rests on a concrete support of the same size.

6

bending
members

6-1   INTRODUCTION

This chapter will cover bending members (beams and girders) loaded principally in bending. The basic behavior of these members will be discussed and will include a discussion as well as definition of compact and noncompact members. A review of allowable stress design (Part 1 of the AISC Specification) and plastic design (Part 2 of the AISC Specification) will be presented. A comparison of both these provisions will afford a more precise view of the Specification provisions.

Allowable stresses and the reason for the various allowable stresses will be developed together with appropriate design examples. The concept of compact and noncompact sections will be accentuated. Deflections, vibrations, and ponding will be discussed toward the end of the chapter. Laterally unsupported beams and box sections will also be reviewed.

In the discussion to follow, ASD and PD stand for allowable stress design and plastic design, respectively. In ASD it is assumed that maximum stress is the controlling factor and that the distribution of stress between the flanges or outermost fibers of a beam is a linear distribution in accordance with the usual elastic assumption. The required section modulus for any beam in simple bending is obtained by simply dividing the maximum moment that occurs on the span (M_{max}) by the allowable bending stress, F_b:

$$S = \frac{M_{max}}{F_b} \qquad (6\text{-}1)$$

The designations of allowable stresses and calculated stresses in the AISC Specification are capital letters and lowercase letters, respectively, which will be used in this text.

The PD procedure is basically the same. The maximum moment that exists at any point in the beam span, referred to as M_p, the plastic moment, is used. It is assumed that the beam has just enough moment capacity to carry this full plastic moment. In the fully plastic condition the distribution of stress is assumed such that maximum tensile stress exists over one portion of the beam cross section while maximum compressive stress exists over another portion of the beam. It is assumed that there is a constancy of these two stresses rather than a linear variation.

The technique for determination of the size of the beam is left exactly in the same form as in ASD. Moment, M_p, is divided by the yield-point stress, F_y, rather than the allowable bending stress, F_b, yielding a required modulus referred to as the plastic modulus, Z:

$$Z = \frac{M_p}{F_y} \qquad (6\text{-}2)$$

Both S and Z values for the various rolled shapes are listed in the Properties for Designing Tables, the ASD Selection Table, and the PD Selection Table, all of which are found in Parts 1 and 2 of the AISC Manual. Knowing either S or Z, a section can be chosen.

Figure 6-1a and b shows the stress distribution for ASD and PD, respectively. Because of the use of the yield-point stress, F_y, in the PD procedure, it is desirable to have a safety factor included in the design somehow. In ASD, this was accomplished by reducing the value of yield stress, F_y, to F_b (a fraction of F_y, usually $0.6F_y$), so that a safety factor of approximately 1.7 was used. The technique is applied a little differently in PD, in that the loads on the beam are factored; that is, the loads are multiplied by the safety factor of 1.7, justifying the use of yield stress rather than a reduced working stress, F_b.

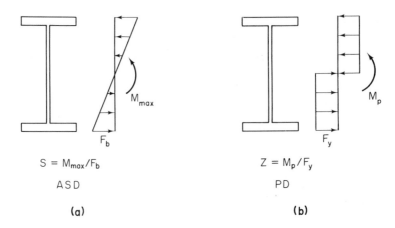

$$S = M_{max}/F_b \qquad\qquad Z = M_p/F_y$$

ASD   PD

(a)   (b)

FIG. 6-1

    Both techniques are basically quite simple, but the question of when each formula is valid is not. In many instances the same beam size is arrived at using the section modulus (ASD) method or the plastic modulus (PD) method; in other instances, the same size member is not arrived at. Obviously, ASD methods would be used if the material were truly an elastic material, and PD methods would be used if the material were truly a plastic material. Unfortunately, none of the materials used fits either of these categories. Every material deviates in some respect from perfect elasticity or perfect plasticity. The question of which method to use depends upon the material's characteristics and the way in which the material is used. In cases where the material is used where impact or fatigue is of importance, the PD procedures are not suggested for use.

    In all structures that are designed, two basic problems should be recognized: the structure and the material must be strong enough and, in addition, the structure has to be stable and the individual elements of the structure have to be stable. In general, stability is the heart of the problem and is much more difficult to handle than the strength problem.

    The usual approach in stability is not to do these problems, because they are so complex. Most of the time the stability problem is converted to a strength problem, which is a manageable design problem. For instance, in the case of columns, the allowable stress is reduced as a function of some slenderness parameter, l/r. This converts the stability problem into a strength problem. The stability problem is not really solved except in a very indirect manner. This is really the purpose of the specification. The specification enables a very difficult problem to be converted into a manageable one.

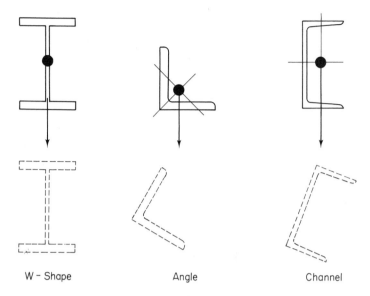

W - Shape   Angle   Channel

FIG. 6-2

6-2   CENTER OF GRAVITY/
        SHEAR CENTER

One of the fundamental assumptions made in all beam problems is that load is applied through the centroid of the cross section. The centroid may be defined as the center of the mass of the cross section. In Fig. 6-2 the solid dots represent the centroid of the section. In the case of the W shape, the centroid is obviously at the center of the cross section. In the case of the angle, it is somewhere between the legs of the section. In the case of the channel, the centroid is somewhere between the flanges on the axis of symmetry.

The question of whether Mc/I, the computed bending stress, is valid for each of the three sections shown in Fig. 6-2 may be asked. The answer to this question would be no. Figure 6-2 also shows how each member will deflect when loads are applied through the centroids. The deflected positions are shown dashed.

In the case of the W shape, when the load is applied through the center of gravity or centroid, it will deflect in the direction of the load, as shown. The angle section, on the other hand, will rotate and deflect. The angle bends about the two principal axes and twists. The channel section will deflect straight down and rotate. The channel bends in the direction of the load and twists.

Loading through the centroid does not ensure simple bending. Only the W shape undergoes simple bending, and therefore only in this case will Mc/I be applicable.

The combined bending and torsion condition is too complex to appear in any specification, because the solutions are generally unique and one should try to prevent torsion or seek an appropriate reference source for a solution to the problem.

For simple bending to occur, load must be applied through the shear center of the cross section. Shear center may be defined as a point through which load must pass so that twisting does not occur. For the angle, the shear center is at the intersection of its legs. For a doubly symmetrical shape, such as an H, W, or S shape, the shear center and center of gravity coincide. Handbooks summarize the location of shear centers for various shapes. If the external forces on a bending member are not directed through the shear center of the cross section, both twisting and bending will result. This behavior is important primarily for members with open-type sections.

In summary, if a cross section contains an axis of symmetry, its shear center lies on that axis. If the cross section is symmetrical about two axes, the shear center is coincident with the centroid. If the cross section comprises only two rectangular elements, the shear center is at the juncture of the elements.

Figure 6-3 shows that, when the load is applied through the shear center of the cross section, the member does not twist, and simple bend-

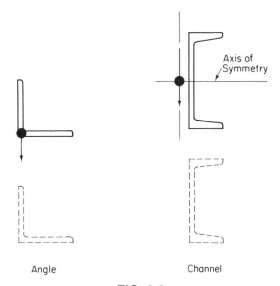

Angle   Channel

FIG. 6-3

ing in the direction of the applied load occurs. Mc/I for this condition is valid. The solid dots in this figure represent the location of the shear center.

Channels can be loaded through their shear center by using an angle as shown in Fig. 6-4a. This is commonly done for lintels and spandrels. Simple bending using Mc/I can be used, and deflection is in the direction of the load. If it is not practical to load through the shear center, torsion can be resisted by bracing or diaphragms, as shown dashed in Fig. 6-4b. The bracing system must resist the twisting moment, Pe.

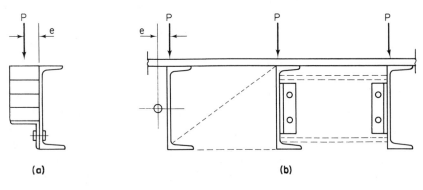

(a)   (b)

FIG. 6-4

6-3   ELASTIC BEHAVIOR / PLASTIC BEHAVIOR

In order at this point is a discussion of the behavior and design of bending members. To understand and properly utilize the design methods and specification provisions for structural steel design, the variables that affect beam behavior will be explored. The principal variables that affect beam load capacity and behavior are material strength, which determines at what level yielding commences; unbraced compression elements; width-to-thickness ratios of plate elements; cross section (or general shape thereof), which gives rise to a parameter termed the shape factor; loading; and support conditions.

It is an unfortunate fact that all beam behavior cannot be represented by a single load-deflection curve. There are a great many variables involved. Figure 6-5 shows five curves, which are supposed to represent the different ways in which a member subject to bending in a similar manner to that shown can behave.

Curve 1 assumes that strain hardening does occur. Curve 2 assumes that the material is ideally plastic. Both curves, as shown, lead to some

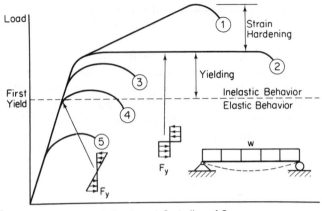

FIG. 6-5   Deflection at Centerline of Span

form of instability collapse, usually represented by the bending down of the curves.

Curves 3 and 4 represent members in which first yield is reached before large deflections occur. By first yield is meant the point at which the outer fibers reach the yield stress. This condition does not represent collapse of the section.

Curve 5 represents a beam that is basically unstable. The yield-point stress at any place on the section is unachievable before some other phenomenon occurs. This usually represents the problem of instability, for example, or local buckling of the web or flanges.

Safe, economical structures can be designed on the basis of any one of these typical curves. Curves 1 and 2 will generally provide the lightest beam, but sometimes the fabrication and detail costs are increased, owing to the necessity of bracing and stiffeners, which afford stability. The proper design is the most economical one, not necessarily the lightest one.

Of course, shear and deflection can also affect design, but in most members with a depth-to-length ratio less than 1/10, bending predominates and shear stresses are usually low. Furthermore, deflection is usually ignored because limitation on the stress is sufficient to prevent deflection from becoming a prime consideration. Deflections are frequently limited directly by depth-to-span limitations on the beam.

Studying these curves in more detail would be of value to see how they relate to the AISC Specification. Curves 1 and 2 will be treated together, because the design provisions are basically the same. Local and lateral buckling have to be controlled until significant yielding takes place. No advantage is taken of the strain-hardening zone as to increased strength. Strain hardening is neglected in curve 2. Curve 2 is an idealization of curve 1, which is more like the actual situation, in which we have a little reserved strength.

A beam behaving like curves 1 and 2 would be a compact section in ASD. A *compact section* is a beam in which advantage is taken of plastic deformations in addition to elastic deformations. When PD approaches are used, this type of behavior must be assumed to occur. Large deformations occur with very little increase in stress.

Curve 1 is really the type of behavior that is closest in mathematical idealizations to the actual behavior of structural steel. The moment varies along the length of the beam, and there is a reserve capacity even at the section that is fully plastic. Reserve strength or capacity as far as PD is concerned relates to cross section. This reserve strength is usually expressed by a term called the *shape factor,* which is a function of the type of cross section involved and varies with this. This shape factor, f, is equal to

$$f = \frac{M_p}{M_y} = \frac{F_y Z}{F_y S} = \frac{Z}{S} \tag{6-3}$$

where M_p is the plastic moment, M_y is the moment at first yield, Z is the plastic modulus, and S is the elastic or section modulus.

6-4 STRONG-AXIS BENDING

In the case of a W shape bent about its x-x axis or strong axis, there is a 12% reserve between the yield-point stress at M_y at the outermost fiber (first yield) and the fully plastic condition M_p, which represents full plasticity through the entire cross section. This condition is shown in Fig. 6-6, which shows moment plotted against deflection. Of course, the as-

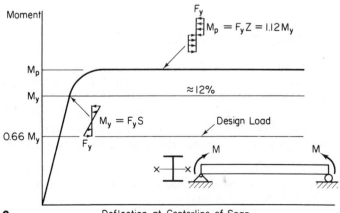

FIG. 6-6   Deflection at Centerline of Span

sumption is made that stability is maintained. Lateral braces are properly spaced; width-to-thickness ratios of flange and web are small enough to avoid local buckling until the entire cross section has yielded.

The fundamental assumption for PD is that stability is maintained. It has been indicated that, on an average, something like a 12% increase in strength over first yield is available for W shapes bending about their strong axis. Shape factors for other cross-sectional shapes are as indicated in Table 6-1.

Table 6-1

Cross-Sectional Shape	f
Rectangle	1.50
Circle	1.70
Diamond	2.00
W-shape y-y bending	1.50

For ASD, the AISC Specification states that the allowable tension and compression on extreme fibers (F_b) of compact hot-rolled or built-up members (except hybrid girders* and members of ASTM A514 steel), symmetrical about and loaded in the plane of their minor axis (provided the other requirements for compact sections are met), is equal to

$$F_b = 0.66F_y \qquad (6\text{-}4)$$

The factor of safety based on full yielding of the cross section can be seen to be $M_p/0.66M_y = 1.12M_y/0.66M_y = 1.7$ (Fig. 6-6). In PD, this factor of safety is called the *load factor, F,* and is equal to $(F_y/F_b)f$.

Almost all the provisions for compact sections in ASD are based on maximum stress, not upon first yield.

6-5   WEAK-AXIS BENDING

Sometimes, as shown in Fig. 6-7a, purlins, girts, or some columns can be subject to loads causing weak-axis bending. For these cases, basically what is involved is a member turned on its side such that what we are dealing with are the two rectangles that constitute the flanges. The resistance to bending of the web when it is on the axis of bending is negligible. The two rectangles of the section (the flanges) each have a shape factor

---

* *Hybrid girders* are beams whose flanges are fabricated from a stronger grade of steel than that in their webs.

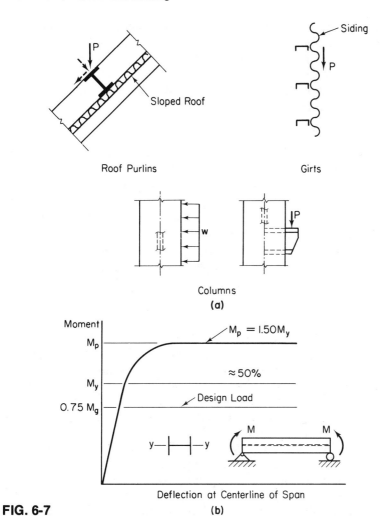

FIG. 6-7

equal to 1.50 (neglecting the web), so there should be an additional 50% reserve strength beyond first yield, as shown in Fig. 6-7b.

Referring to Fig. 6-7b, and from the AISC Specification, the allowable tension and compression on extreme fibers, F_b, of doubly symmetrical I- and H-shape members having their flanges continuously connected to the web or webs and having the proper width-to-thickness ratio of unstiffened compression elements (e.g., half of the W-shape flange) is equal to

$$F_b = 0.75F_y \qquad (6\text{-}5)$$

This formula does not apply to A514 steel. This allowable stress may be used for solid round and square bars and solid rectangular sections bent about their weak axis.

The factor of safety based on full yielding of the cross section is $M_p/0.75M_y = 1.50M_y/0.75M_y = 2.0$. The factor of safety for this case has risen in spite of an increase in allowable stress.

6-6   BIAXIAL BENDING

When a member is subject to biaxial loading, it is suggested to use an interaction formula. If load applied to a member is not parallel to the member's principal axes, the load should be resolved into components along the two principal axes and the allowable stresses for strong- and weak-axis bending should be used. Figure 6-7a (roof purlins) illustrates the resolving of the nonparallel-to-the-principal-axis load, P, into parallel components. The suggested interaction formula is

$$\frac{M_x/S_x}{0.66F_y} + \frac{M_y/S_y}{0.75F_y} \le 1.0 \qquad (6\text{-}6)$$

where $F_{bx} = 0.66F_y$ and $F_{by} = 0.75F_y$.

Referring again to Fig. 6-7a, Roof Purlins, to be more exact, the vertical load, P, should be resolved into the components shown, except that the component parallel to the roof slope can be shown going through the purlin centroidal axis (x-axis) with a twisting counterclockwise moment equal to $P_x(d/2)$, where d is the depth of the purlin. Sag rods spanning in the direction parallel to the roof slope from purlin to purlin are usually provided to support the purlins in the weak axis direction and also to resist the P_x force. The sag rod is merely a tension rod designed to resist P_x or the summation of P_x forces below the roof ridge on one side of the roof.

A more detailed discussion of biaxial bending when the plane of loading does not coincide with either of the principal axes of the cross section may be found in *Structural Steel Design*, by L. S. Beedle and others, The Ronald Press Company, New York, 1974, and *Steel Design Manual*, by R. L. Brockenbrough and B. G. Johnston, United States Steel Corporation, 1968.

6-7   PLASTIC DESIGN

In PD, a load factor, F, of 1.7 is used for gravity loads regardless of the type of cross section. Some slight local yielding at working load is of no concern. This will always occur due to residual stresses, stress concentrations, erection stresses, and so on. Furthermore, if we ever load into the maximum stress zone, upon loading and unloading, further response will be elastic.

In PD, bending strength is based on a full yield of cross section. Maximum strength is assumed without accounting for strain hardening.

For the remainder of this chapter, PD will be discussed along with ASD for each of the subjects discussed. The AISC Specification provisions for both means of design will be given. Design examples will illustrate both design methods. The discussion of the behavior of indeterminate beams relates PD behavior assumptions, which are the basis for many of the elastic AISC Specification provisions.

The following design example illustrates the design procedures for ASD and PD, the use of the AISC Manual's ASD and PD Selection Tables, and the fact that ASD and PD yield the same answers (size of member) for all determinate beams.

EXAMPLE 6-1

Design the most economical A36 section for a continuously braced, 22-ft. single span simply supported, 2.0 kips/linear ft. uniformly loaded, determinate beam. Use **(a)** ASD and **(b)** PD methods to determine the section (Fig. 6-8). Assume strong-axis bending.

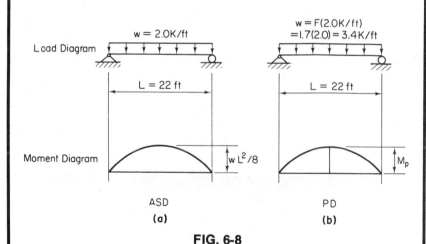

FIG. 6-8

solution:

(a) ASD:

$M = wL^2/8 = 2.0(22)^2/8 = 121$ kip-ft. Assume that $F_b = 0.66F_y = 24$ ksi. $S_{req} = M/F_b = 121(12)/24 = 60.5$ in.^{3}. From the AISC Manual, ASD Selection

Tables, *try* W 16 × 40 (S_x = 64.7 in.^{3}) (the most economical section is in boldface type; and F_y = 36 ksi
< F_y' and F_y'', so the section is compact and our orig-
inal assumption of F_b is correct). Web shear check:
V_{max} = $wl/2$ = 2.0(22)/2 = 22 kips, V_{allow} = 70 kips,
(from the bottom of the Beam Load Tables, V_{allow} =
$F_v dt$). *Use* W 16 × 40. (*Note:* For a discussion of F_y'
and F_y'', see Sections 6-8 and 6-9, respectively.)

(b) PD:
M_p = $wL^2/8$ = 3.4(22)^2/8 = 206 kip-ft. Z_{reg} = M_p/F_y =
206(12)/36 = 68.7 in.^{3}. From the AISC Manual, PD Selection Tables, *use* W 16 × 40 (Z_x = 72.9 in.^{3}) (the
most economical section is in boldface type).

The two solutions in Example 6-1 are based upon the idealized
behavior, curve 2 of Fig. 6-5. To achieve this behavior, lateral buckling
and local flange and web buckling must be controlled. One or both types
of buckling will always eventually cause failure of the member, but only
after the structure becomes useless because of excessive deflection. Sections that satisfy the width-to-thickness and bracing requirements are
called compact sections.

6-8   LOCAL FLANGE BUCKLING

Figure 6-9 shows how a member can fail by local flange buckling. When
a beam is loaded with significantly large loads, the steel does not break.
Yielding is not a big problem—stability is the problem.

There are two different cases of flange buckling. The first case is
an element supported only along its back edge, such as the compression
flange of a W shape, which can buckle. We term this an *unstiffened
compression element*. It is defined in Section 1.9.1 of the AISC Speci-

FIG. 6-9 Local flange buckling.

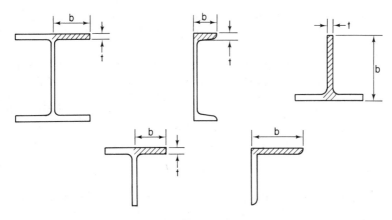

FIG. 6-10

fication. In addition, Section 1.9.1 presents the limiting width-to-thickness ratios for various sections. The thickness to be used for the flanges of channels or S shapes is the average thickness. When these limiting values in Section 1.9.1 are exceeded, Appendix C of the AISC Specification is to be used. Figure 6-10 shows some examples of unstiffened (projecting) compression elements.

The other case is that of a *stiffened compression element* (e.g., the compression flange of a box section). It is defined in Section 1.9.2 of the AISC Specification. The b/t requirements are relaxed about 50% when compared to unstiffened elements. Section 1.9.2 also presents limiting width-to-thickness ratios for various shapes. Appendix C of the AISC Specification should be used when these parameters are exceeded. Figure 6-11 illustrates some examples of stiffened (restrained) compression elements. So these are restrained and unrestrained compression elements, and there is a large difference in their carrying and buckling capacity.

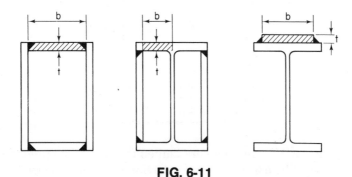

FIG. 6-11

The stiffnesses of the two members are different, and the buckling load, which we get on a projecting element, is much less than that on a restrained element.

The restrictions (limitations) on width-to-thickness ratios of unstiffened and stiffened compression flanges for ASD and PD are given in the AISC Specification. For compact sections, the AISC Specification states that the limiting width-to-thickness ratio of unstiffened projecting elements of the compression flange is

$$\frac{b}{t} \leq \frac{65}{(F_y)^{1/2}} \qquad (6\text{-}7a)$$

For stiffened elements of the compression flange, the AISC Specification states the limiting width-to-thickness ratio to be

$$\frac{b}{t} \leq \frac{190}{(F_y)^{1/2}} \qquad (6\text{-}7b)$$

Equations 6-7a and 6-7b are for ASD.

For PD, the limiting width-to-thickness ratios of the compression flanges of W shapes are given in tabular form in the AISC Specification. These are shown in Table 6-2 for different values of F_y. The limiting values for PD for stiffened compression elements are the same as for ASD (Eq. 6-7b). The limiting b/t values for ASD are given in Table 6-2 for comparison with the PD values. These values are also given in Appendix A of the AISC Specification. The difference between the ASD and PD values is due to the necessity of larger rotational capacity for PD. Local buckling for PD is more critical.

Table 6-2

	Limiting b/t	
F_y	PD	ASD
36	8.5	10.8
42	8.0	10.0
45	7.4	9.7
50	7.0	9.2
55	6.6	8.8
60	6.3	8.4
65	6.0	8.1

Experimental data to date are limited for very high strength steels. Because of this, the use of compact behavior and plastic design is limited to steel strength levels up to $F_y = 65$ ksi.

The AISC Manual lists values of F_y' in the ASD Selection Table for Shapes Used as Beams (Part 2 of the Manual) and in the Properties for Designing Table for various shapes. F_y' is derived from Eq. 6-7a. If the equation is solved for F_y, F_y' is arrived at. F_y' is the hypothetical yield stress of the material for which b/t is just satisfactory. If the value of the actual yield stress is beyond F_y', the section is noncompact and local flange buckling is not controlled. F_y' is defined in Eq. 6-7c for sections symmetrical about their minor axis.

$$F_y' = \left(\frac{65}{b_f/2t_f} \right)^2 \qquad (6\text{-}7c)$$

The AISC Specification relating to the limiting width-to-thickness ratios for stiffened and unstiffened compression elements subject to compression due to bending is equally applicable to axial compression (Chapter 5). Further, circular tubular elements subject to axial compression are fully effective when the outside diameter-to-wall thickness ratio $\leq 3300/F_y$, and when $3300/F_y <$ outside diameter-to-wall thickness ratio $< 13,000/F_y$, Appendix C of the AISC Specification is applicable. For hollow circular tubular compact bending members, the diameter-thickness ratio $\leq 3300/F_y$.

6-9   WEB BUCKLING

Figure 6-12 shows a beam that has exhibited a stability failure—in this case, failure by buckling of the web. Web buckling depends on the stress distribution in the web and the presence of axial force in addition to moment. The latter alters the stress in the web. The compact section criteria spelled out in the AISC Specification include the effect of axial stress. These slenderness requirements for compact sections try to ensure that web yielding occurs before web buckling commences.

The AISC Specification for compact sections limits the web depth-to-thickness ratio as follows: when $f_a/F_y \leq 0.16$,

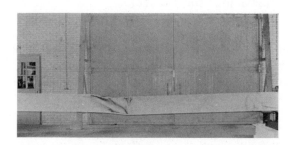

FIG. 6-12 Web buckling.

$$\frac{d}{t} \le \frac{640}{(F_y)^{1/2}} \left[1 - 3.74 \left(\frac{f_a}{F_y} \right) \right] \qquad (6\text{-}8a)$$

[When $f_a = 0$, as is the case in most beams, then Eq. 6-8a becomes $d/t \le 640/(F_y)^{1/2}$.] And when $f_a/F_y > 0.16$,

$$\frac{d}{t} \le \frac{257}{(F_y)^{1/2}} \qquad (6\text{-}8b)$$

where f_a is the computed axial stress, d is the nominal depth or length of the element, and t is the thickness of the element.

From Eq. 6-8a, when axial load is zero, the d/t limitation becomes $640/(F_y)^{1/2}$. In this case, half the web is in tension and the other half is in compression. It is worth noting that this parameter is higher than Eq. 6-7a for projecting flange compression elements and Eq. 6-7b for unstiffened flange compression elements. The reason for this is that stress does not exist over the entire section and that the flanges offer restraint.

If axial load is on the member, the axial load is a compressive load, which lowers the neutral axis of the section so that more of the cross section is in compression, less in tension, and a more severe d/t restriction is to be expected.

The upper limit would be when the neutral axis is completely out of the beam and full plasticity is developed. Uniform stress exists over the entire depth of the member. For this case, $f_a/F_y > 0.16$, and Eq. 6-8b is applicable. The entire web has a uniform compressive stress distribution at ultimate load. These requirements (Eqs. 6-8a and 6-8b) are for ASD.

For PD, the AISC Specification web depth-to-thickness limiting ratio is as follows: when $P/P_y \le 0.27$,

$$\frac{d}{t} \le \frac{412}{(F_y)^{1/2}} \left[1 - 1.4 \left(\frac{P}{P_y} \right) \right] \qquad (6\text{-}8c)$$

and when $P/P_y > 0.27$, Eq. 6-8b is used, P is the applied axial load, and P_y is the plastic axial load equal to AF_y.

Two terms are introduced in ASD: F_y'' and F_y'''. F_y'' is the theoretical maximum yield stress in kips per square inch based on the depth-to-thickness ratio of the web, beyond which a particular shape is noncompact. F_y'' is only applicable to cases of pure bending (e.g., $f_a = 0$). F_y'' is derived from Eq. 6-8a when $f_a = 0$, as seen from Eq. 6-8d.

$$F_y'' = \left[\frac{640}{d/t} \right]^2 \qquad (6\text{-}8d)$$

F_y''' is the theoretical yield stress in kips per square inch based on depth-to-thickness ratio of the web, beyond which a particular shape is non-compact. F_y''' is for any condition of combined bending and axial stress. F_y''' is derived from Eq. 6-8b when $f_a/F_y > 0.16$, as seen from Eq. 6-8e.

$$F_y''' = \left[\frac{257}{d/t} \right]^2 \qquad (6\text{-}8e)$$

Values of F_y', F_y'', and F_y''' are listed in the AISC Manual Properties for Designing Tables for rolled shapes symmetrical about their minor axis. F_y' and F_y'' values are also listed in the ASD Selection Table for Shapes Used as Beams.

F_y', F_y'', and F_y''' are valuable parameters to the designer. At a glance the designer, knowing the yield stress of the material used and assuming the beam is properly braced laterally, can see if a section is compact ($F_y \le F_y'$ and F_y'' and F_y'''). If no axial load is present, and $F_y \le F_y''$, the web is compact. When axial load is present, and $F_y \le F_y'''$, the web is compact. If F_y is somewhere between F_y'' and F_y''', Eq. 6-8a must be checked.

Example 6-2 illustrates the use and value of F_y', F_y'', and F_y'''.

EXAMPLE 6-2

Determine whether a W 36 × 135 is compact for A36 steel. Assume the section to be continuously braced.

solution:

From W-Shapes Properties for Designing Tables, Part 1 of the AISC Manual:

$F_y = 36$ ksi $< F_y' = -$[projecting compression flange element, $b/t = b_f/2t_f = 7.6 < 65/(F_y)^{1/2} = 10.8$; and therefore this meets the compactness criteria]. A dash in the table signifies that $F_y' > 65$ ksi (all allowable grades of steel for compact use).

$F_y = 36$ ksi $< F_y'' = -$[web $d/t = 59.3 < 640/(F_y)^{1/2} = 107$ and meets the compactness criteria when $f_a = 0$]. Again, a dash in the table signifies that $F_y''' > 65$ ksi (all allowable grades of steel for compact section use).

$F_y = 36$ ksi $> F_y''' = 18.8$ ksi. Therefore, when combined bending and axial load is encountered, Eq. 6-8a must be checked with the appropriate f_a/F_y value for the limiting d/t ratio (since $F_y'' > 36$ ksi $> F_y'''$).

Since $d/t = 59.3$, from Eq. 6-8a: $59.3 = 640/(36)^{1/2}$ [1

FIG. 6-13   Lateral buckling.

$- 3.74(f_a/36)]$, $f_a = 4.29$ ksi, $A = 39.7$ in.^{2}, and $P = f_a A = 170$ kips.

If $P \leq 170$ kips, W 36 × 135 is compact (the d/t compactness criteria are met).

6-10   LATERAL BUCKLING

The problem of lateral buckling must also be dealt with. Lateral buckling is affected by the steel strength, the unbraced length of the compression flange, and the moment gradient along the member. Figure 6-13 shows a beam that has buckled laterally. Bracing has to be spaced close enough to prevent lateral buckling from significantly affecting the idealized plastic behavior. L_c is the maximum bracing interval of the compression flange at which the allowable bending stress, F_b, is $0.66F_y$ or the value obtained by Eq. 6-11, whichever is applicable. Equation 6-11 is discussed in Section 6-12.

The AISC Specification provisions for compact sections (ASD) include the provision that the compression flange be laterally supported at a limiting interval as expressed by Eqs. 6-9a and 6-9b (for other than box or circular members). L_c is the lesser of the two values (Eqs. 6-9a and 6-9b). L_b, the bracing interval, should be less than or equal to L_c.

$$L_c = \frac{76.0b_f}{(F_y)^{1/2}} \qquad (6\text{-}9a)$$

$$L_c = \frac{20,000}{(d/A_f)F_y} \qquad (6\text{-}9b)$$

where b_f is the flange width, d is the depth of section, and A_f is the area of the flange. L_c values for various rolled sections are listed in the ASD Selection Tables and the Allowable Uniform Load Beam Tables in Part 2 of the AISC Manual.

The PD rules for limiting lateral bracing intervals, as found in Part 2 of the AISC Specification, are as expressed in Eqs. 6-10a and b as L_b. When $+1.0 > M/M_p > -0.5$,

$$L_b \le r_y \left[\frac{1375}{F_y} + 25 \right] \qquad M_p \qquad (6\text{-}10a)$$

When $-0.5 \ge M/M_p > -1.0$,

$$L_b \le r_y \frac{1375}{F_y} \qquad M_p \qquad (6\text{-}10b)$$

where M is the lesser of the moments at the ends of the unbraced segment and M/M_p, the end-moment ratio. M/M_p is positive for reverse curvature and negative for single curvature bending.

Lateral buckling control is not completely understood to date, as is evidenced by the vast difference in appearance between ASD (Eqs. 6-9a and b) and PD (Eqs. 6-10a and b) lateral bracing requirements.

The two ASD equations measure two different kinds of distortion: distortion of the compression flange and rotation of the entire section. Both equations must be satisfied. The ASD equations make no distinction between uniform and moment gradient as PD equations do. For PD, Eq. 6-10b, which is for uniform moment, is more severe in terms of loading than Eq. 6-10a, which is for moment gradient, so that the requirements for bracing are more severe. The uniform-moment case in design is unusual.

In general, the PD provisions usually permit larger unbraced lengths. Only for the case of uniform moment will PD require a shorter bracing interval than for ASD.

6-11   BEHAVIOR OF INDETERMINATE BEAMS

This section will discuss the behavior of indeterminate beams when adequately braced compact sections are used. In effect, this means that the unbraced length is less than L_c (Eqs. 6-9a and b).

If a beam is loaded and fixed at the ends, such as the case where the restraint of the columns to which the beam frames is such that practically full fixity occurs at the ends, moments are developed at the end and at the midspan. If the stiffness of the columns is sufficiently great, the moment at the ends will be the controlling value. The moment at the end would be $wl^2/12$, and the moment at midspan would equal $wl^2/24$.

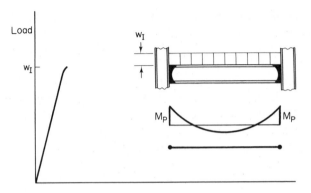

Deflection at Centerline of Beam Span

(a)

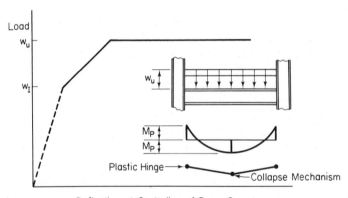

Deflection at Centerline of Beam Span

(b)

FIG. 6-14

In accordance with ASD, the member is loaded and elastic behavior exists up until the time first yield occurs. The limit of usefulness is based on the yielding of the cross section at one point only. First yield would occur at the outermost fibers at the columns, and this would represent the limiting condition. This condition is represented in Fig. 6-14a as W_1. This does not really represent collapse of the beam. Continuing to load beyond point W_1 causes a plastic penetration into the cross section toward the neutral axis at the column faces. The plastic penetration means that the yield-point stress at other than the outermost fibers is developed. At the point at which complete yielding of the cross section at the face of the column occurs, the beam has still not failed. It has generated into a beam in effect that has pins at the ends with applied moments equal to the full plastic moments, M_p, at the ends.

When the beam load increases, the magnitude of stress at midspan is gradually increased, and a redistribution in the moment diagram will occur such that the condition changes from M_p at the ends and something less at midspan to one in which full plasticity occurs at the ends as well as at midspan. The moments at the ends were limited to the bending strength of the cross section, $M_p = ZF_y$, and plastic hinges then form and there is a redistribution of stress to the center portion of the beam until it also yields and forms a plastic hinge at the center. Moment redistribution has occurred and a plastic mechanism has developed. The beam is now in a state of collapse.

The basic principle of PD is that of a collapse mechanism, as shown in Fig. 6-14b, with hinges not only at the columns, but also at the midspan for collapse to occur. This load, W_u, as shown in Fig. 6-14b, represents one where very large deflections with minimal increase in load are encountered. This is PD, which uses W_u as the limit of usefulness. This gives rise to simple design procedures that are easier than ASD.

ASD attempts to take advantage of the same sort of moment redistribution by using $W_1 + 10\%$ in an attempt to approach W_u. ASD does this in an indirect manner by reducing the moment at the support by 10% and increasing the magnitude of the midspan moment by 10%. This principle is expressed in the AISC Specification.

As a practical matter, with a load of $W_u/1.7$, all behavior will really be more like the elastic. Plastic hinges do not occur at working load. Figure 6-15 shows a photograph of a plastic hinge. The area shown has been forced much beyond anything that would be tolerated in a real design situation. The beam has been whitewashed to bring out the movement.

Example 6-3 is a design example illustrating ASD procedures (the 10% redistribution rule), while Example 6-4 illustrates PD procedures.

EXAMPLE 6-3 (ASD)

Given the continuously braced (laterally) two-span continuous uniformly loaded A36 beam of uniform cross section as shown in Fig. 6-16, design the beam for the most economical section using the AISC Specification ASD procedures.

The moments shown in Fig. 6-16b are calculated by the usual elastic procedures, $wL^2/8$ at the interior support and $9wL^2/128$ in the span. These values are computed by any of the usual analyses (slope deflection, moment dis-

FIG. 6-15 Plastic hinge.

tribution, etc.) or they can be gotten from Part 2 of the AISC Manual (Beam Diagrams and Formulas).

The solid line shown in Fig. 6-16c represents the moment diagram in Fig. 6-16b. The AISC Specification says that the moment over the support (for compact sections that are continuous) can be reduced by 10% provided that we increase the positive moment by 10%, the average decrease at the two ends of the span (10% of the average negative moment). This is shown as the dotted line in Fig. 6-16c. The 10% redistribution rule can be used only if the support moment is larger than the maximum positive moment.

solution:

Design moment = 141 kip-ft, $S_{req} = M/0.66F_y = 141(12)/24 = 70.5$ in.^{3}. From the AISC Manual, ASD Selection Table: entering with $S_{req} = 70.5$ in.^{3}, *try* W 21 × 44 ($S = 81.6$ in.^{3}).

Checking F'_y and F''_y values in the table, we find a dash indicated, which means that both values are larger than the highest strength of steel permitted for the compact section assumptions (65 ksi). Since F'_y and $F''_y > F_y = 36$ ksi, the trial section is compact and F_b does equal $0.66F_y$ and the redistribution value is valid (the redistribution of moment from supports to midspan is only valid for compact continuous sections).

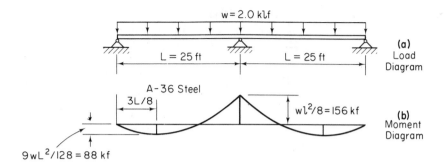

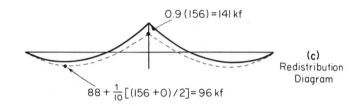

FIG. 6-16

Use W 21 × 44.

It is of value to note that, if the advantage of this re-distribution of moment idea was not used, a larger section would be required in many cases.

EXAMPLE 6-4 (PD)

Given the three-span uniform cross section adequately braced uniformly loaded A36 beam shown in Fig. 6-17, de-sign the cross section and solve for the most adequate spacing of supports (position the two interior supports so that the lightest or least-weight member is obtained). The uniform load is 2.4 klf.

In one method of plastic analysis, internal work is equated to external work at the assumed collapse load. At collapse load, M_p, the full plastic moment, is at each plastic hinge. Notice that only two plastic hinges are assumed in the end span because there already exists a real hinge at the exterior support. A plastic hinge is also assumed at the center of end span L_1 (Fig. 6-17b). Because of symmetry, the two end spans behave similarly. The assumption of a plastic hinge at $L_1/2$ is the same as assuming maximum moment at midspan. The actual location is $0.41 L_1$ from the

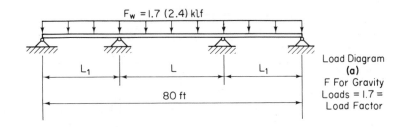

Load Diagram
(a)
F For Gravity
Loads = 1.7 =
Load Factor

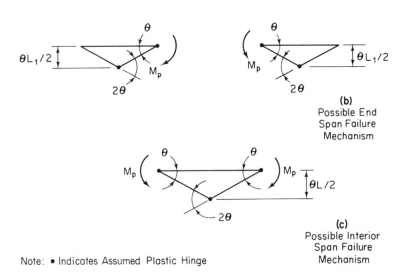

(b)
Possible End
Span Failure
Mechanism

(c)
Possible Interior
Span Failure
Mechanism

Note: • Indicates Assumed Plastic Hinge

FIG. 6-17

end support. The midspan approximation causes only a 3% error in the design moment.

solution:

For the end-span failure mechanism (Fig. 6-17b), internal work is simply $M_p\theta$ (θ is the angle of rotation) at the right end of the end span plus $M_p(2\theta)$ at the interior hinge. So the internal work equals $M_p\theta + M_p(2\theta)$ or $3M_p\theta$. M_p is the plastic moment that exists at each assumed hinge location.

External work is to say we have a load, F_w, moving along the span through a deflection that varies along the span. At the center of the end span, $\tan\theta = \Delta/(L_1/2)$ ($\Delta =$

vertical deflection) or, for small angles, tan $\theta = \theta$ an
$\theta L_1/2$. We multiply load times L_1 by Δ, or

$$F_w L_1 \left(\frac{\theta L_1}{2}\right)\left(\frac{1}{2}\right) = \frac{F_w \theta L_1^2}{4}$$

This is the external work, which is actually the area under the deflected shape multiplied by F_w.

Equating internal work and external work and solving for M_p,

$$M_p = \frac{F_w L_1^2}{12} \tag{A}$$

For the interior-span failure mechanism, the other possible failure mechanism (Fig. 6-17c), a plastic hinge occurs over each of the supports and at midspan. Again, equating internal and external work (M_p at both ends and at center and θ at both ends and 2θ at center),

$$\text{internal work} = \text{external work}$$

$$4 M_p \theta = F_w \left(\frac{\theta L}{2}\right)\left(\frac{L}{2}\right)$$

Solving for M_p,

$$M_p = \frac{F_w L^2}{16} \tag{B}$$

Following the original requirement for the most efficient design, a solution is sought after where the two collapses occur simultaneously. The assumed end-span failure mechanism and the assumed interior-span failure mechanism will occur at the same time with the same required M_p value.

Equating M_p values for end span and interior span,

$$\frac{F_w L_1^2}{12} = M_p = \frac{F_w L^2}{16}$$

Solving for L_1: $L_1 = 0.866L$. *Use* $L_1 = 25$ ft and $L = 30$ ft.

$M_{p\,req} = F_w L^2/16 = 1.7(2.4)(30)^2/16 = 230$ kip-ft
$Z_{req} = M_p/F_y = (12)230/36 = 76.7$ in.^{3}

From the PD Selection Table in the AISC Manual, *try* W 18 × 40 ($Z = 78.4$ in.^{3}).

The section meets all width or depth thickness requirements for flange and web: $b_f/2t_f = 5.7 < 8.5$; $d/t_w = 56.8 < 640/(F_y)^{1/2} = 107$. *Use* W 18 × 40.

Referring to Fig. 6-5, curve 3 shows a beam where lateral buckling is controlled but where flange or web slenderness ratios exceed the compactness limits. PD would not be permitted in this case.

6-12   *NONCOMPACT ALLOWABLE STRESSES*

ASD currently permits a gradual change in allowable bending stress when the compactness limits for unstiffened projecting compression elements are exceeded. F_b is somewhere between $0.66F_y$ and $0.60F_y$ for strong-axis bending. For members that meet all the compact requirements with the exception of Eq. 6-7a and $65/(F_y)^{1/2} < b/2t < 95/(F_y)^{1/2}$, F_b should be computed as in Eq. 6-11. This equation (Section 1.5.1.4.2, AISC Specification) is for values of F_y between 36 and 65 ksi and does not apply to hybrid girders.

$$F_b = F_y \left[0.79 - 0.002 \left(\frac{b_f}{2t_f} \right) (F_y)^{1/2} \right] \qquad (6\text{-}11)$$

where b_f and t_f are the flange width and thickness, respectively. This is a linear transition of allowable stress between compact and noncompact members. Table 6-3 shows these values for various sections when $b_f/2t_f > F'_y$ but is $< 95.0/(F_y)^{1/2}$. Appendix C of the AISC Specification presents a procedure for very large b/t ratios. Since large b/t ratios are relatively infrequent and are not encountered in any current rolled sections, a further discussion of the Appendix will not be made. Appendix C provisions are used when $b_f/2t_f > 95.0/(F_y)^{1/2}$ for such fabricated sections.

ASD also currently permits a gradual change in allowable bending stress for doubly symmetrical I- and H-shape members bent about their weak axis when the compact requirement of the flanges being continuously connected to the web or webs is met, but $65/(F_y)^{1/2} < b_f/2t_f < 95.0/(F_y)^{1/2}$. F_b values vary between $0.75F_y$ and $0.60F_y$. F_b (Section 1.5.1.4.3, AISC Specification) is computed as follows:

$$F_b = F_y \left[1.075 - 0.005 \left(\frac{b_f}{2t_f} \right) (F_y)^{1/2} \right] \qquad (6\text{-}12)$$

This equation is for values of F_y between 36 and 65 ksi and does not apply to hybrid girders.

Table 6-3

NONCOMPACT "F_b" STRESSES									
			F_y						
SECTION	$b_f/2t_p$	F'_y	36	42	45	50	55	60	65
W 14 × 109	8.5	58.6						39.5	42.4
W 14 × 99	9.3	48.5				32.9	35.9	38.8	41.6
W 14 × 90	10.2	40.4		27.6	29.4	32.3	35.1	37.9	40.7
W 12 × 79	8.2	62.6							42.8
W 12 × 72	9.0	52.3					36.1	39.1	41.9
W 12 × 65	9.9	43.0			29.6	32.5	35.4	38.2	41.0
W 10 × 54	8.1	63.5							42.8
W 12 × 53	8.7	55.9						39.3	42.3
W 10 × 49	8.9	53.0					36.2	39.2	42.0
W 16 × 36	8.1	64.0							42.8
W 8 × 35	8.1	64.4							42.8
W 10 × 33	9.1	50.5					36.0	38.9	41.8
W 8 × 31	9.2	50.0					36.0	38.8	41.7
W 14 × 30	8.7	55.3						39.3	42.3
W 12 × 26	8.5	57.9						39.5	42.4
W 8 × 24	8.1	64.1							42.8
W 6 × 20	8.3	62.1							42.6
W 6 × 15	11.5	31.8	23.5	26.9	28.6	31.4	34.0	36.7	39.3
W 12 × 14	8.8	54.3					36.2	39.2	42.1
W 10 × 12	9.4	47.5				32.9	35.8	38.6	41.5
W 8 × 10	9.6	45.8				32.7	35.6	38.5	41.3
W 6 × 9	9.2	50.3					36.0	38.8	41.7
M 8 × 34.3	8.7	55.6						39.3	42.2
M 8 × 32.6	8.7	56.5						39.4	42.3

Courtesy of American Institute of Steel Construction.

6-13 BOX SECTIONS/CIRCULAR SECTIONS

Curve 4 of Fig. 6-5 is typical of sections with noncompact webs, such as welded plate girders, in general, where thin webs are used. It is also typical of box girders that are unbraced laterally. Box girders usually have thin webs also.

The allowable tension and compression, F_b, on extreme fibers of box-type flexural members, whose compression flange or web width-to-thickness ratio exceeds Eqs. 6-7b, 6-8a, and 6-8b, respectively (compact requirements), but does conform to the width-to-thickness ratio requirements of Section 1.9 of the AISC Specification, is $0.60F_y$.

The Specification for box sections is somewhat less severe than that for W shapes because the box girder is very stable laterally and torsionally stiff. Lateral torsional buckling, according to the AISC Specification, is not a problem when the depth of the box section is less than 6 times its width. Beyond this depth-to-width ratio, the lateral support requirements must be determined by special analysis.

In summary, when $190/(F_y)^{1/2} < b_f/t_f < 238/(F_y)^{1/2}$ and $d \leq 6b_f$, $F_b = 0.60F_y$.

A subparagraph in the AISC Specification compactness section covers the case of compact box sections. When $d \leq 6b_f$, $t_f \leq 2t_w$ and the laterally unsupported length meets the condition of Eq. 6-13, $F_b = 0.66F_y$. This does not apply to hybrid girders or A514 steel.

$$l_b \leq \frac{[1950 + 1200(M_1/M_2)]b_f}{F_y} \qquad (6\text{-}13)$$

except that l_b need not be less than $1200(b_f/F_y,)$ where b_f is the flange width, t_f the flange thickness and t_w the web thickness, and M_1 is the smaller and M_2 the larger bending moment at the ends of the unbraced length, l_b. Further, the end-moment ratio M_1/M_2 is positive for reverse curvature bending and negative for single curvature bending. The bracing requirement expressed in Eq. 6-13 takes moment gradient into account.

For flexural box sections bent about their weak axis, continuously welded to its webs, and that meet Eqs. 6-7b, 6-8a, and 6-8b, $F_b = 0.66F_y$.

The AISC Specification compactness section also covers hollow circular sections. When the diameter-to-thickness ratio $\leq 3300/F_y$, $F_b = 0.66F_y$.

6-14   DESIGN AIDS

The AISC Manual has valuable tables and charts that greatly aid the design. These charts and tables, many of which have already been mentioned and discussed, make computations easier. The *AISC Engineering Journal* (a quarterly technical journal) and *Civil Engineering's Engineer's Notebook* (an American Society of Civil Engineers' monthly periodical) often give additional design aids that simplify and quicken design computation.

6-14.1   Beam Selection Tables

The AISC Manual contains ASD and PD Selection Tables in which are listed, for shapes in descending order of strong-axis section modulus, S_x, for ASD, or strong-axis plastic modulus, Z_x, for PD, many valuable design parameters, such as L_c, L_u, M_R, F'_y, A, d/t_w, r_x, r_y, M_p, and P_y for 36- and 50-ksi steels. These tables allow the selection of bending members on the basis of ASD or PD rules as found in the AISC Specification. The parameters are as previously defined or as in the definitions that follow.

L_u is the largest unbraced length of the compression flange in feet for which $F_b = 0.60F_y$.

L_u is derived from solving for l in feet from Eq. 8-7c using $C_b = 1.0$ and $F_b = 0.60F_y$. For some shapes L_u is derived from the defining l/r_T lower-bound condition for Eq. 8-7a by solving for l in feet and using a value of $C_b = 1.0$. For the latter case, $L_u = \{[102(10^3)r_T^2]/144F_y\}^{1/2}$. The more liberal value of the two aforementioned means of arriving at L_u should be used.

M_R is the beam resisting moment $= F_b S_x/12$, where F_b is $0.66F_y$ for compact sections; it is computed by Eq. 6-11 for noncompact sections (when $F_y > F'_y$), and is $= 0.60F_y$ for noncompact sections (when $F_y > F''_y$).

$M_p = F_y Z_x$ (in kip-ft).
$P_y = F_y A$ (in kips).

Knowing S_x or M_R, enter the ASD Selection Tables and select a shape having values equal to or greater than S_x or M_R. That shape and all shapes above it are strong enough for the encountered loading. The first shape appearing in boldface type adjacent to or above the S_x or M_R is the most economical (lightest) beam for A36 steel. For 50-ksi steel, the tables must be scanned in the vicinity of S_x or M_R for the lightest section. The lateral bracing interval should be checked, depending on the allowable stress assumed (e.g., $0.66F_y$, Eq. 6-11, or $0.60F_y$). When a blank is shown under the L_c column, the maximum bracing interval will be L_u. For beams of greater bracing interval, the procedures described in Chapter 8 should be used. Finally, a check on the web shear capacity should be made (Eq. 4-2).

For fully laterally supported, uniformly loaded W, M, S, C, MC, and L shapes, simply supported (or some equivalent symmetrical loading), the ASD uniform load constants for beams tables can be utilized. These tables are also found in the AISC Manual. They list constants for the direct computation of the total maximum allowable uniform load for laterally supported beams for $F_y = 36$ ksi steel. Tables for $F_y = 50$ ksi steel are also given for W and M shapes.

The listed uniform load constant W_c is computed from the moment stress relationship for a uniformly loaded, simply supported beam and is equal to a value of $2S_xF_b/3$ kip-ft. The total allowable uniform load is $W = W_c/L$ in kips.

For symmetrical shapes (W, M, and S shapes), the load constants are based upon L_b (the bracing interval) $\leq L_c$. When $L_c < L_b < L_u$, the tabulated constant must be multiplied by $0.60F_y$ divided by the allowable stress used to compute its capacity. For noncompact shapes the allowable stress is $0.60F_y$ or a value between $0.60F_y$ and $0.66F_y$, depending on $b_f/2t_f$ (Eq. 6-11). The F_b value from Eq. 6-11 is flagged by a footnote to the tables referring to AISC Specifications Section 1.5.1.4.2.

When $L_b > L_u$, the charts should be used as described in Chapter 8.

For PD, the PD Selection Tables list Z_x in descending order for W and M shapes, which are appropriate for PD for 36- and 50-ksi steel when no axial load is present (except that for 50-ksi steel, where values of M_p and P_y do not appear). The other parameters listed are useful for the appropriate PD formulas included in the AISC Specification, Part 2.

6-14.2   Allowable Beam-Moment Curves

The allowable moment curves found in Part 2 of the AISC Manual provide a simple design solution, since the allowable stress varies almost continuously with $b_f/2t_f$ (for noncompact sections; Eq. 6-11), and without these charts, the allowable stress would have to be guessed at to design the beam.

The ordinate to the curves is the total allowable moment in kip-feet, and the abscissa, the unbraced length, L_b in feet. The charts give the total allowable moment of various sections for all unbraced lengths (e.g., $L_b < L_c$, $L_b > L_c$, and L_b between L_c and L_u). This design aid will be treated in more detail in Chapter 8.

For the purpose of fully braced beams, the chart can be entered with an abscissa value of $L_b = 0$, and knowing the maximum moment in the beam, a section can be chosen. The curve automatically accounts for the proper bending stress as a function of $b_f/2t_f$. These charts are handy to select beams that are braced but not compact.

6-14.3   Dimensions for Detailing and
Properties for Designing Tables

Part 1 of the AISC Manual provides tables that list the various dimensions for detailing of W, M, S, HP, C, and MC shapes, as well as the design properties for these shapes and L shapes. Additional property tables are given in Part 1 of the AISC Manual for double angles, structural tees, and tubing.

The Properties for Designing Tables provides values of strong- and weak-axis moments of inertia, section moduli, and radii of gyration, as well as F_y', F_y'', and F_y'''. Dimensions and properties are representative, for the most part, of all producers offering the particular sections. The exceptions are listed in the AISC Manual in the text preceding the tables.

6-14.4 Beam Diagrams and Formulas

Part 2 of the AISC Manual provides beam diagrams and formulas for cases frequently encountered in design. Various loading diagrams of beams with particular end-support conditions are given, together with shear and moment diagrams. Formulas for reactions, shears, and moments are also provided for these cases.

6-14.5 Camber and Deflection Tables

The AISC Manual, Part 2, also provides tables that give camber and deflection coefficients for simply supported spans uniformly loaded. These coefficients enable the designer to find the center of span deflection (in inches) by multiplying the span length (in feet) by the tabulated coefficients. The span length, depth of section, and maximum design unit bending stress (ksi) must be known.

6-14.6 Quick Method for Estimating Beam Size

This rapid method of estimating beam size of simply supported, uniformly loaded, laterally supported beams with an allowable stress of $F_b = 24$ ksi was recognized by a designer and subsequently published in a technical magazine.* It is a useful and simple approach that can be utilized for estimating, preliminary design, or checking.

The hypothetical required section depth in inches is $d = \frac{1}{2}$ (span in feet). The hypothetical required section weight in pounds per foot is $w = (5W/4)$, where W is the total load in kips. The hypothetical deflection of the required section in inches is $\frac{1}{10} d$.

EXAMPLE 6-5

Without the aid of the AISC Manual, determine the hypothetical required W shape and its maximum deflection for

---

* Adrian J. Perez, Estimating Beam Size, *Civil Engineering,* December 1971, p. 67.

an A36 compact section that is simply supported and fully laterally supported. The span length is 20 ft and the span is uniformly loaded with a load of 4 klf.

solution:

(a) $W = 4(20) = 80$ kips.

(b)   Required beam depth, d: $d = (\frac{1}{2})20 = 10$ in.

(c)   Required beam weight, w: $w = (\frac{5}{4})80 = 100$ plf.

(d)   Maximum deflection, Δ: $\Delta = (\frac{1}{10})10 = 1.0$ in.

(e)   Hypothetical section: W 10 × 100.

Try W 10 × 100: $S_x = 112$ in.³, $M_{max} = wl^2/8$ (kip-ft), $F_y = 36 < F_y', F_y''$, and F_y''', $F_b = 24$ ksi, $I = 623$ in.⁴.

   $F_b S_x = 12M_{max} = 12wl^2/8$.

   $w = 8F_b S_x/12l^2 = 8(24)112/12(400) = 4.48$ k/f.

   $wl = W_{allow} = 4.48(20) = 89.6$ kips > 80 kips.

   Use W 10 × 100.

   $\Delta = 5W_{allow}l^3/384EI = 5(89.6)(240)^3/384(29,000)623 = 0.89$ in. ≈ 1.0 in.

For illustrative purposes, assume that there is no size of section close to the hypothetical section. Therefore, go to the next deeper section.

   Try W 12: $w = (10/12)100 = 84$ p/f, $\Delta = (10/12)1.0 = 0.833$ in.

The closest real shape is W 12 × 87.

From the AISC Manual, Uniform Load Constants for Beams Table: for a W 12 × 87, span = 20 ft, A36 steel, $W_c = 1890$ kip-ft and $W = W_c/L = 94$ kips, and $\Delta = 0.79$ in.* $D_c L^2/1000 = 2.0(4.00)/1000 = 0.80$ in.

This method can be extended to other loading conditions as follows:

(1)   (For total concentrated load at midspan, double beam weight as computed above and take 80% of computed deflection.) $W =$ concentrated load at midspan: $w' = 2w$; $\Delta' \simeq 0.80\Delta$.

(2) $W/2$ at third points of span: $w' = 1.33w$; $\Delta' \simeq \Delta$ (slightly larger than Δ). (For total load divided into two equal concentrated loads at the third points of the span, increase the computed beam weight by one third. The deflection will be negligibly larger.)

---

* Section 8-3.3.

The derivation of the method is as follows: for any rolled W-shape, $dw \approx 10S$. Assuming d in inches = $L/2$ (L = span length in feet), since max $M = WL/8$ for uniform loading and F_b = 24 ksi for A36 steel (compact section):
$S = 12M/F_b = 12(WL/8)/24 = WL/16$.

Since $S \approx dw/10$ and $d \approx L/2$, then $Lw/20 \approx WL/16$ and $w \approx 1.25W$.

Further, from the AISC Manual, Part 2, Δ_{max} at the center of a simply supported beam uniformly loaded = $(5/384EI)wl^4 = 0.02483L^2/d$ or $\Delta \approx 0.025L^2/d$ and for $d \approx L/2$, $\Delta \approx d/10$.

6-15 BEARING-PLATE DESIGN/PREVENTION OF WEB CRIPPLING DESIGN

Just as a column base plate (Section 5-8) transmits and distributes loads to masonry supports or footings, when a beam is supported by a masonry wall, a bearing plate serves to distribute the beam reaction over an area of the masonry sufficiently large enough to sustain the pressure on the masonry within some allowable value. The AISC Specification, in the absence of any other code regulation, suggests various allowable bearing pressures on supports, F_p.

The AISC Manual recommends the following design methods for determining the end bearing plate. From Fig. 6-18, R, the beam reaction in kips, is assumed to be uniformly distributed on the bearing plate area $2kN$. The bearing plate, in turn, is assumed to distribute this load uniformly over the masonry support area.

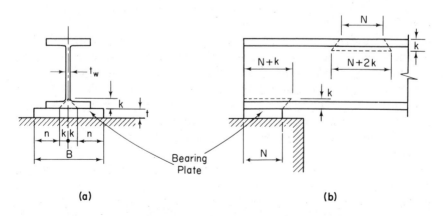

(a)   (b)

FIG. 6-18

(1) Establish values for F_b and F_p. F_b, the allowable bending stress in the bearing plate, is equal to $0.75F_y$ (weak-axis bending for a plate as per AISC Specification). F_p, the allowable bearing pressure on the support, is determined from the AISC Specification and is dependent on the type of support material (e.g., sandstone, limestone, brick in cement mortar, concrete). In the latter case (concrete), the allowable bearing pressure depends on whether the full area or less than the full area is used for bearing. In this case, the allowable bearing pressure is expressed as a fraction of f_c', the specified compression strength of the concrete (Eqs. 5-12 and 5-13).

(2) Find the required area of bearing plate, A. $A = R/F_p$.

(3) Assume a value for N (in whole inches), the length of the bearing plate. It is usually governed by a code or specification requirement and is dependent on the wall thickness used. $N_{min} = (R/0.75F_y t_w) - k$. Find $B = A/N$. Round B off to the next highest inch.

(4) Find the actual bearing pressure, f_p. $f_p = R/BN$.

(5) Determine n and solve for the required bearing plate thickness, t, by using the equation $t = (3f_p n^2/F_b)^{1/2}$. This equation is derived in a similar manner to that illustrated for column-base-plate design in Section 5-8.

(6) The check for web crippling as per the AISC Specification (Section 1.10.10) on length $N + k$ (N_{min} in step 3) involves

$$\frac{R}{(N + k)t_w} \leq 0.75F_y \qquad (6\text{-}14)$$

The AISC Specification suggests that the webs of the beams should be of such size that the compressive stress at the web toe of the fillets (distance k from the outermost surface of the beam flange) shall not be more than $0.75F_y$. The compressive stress can be created by concentrated loads not supported by bearing stiffeners. If Eq. 6-14 is not met, bearing stiffeners should be used or the length of bearing increased.

Figure 6-18b also shows the case where a concentrated load is transferred to the top of the upper beam flange over a length N. The stress at the toe of the upper flange fillet, due to some interior load, is assumed distributed over a length of bearing plus $2k$ or $(N + 2k)$ and to prevent web crippling as per the AISC Specification:

$$\frac{R}{(N + 2k)t_w} \leq 0.75F_y \qquad (6\text{-}15)$$

The length of bearing or length of concentrated interior load should not be less than k for end reactions. In both cases the compression flange of the beam must be properly braced laterally.

EXAMPLE 6-6

Design an end bearing plate for a W 16 × 50 beam supported on a pocket in a brick-and-cement mortar wall. Assume A36 steel, that the beam has an end reaction of 20 kips, and that the length of bearing is 6 in.

solution:

(a) $F_b = 0.75F_y = 27$ ksi (AISC Specification, Section 1.5.1.4.3), $F_p = 0.25$ ksi (AISC Specification, Section 1.5.5).

(b) $A = R/F_p = 20/0.25 = 80$ in.^{2}.

(c) $N > N_{min} = 6$ in. (given); $B = A/N = 80/6 = 13.3$. Use $B = 14$ in.

(d) $f_p = R/BN = 20/14(6) = 0.238$ ksi.

(e) $n = (B/2) - k = 7 - 1.31 = 5.69$ in. ($k = 1\frac{5}{16}$ in. from Dimension for Detailing for W-Shapes, Part 1 of the AISC Manual). $t = (3f_p n^2/F_b)^{1/2} = [3(0.238)(5.69)^2/27]^{1/2} = 0.925$ in. Use 1 in.

(f) $R/(N + k)t_w \le 0.75F_y = 20/(6 + 1.31)0.375 \le 27$ ksi $= 7.30 \le 27$ ksi. Use PL6 × 1 × 1 × 1' − 2.

Steel bearing plates are usually shipped separately to the construction site and grouted in place before the beam is put in place. The beam need not be attached to the bearing plate, but the beam should be properly attached to the wall by angle wall anchors, government anchors, or similar connecting devices. Suggested details are provided in Part 4 of the AISC Manual.

When high allowable bearing pressures or relatively light loads are encountered that yield small required bearing areas, and/or bearing-plate thicknesses, the beam may rest (without end bearing plates) directly on the support. In this case, the beam's bottom flange should be investigated for bending such that $t \ge (3F_p n^2/F_b)^{1/2}$, where F_p is the allowable bearing pressure on the support; $n = (b_f/2) - k$; $F_b = 0.75F_y$, which is the allowable bending stress in the flange, which is in this case a bearing plate.

6-16   DEFLECTION

Any discussion relative to the design of bending members used in floor or roof framing would be incomplete without something being said about deflection. Deflection, together with vibration and ponding, which will be discussed in the next two sections, are more design criteria, as opposed to stress criteria. That is, when considering deflection limitations, the sound judgment and experience of the designer with the behavior of similar structures are the governing factors. The only precise limits enumerated in the AISC Specification (Section 1.13.1) are the 1/360 of the span live-load deflection for beams supporting plaster ceilings and the ponding formulas to be checked for flat roofs. The ponding formulas will be dealt with in Section 6-18.

The Commentary to the AISC Specification gives as a guide the following: For fully stressed floor beams and girders, the depth, if practicable, shall not be less than $(F_y/800)L$, where L is the span length. If members of lesser depth are used, the unit stress in bending should be decreased in the same ratio as the depth is decreased from this.

For fully stressed roof purlins (except for flat roofs), the depth, if practicable, should not be less than $(F_y/1000)L$, where L is the span length.

Table 6-4 shows the depth-to-span ratios for various yield-stress levels arrived at from these formulas.

Table 6-4
MINIMUM DEPTH-TO-SPAN RATIOS

	F_y								
	36	42	45	50	55	60	65	90	100
Floor beams	$\frac{1}{22}$	$\frac{1}{19}$	$\frac{1}{18}$	$\frac{1}{16}$	$\frac{1}{15}$	$\frac{1}{13}$	$\frac{1}{12}$	$\frac{1}{9}$	$\frac{1}{8}$
Roof purlins	$\frac{1}{28}$	$\frac{1}{24}$	$\frac{1}{22}$	$\frac{1}{20}$	$\frac{1}{18}$	$\frac{1}{17}$	$\frac{1}{15}$	$\frac{1}{11}$	$\frac{1}{10}$

6-17   VIBRATION

Large open floor areas in a building, free of floor-to-ceiling (permanent) partitions or other sources of damping, may be susceptible to undesirable transient vibration caused by the footfall of pedestrian traffic. Although there are design methods available to check a floor system for vibration susceptibility, they necessarily involve trying to evaluate the difficult problem of human perception of vibration. The AISC Commentary (Section 1.13.2) recommends, as a guide, that the depth-to-span ratio of a steel beam be not less than $\frac{1}{20}$ to avoid a problem of perceptible transient vibration for large open areas, free of permanent (floor-to-ceiling) partitions. This ratio applies to the bare steel beam, regardless of whether the beam is composite or noncomposite with the slab. It also applies to any type of cross section of a bending member.

6-18 PONDING

Ponding refers to the retention of water on a roof surface resulting solely from the deflection of flat roof framing. The amount of retention is dependent upon the flexibility of the framing. If the framing lacks stiffness, the accumulation of water can conceivably cause a failure of the roof.

The AISC Specification (Section 1.13.3) proper presents a rational approach to this analysis, which assures stability when a roof surface is not provided with sufficient slope toward points of free drainage or with adequate individual drains that would prevent the accumulation of water. The Commentary to the AISC Specification provides a more exact method, which should be used when the Specification check is not satisfied. The procedure used in the Commentary was adopted from an *AISC Engineering Journal* (July 1966) paper by Frank J. Marino, Ponding of Two Way Roof Systems. This procedure will not be discussed, in favor of the Specification approach.

The roof-framing system is considered to be stable with no further stability investigations to be performed if Eqs. 6-16 and 6-17 are not violated.

$$C_p + 0.9C_s \le 0.25 \qquad (6\text{-}16)$$

$$I_d \ge \frac{25\,S^4}{10^6} \qquad (6\text{-}17)$$

where C_p and C_s represent stiffness factors for primary and secondary members in the framing of a flat roof, respectively, and

$$C_p = \frac{32L_s L_p^4}{10^7 I_p} \quad \text{and} \quad C_s = \frac{32SL_s^4}{10^7 I_s}$$

where L_p = length of the primary member, ft (the column spacing in the direction of the girder)

L_s = length of the secondary member, ft (the column spacing normal to the direction of the girder)

S = spacing of the secondary member, ft

I_p = moment of inertia, in.^{4}, for the primary member

I_s = moment of inertia, in.^{4}, for the secondary member

I_d = moment of inertia, in.^{4} per foot of the steel deck (when a steel deck is used), supported on secondary members

A typical roof framing plan is illustrated in Fig. 6-19.

The Specification goes on to state that I_s is reduced by 15% in the applicable equation for trusses and open web joists. Steel deck is consid-

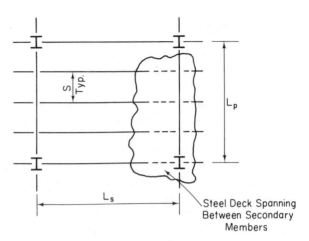

FIG. 6-19

ered a secondary member when directly supported by the primary members. Finally, the total allowable bending stress, F_b, due to dead loads, gravity live loads (if any), and ponding shall not exceed $0.80F_y$ for primary and secondary members.

In an article by Lewis B. Burgett, Fast Check for Ponding, published in the *AISC Engineering Journal,* January 1973, the equations for C_p and C_s are modified and used to develop Eq. 6-16 in a different form. Two graphs are developed to enable the designer to evaluate this new equation directly when knowing L_s, L_p, and S. A third graph is developed to check the stiffness of steel deck used and supported on the secondary members. The third graph is entered with I_d and S, and the intersection point shows when the deflection is satisfactory or unsatisfactory, completing the check on whether the roof system is stable or unstable. The article by Burgett introduces a graphical, rapid, and simple solution to Eqs. 6-16 and 6-17.

Ponding becomes increasingly critical as the lengths of supporting members increase. The designer should be particularly aware of the possibility of ponding when bay sizes (center line to center line of column) are large.

PROBLEMS

(6-1) Using ASD, choose the most economical W shapes (A36 steel) for the beams and girders in Problem 2-2. Assume full lateral support of the compression flange.

(6-2) Assuming full lateral support of the compression flange for each of the simply supported members shown, choose the most economical W shapes (A36 steel) for each beam and girder. Use ASD in all cases. Assume 60 psf live load and 3-in. concrete deck (150 pcf). In all cases, be sure to indicate a check for (a) dead load of chosen beam, (b) shear, (c) deflection.

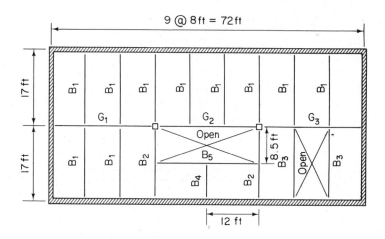

(6-3) Select the least-weight W shape for each of the following conditions. Use ASD, A36 steel, and assume full lateral bracing of the compression flange.

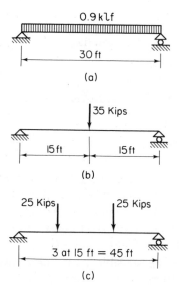

(6-4) Select the most economical beam to carry a uniform load (in-cluding the beam weight) of 5 kips per foot on a span of 25 ft, if the steel is
(a)   A36 steel ($F_y = 36$).
(b)   A588 steel ($F_y = 50$).
Use ASD. Full lateral support.

(6-5) Select the appropriate section (ASD) for the beams shown. The section should be the following:
(a)   Least weight (A36 steel).
(b)   Least weight (A588 steel).
(c)   Least weight with a depth not to exceed 21 in. (A36 steel).
(d)   Least weight with a depth not to exceed 15 in. (A36 steel).

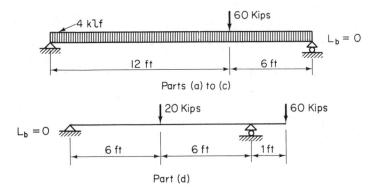

Parts (a) to (c)

Part (d)

(6-6) Select the W shape to carry the loads shown assuming full lateral support of the compression flange. Use ASD and A36 steel.

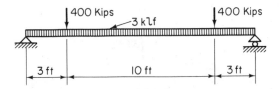

(6-7) What is the allowable concentrated *live* load that can be carried by a W 21 × 68 used as a simple span beam. The beam span is 26 ft and the load is applied as a single midspan concentrated load. The beam is supported laterally. A plastered ceiling is below the beam. Use ASD and A36 steel.

(6-8) Determine the permissible midpoint P (assume concentrated). Assume full lateral support of compression flange and A36 steel.

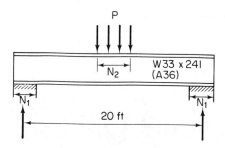

(6-9) Using the diagram of Problem 6-8, determine the minimum required bearing dimensions N_1 and N_2 to prevent web crippling on
the beam.

(6-10) If the compression flange is fully braced, select the most economical W shape using PD. $P_{design} = 20$ kip. Use A36 steel.

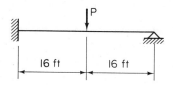

(6-11) Assuming the compression flange is supported laterally along the
full length, choose the most economical W shape (A36 steel) using
(a) ASD, and (b) PD.

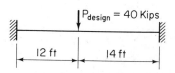

(6-12) Using PD, choose the most economical W shape ($F_y = 50$ ksi)
for the loading shown on the continuous beam. Assume the compression flange is supported laterally along its entire length. Compare the results with those obtained using simple ASD.

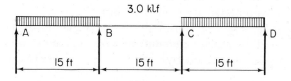

7

plate girders
and
rolled beams

7-1   INTRODUCTION

A brief description of plate girders was presented in Chapter 2. However, a brief review and additional discussion will be made in this chapter.

Plate girders are flexural members, usually having the same shape as a rolled W-shape member. They are used where heavy loads and/or long spans make the use of standard rolled shapes impossible or uneconomical. Since the section is fabricated to the desired geometry (depth and thickness of web plate, width and thickness of flange plate, etc.), individual elements are used having the most economical dimensions. For example, web plates are deep and thin.

For clarity, Fig. 7-1 shows the basic elements of a typical plate girder fabricated from plate elements and of constant depth. These plate elements may be connected by rivets, bolts, or welds. Sometimes the

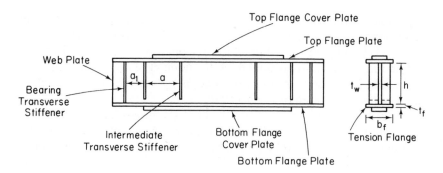

FIG. 7-1

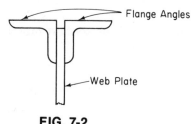

FIG. 7-2

flanges of plate girders can be made from angles as shown in Fig. 7-2. Hybrid plate girders may also be used. They are girders whose flanges are fabricated from a stronger grade of steel than that used in the web. The flanges of the hybrid girder at any section are required to have the same cross-sectional area, as well as to be of the same grade of steel. Tapered plate girders may also be employed as structural members (see Fig. 7-3).

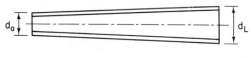

FIG. 7-3

7-2   PROPORTIONS

The AISC Specification recommends that plate girders in general be proportioned by the moment of inertia of the gross section. Holes for fasteners in either flange need not be deducted except where the reduction in area of either flange caused by these holes exceeds 15% of the gross flange area. In this case, the excess is to be deducted. Reduction of the area of flange caused by holes should be calculated as discussed in Section 3-3. Hybrid girders acted upon by an axial force equal to or less than $0.15F_yA$ may also be proportioned by the moment of inertia of their gross section subject to the provisions of Section 1.10 of the AISC Specification (A is the gross area of the section).

As a review, the moment of inertia of the gross-section method of proportioning will be illustrated by Example 7-1. Several of the design aids found in the AISC Manual will be used for ease of computation.

EXAMPLE 7-1

Given a constant-depth all-welded plate girder whose dimensions are web plate, $\frac{7}{16}$ in. $\times$ 60 in.; two flange plates, 1 in. $\times$ 20 in. Find the section modulus furnished by the section.

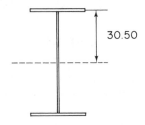

30.50

solution:

Section	A (in.²)	y (in.)	Ay² (in.⁴)	I (in.⁴)	I_gr (in.⁴)
Web $\frac{7}{16}$ × 60	26.25				7,875[a]
Flange 1 × 20	20.00 ⎫				37,222
Flange 1 × 20	20.00 ⎬	30.5	37,220[b]	2	
		Moment of inertia = 45,097			

AISC Manual design aids: [a]Part 2, Plate Girders Moment of Inertia of One Plate About Axis x-x.

[b] Part 2, Plate Girders, Values of 2y² for Computing Moment of Inertia of Areas About Axis x-x [ΣAy^2 = area of one flange $(2y^2)$]. Enter table with $2y = 61.0$ and read $2y^2 = 1861$. 1861 (area of one flange) = $37,220 = \Sigma Ay^2$.

$$S = 45,407/31.0 = 1465 \text{ in.}^3.$$

7-3   WEB/WEB CRIPPLING

To ensure that the compression flange is laterally supported in the plane of the web and will not fail by vertical buckling (pushing) into the web when subjected to bending, the h/t_w ratio must be limited, as follows, when transverse stiffeners are provided and are not spaced more than $1\frac{1}{2}$ times the girder depth:

$$\frac{h}{t_w} \leq \frac{14,000}{[F_y(F_y + 16.5)]^{1/2}} \qquad (7\text{-}1a)$$

except that

$$\frac{h}{t_w} \leq \frac{2000}{(F_y)^{1/2}} \qquad (7\text{-}1b)$$

where h is the clear distance between the flanges, t_w is the web thickness (Fig. 7-1), and F_y is the yield stress of the compression flange.

To prevent web crippling due to end reactions and/or interior concentrated loads, Eqs. 6-14 and 6-15 should be checked. If these equations are violated, bearing stiffeners should be provided.

Because plate girder webs are thin, the AISC Specification has a further limitation on the amount of load that may be applied directly to the girder flange on the compression edge of the web plate between stiffeners when bearing stiffeners are not utilized. This limitation of load is expressed as a limitation of the sum of compression stresses caused by concentrated and distributed loads. For the case where the compression flange is restrained against rotation about its longitudinal axis (for example, when it is embedded or in contact with concrete slab),

$$\Sigma f_1 + \Sigma f_2 \leq \left[5.5 + \frac{4}{(a/h)^2} \right] \frac{10,000}{(h/t_w)^2} \qquad (7\text{-}2)$$

and for the case where the flange is not so restrained:

$$\Sigma f_1 + \Sigma f_2 \leq \left[2 + \frac{4}{(a/h)^2} \right] \frac{10,000}{(h/t_w)^2} \qquad (7\text{-}3)$$

where h is the girder web depth, a is the clear spacing between stiffeners, h is the clear distance between flanges; the larger of P/ht_w or $P/at_w = f_1$ and $f_2 = w/t_w$, where P is a concentrated load in kips and w is a distributed load in kips per linear inch of length.

Often light beams or girders with slender webs rest atop columns and are supported laterally only in the plane of their top flanges, as in Fig. 7-4a. This case is unstable because of the lack of lateral support in the plane of the bottom flange (bracing or beam-column-moment connec-

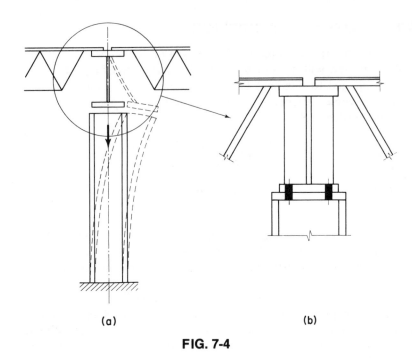

(a)   (b)

FIG. 7-4

tion). Lateral displacement at the top of the column along with distortion ot the slender web (shown dotted) can lead to an instability collapse. This condition can be vastly improved through the use of web stiffeners coupled with an appropriate moment connection at the bottom flange, as shown in Fig. 7-4b

7-4   FLANGES

Referring to Fig. 7-1, flanges of welded plate girders can vary in thickness or width by using cover plates. A series of plates may be spliced to accomplish this end. The area of cover plates of riveted plate girders may not be larger than 70% of the total flange area (AISC Specification).

The outstanding (unstiffened) compression elements of the flanges must conform to the width-to-thickness ratio limitations of the AISC Specification (Section 6-8), $b/2t_f \leq 95.0/(F_y)^{1/2}$. Owing to compressive bending stresses above the neutral axis of a thin girder web, lateral deformation (buckling) of the web takes place. The web tends to be held straight by bending tension stresses that occur below the neutral axis of the section, while for a distance down from the top flange, about $30t_f$, the flange restrains the web plate from buckling. The effect of this lateral displacement of the web (above the neutral axis to a distance $30t_f$ down from the top flange) is to reduce the bending resistance furnished by the portion of the web in compression. As a matter of convenience, this web deficiency can best be provided for in designing the compression flange by using an adjusted allowable bending stress, F_b', which is lower than would otherwise be permitted if no loss of bending resistance due to web buckling would occur. For most cases, this reduction in allowable flange bending stress amounts to only a small percentage. No reduction in allowable flange stress is required when $h/t_w \leq 760/(F_b)^{1/2}$. For cases where this h/t_w ratio is exceeded, Eq. 7-4 is applicable.

$$F_b' \leq F_b \left\{ 1.0 - 0.0005 \frac{A_w}{A_f} \left[\frac{h}{t_w} - \frac{760}{(F_b)^{1/2}} \right] \right\} \qquad (7\text{-}4)$$

where F_b is the applicable allowable bending stress, A_w is the area of the web at the section under investigation, and A_f is the area of the compression flange.

In hybrid girders (girders with flanges of higher strength than their webs), the maximum allowable stress in either flange must not exceed Eq. 7-4 or 7-5.

$$F_b' \leq F_b \left[\frac{12 + (A_w/A_f)(3\alpha - \alpha^3)}{12 + 2(A_w/A_f)} \right] \qquad (7\text{-}5)$$

where $\alpha = (F_y)_{\text{web}}/(F_y)_{\text{flange}}$. Connectors that join flange to web or cover plate to flange should be proportioned to resist the horizontal shear resulting from applied bending forces. The distribution of these connectors along the span length should be in proportion to the magnitude of the horizontal shear, with the longitudinal spacing limited as per the AISC Specification for compression and tension members (Section 1.18.2.3 for compression members and Section 1.18.3.1 for tension members). Horizontal shear is computed in a manner similar to that discussed in Chapter 4. In addition, the connectors joining flange to web must be capable of transmitting to the web any loads directly applied to the flange when no provision is made to transmit these loads by direct bearing.

When partial-length cover plates are used, they should be extended beyond their theoretical cutoff point. This extended portion of cover plate must be developed by enough connectors to accommodate its portion of flexural stresses that it would have received if it had been extended the full length of the member. This ensures the partial-length cover plate functioning as a part of the beam. The cover-plate force to be developed by the plate extension fasteners is MQ/I, where M is the moment at the theoretical cutoff point, Q is the statical moment of the area of the cover plate about the neutral axis of the cover-plated section, and I is the moment of inertia of the cover-plated section.

For the welded cover-plated condition, an additional requirement is provided by the AISC Specification. The welds connecting the end of the cover plate to the beam in length a' should be sufficient at the allowable stresses to develop the cover plate's portion of the flexural stresses in the beam at the distance a' from the cover-plate end (the amount of stress carried by the cover plate at a distance a' in from its end shall not exceed the capacity of the terminal welds deposited along the plate's edges and optionally across its end within the distance a'). The length, a', measured from the end of the partial-length cover plate, will be as defined in Fig. 7-5.

(a) $a' = b$ When Continuous Weld $\geq \frac{3}{4}$ Cover Plate Thickness as Indicated

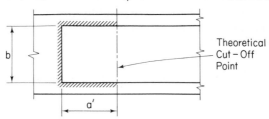

(b) $a' = 1.5\,b$ When Continuous Weld $< \frac{3}{4}$ Cover Plate Thickness as Indicated

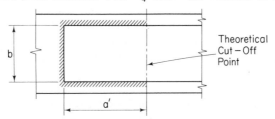

(c) $a' = 2b$ When No Weld Across End of Cover Plate but Continuous Welds Along Both Sides of Cover Plate

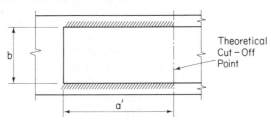

FIG. 7-5

7-5   *BEARING AND INTERMEDIATE STIFFENERS*

To take advantage of the most effective use of material in a plate girder, the designer must recognize a basic concept long in use in the design of the metal skin of aircraft. This concept, called *tension field action,* takes place in the panels of a plate girder web. These panels are bounded by the girder flanges and transverse stiffeners. The limit of structural usefulness of a girder web is no longer assumed to be reached when the web attains the level of stress causing buckling of the web. This would be true for columns, but the panels of a plate girder web are capable of carrying loads in excess of the web buckling load.

Properly spaced and sized intermediate stiffeners act as compression struts, and membrane stresses created by vertical shear forces greater than the theoretical web buckling load form diagonal tension fields. The diagonal tension stresses in the web plate, acting in conjunction with compressive stresses in the intermediate stiffeners and compression, tension, and cross-bending stresses in the flange, form a structural system likened to a Pratt truss. Figure 7-6 illustrates this action. The vertical component of the tension field stresses is resisted by compression stress in the stiffeners. The horizontal component of the tension field stress is transferred to the flanges in shear.

The stiffener spacing in the end panels, a_1 in Fig. 7-1, should be small enough for girders designed on the basis of tension field action. This is assured by limiting the largest computed average web shear value to the allowable shear-stress value in Eq. 7-6. This limitation is also valid at panels containing large holes and panels adjacent thereto.

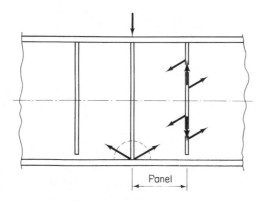

Panel

FIG. 7-6

$$F_v = \left(\frac{F_y}{2.89} \right) C_v \leq 0.4F_y \qquad (7\text{-}6)$$

where C_n = $45{,}000k/F_y(h/t_w)^2$ when $C_v < 0.8$
      = $[190/(h/t_w)](k/F_y)^{1/2}$ when $C_v > 0.8$
  k = $4.00 + 5.34/(a/h)^2$ when $a/h < 1.0$
      = $5.34 + 4.00/(a/h)^2$ when $a/h > 1.0$
  h = clear distance between flanges, in., at the section under investigation

The terms t_w, a, and a_1 are in inches and are as defined in Fig. 7-1.

The largest average web shear value computed for any possible loading condition, partial or full, must not exceed the allowable shear-stress value in Eq. 7-6, except that, for all girders other than hybrid when intermediate stiffeners are used and spaced as required and $C_v \leq 1.0$, the allowable shear stress as computed by Eq. 7-7 may be used in lieu of the value obtained from Eq. 7-6.

$$F_v = \left[\frac{F_y}{2.89} \right] \left[C_v + \frac{1 - C_v}{1.15[1 + (a/h)^2]^{1/2}} \right] \leq 0.4F_y \qquad (7\text{-}7)$$

Equation 7-7 recognizes the contribution of tension field action, whereas Eq. 7-6 does not. Use of Eq. 7-7 for hybrid girders is not recommended, pending further investigation. Both formulas assure the usual factor of safety relative to the yield stress.

Bearing stiffeners are used in pairs (one each side of the web) at unframed ends in plate girders and rolled beams and at points of concentrated loads where required (Eq. 6-15). These bearing stiffeners should bear closely against the flanges or flange through which they receive their loads or reactions. They should extend close to the edge of the flange of the plate girder. They should be designed as if they were columns, and the assumed column section shall be as in Fig. 7-7a or Fig. 7-7b, as applicable.

In computing the l/r ratio of the stiffeners, $l \geq 75\%$ of the actual stiffener length. The effective bearing width of the stiffeners will be considered to be the width clear of the flange angle fillet or the flange/web weld, as shown in Fig. 7-8a and b.

Intermediate stiffeners are not required when $h/t_w < 260$ and $f_v < F_v$ as computed by Eq. 7-6 (no reliance on tension field action) provided the provisions of Eqs. 7-1a and 7-1b are met. The spacing of intermediate stiffeners, when required, shall be limited such that $f_v < F_v$, as computed by Eq. 7-6 or 7-7, as applicable, and $a/h \leq [260/(h/t_w)]^2$ and ≤ 3.0; a/h, the panel *aspect ratio*, is limited to these values to facilitate handling during fabrication and erection.

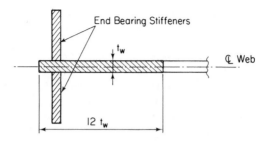

(a) For End Bearing Stiffeners at Reaction

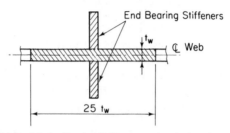

(b) For Interior Bearing Stiffeners at Interior Loads

FIG. 7-7

The moment of inertia of a single stiffener or a pair of intermediate stiffeners about the center line of the web (Fig. 7-7b) should be equal to or larger than $(h/50)^4$ to provide adequate lateral support to the web. Equation 7-8 assures that the area of the stiffeners acting as struts is sufficient to resist the compression stresses necessary to balance the vertical component of the tension field (Fig. 7-6).

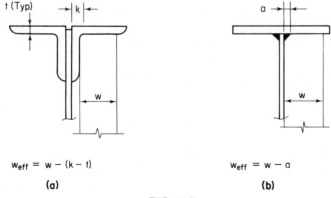

$$w_{eff} = w - (k - t)$$

(a)

$$w_{eff} = w - a$$

(b)

FIG. 7-8

$$A_{st} = \frac{1 - C_v}{2} \left[\frac{a}{h} - \frac{(a/h)^2}{[1 + (a/h)^2]^{1/2}} \right] YDht \qquad (7\text{-}8)$$

where C_v, a, h, and t_w are as defined in Fig. 7-1 or as for Eq. 7-6, and

A_{st} = gross area of intermediate stiffeners spaced as required by Eq. 7-7, in.^{2} (total area for a pair and total area for one stiffener to compensate for the eccentric loading)

Y = $(F_y)_{web}/(F_y)_{stiffeners}$

D = 1.0 for a pair of stiffeners

  = 1.8 for a single angle stiffener

  = 2.4 for a single plate stiffener

When the largest computed shear stress in a panel, f_v, is less than that permitted in Eq. 7-7, F_v, A_{st} may be reduced in equal proportion.

In general, a nominal weld from stiffener to web plate is required to transfer the tension field stress from web to stiffeners. However, the AISC Specification says that the stiffener must be connected so as to be capable of total shear transfer not less than f_{vs} calculated as follows:

$$f_{vs} = h \left[\left(\frac{F_y}{340} \right)^3 \right]^{1/2} \qquad (7\text{-}9)$$

where F_y = yield stress of the web

f_{vs} = shear between transverse stiffeners and web, kips per linear inch of single or pair of stiffeners

In general, f_{vs} is so small that the nominal connector of stiffener to web takes care of this. f_{vs} may be reduced in the same proportion that the largest shear stress, f_v, in the adjacent panels is less than F_v as computed by Eq. 7-7. However, connectors in intermediate stiffeners required to transfer applied concentrated loads or reactions to the web must be proportioned for not less than that applied concentrated load or reaction. When tension field action acts together with an applied concentrated load or reaction, the shear transfer required to be transferred is the larger of the two. Where bearing is not required to transmit applied concentrated loads or reactions, the intermediate stiffener may be stopped short of the tension flange. The stiffener-to-web weld shall be terminated a distance $\geq 4t_w$ and $\leq 6t_w$ from the near toe of the flange-to-web weld, as shown in Fig. 7-1.

When single plate-type stiffeners are used, they should be attached to the compression flange to resist any potential uplift created by torsion on the plate. Whenever a lateral brace is connected to the stiffener or pair of stiffeners, the stiffeners should be connected to the compression flange. This connection should be capable of transmitting 1% of the total flange stress. This latter requirement is unnecessary when the flange is composed solely of angles.

When intermittent fillet welds are used as connectors of stiffeners to the girder web, the clear distance between welds must not be greater than $16t_w$ or 10 in. When rivets are used as connectors, their maximum spacing should be 12 in.

7-6 COMBINED SHEAR AND TENSION STRESS

When plate girders are not designed on the theory of tension field action, no stress reduction is required for cases of combined shear and tension (bending) stress. When tension field action is assumed, in the area of the web marked by a dashed arc envelope, as shown in Fig. 7-6, a combination of tension field stresses and shear stresses occurs. The web area so marked may be subject to near maximum tension stress. However, it has been shown that such webs are capable of resisting large bending stresses with no loss in shear strength or of resisting large shear stresses with no loss in bending strength. The webs can be proportioned on the basis of either

(1) F_b when the simultaneously applied $f_v \leq 0.6F_v$.
(2) F_v when the simultaneously applied $f_b \leq 0.75F_b$.

If rule (1) or (2) is violated, Eq. 7-10 (a linear interaction equation) should be used.

$$F_b \leq \left[0.825 - 0.375\left(\frac{f_v}{F_v}\right) \right] F_y \leq 0.6F_y \qquad (7\text{-}10)$$

where F_b = allowable bending tensile stress due to moment in the plane of the plate girder web
   F_v = allowable web shear computed by Eq. 7-7
   f_v = calculated web shear stress = V/ht_w

For plate girders with flanges and webs of A514 steel, the use of the tension-field-action assumption is limited to web regions where the flange $f_b \leq 0.75F_b$. When $f_b > 0.75F_b$, F_v shall be limited to the value computed by Eq. 7-6 (no tension field action).

7-7   SPLICES

The AISC Specification requires that groove-welded splices in plate girders develop the full strength of the smaller spliced section while any other splice type develop the strength required by the stresses at the section of the splice.

7-8   HORIZONTAL FORCES

The flanges of plate girders that support cranes or other moving loads should be designed to resist any horizontal forces produced by such loads. For example, the horizontal lateral forces produced on the flange of a crane runway plate girder, as shown in Fig. 7-9, should be 20% of the sum of the lifted load and the crane trolley (exclusive of other parts of the crane). This force should be applied to the top of the rails, distributed with due regard for lateral stiffness of the structure supporting the runway rails, on each side of the runway. They may be considered to act in either direction normal to the rail. An additional horizontal longitudinal force applied to the top of the rail should be 10% of the maximum wheel loads of the crane. These horizontal forces are as specified in the AISC Specification in the section on crane runway horizontal forces and can be used in lieu of any other governing requirements.

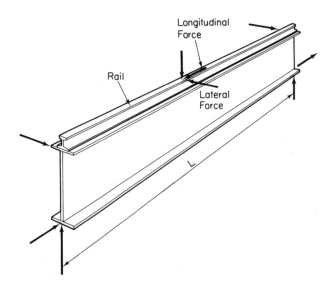

FIG. 7-9

7-9 DESIGN AIDS AND TABLES

Although when the rules for the design of plate girders first are encountered they may seem to be involved, they really are not. With the aid of Appendix A of the AISC Specification, coupled with the many design aids and tables found in Part 2 of the AISC Manual, the design of plate girders is no more complex than the design of any other structural element. This section will convey to the reader the methods by which the aforementioned can be utilized to minimize necessary design time of plate girders. The actual design will be illustrated by design examples that follow.

The table of *Dimensions and Properties of Welded Plate Girders* is valuable for the selection of welded-plate-girder dimensional geometrical proportions to yield an economical design. The tables provide dimensions and properties for a range of sections with nominal depths (d = overall nominal depth) ranging from 45 to 92 in. Although only one width of flange plate is given for each group of girders of a nominal depth (flange plate thickness varies), the substitution of a wider and thinner flange plate, equivalent in area to the one tabulated, yields a section modulus slightly reduced.

The thinnest flange plate listed in the tables conforms to the width-to-thickness ratio requirement for A36 steel. If stronger steels are used or a thinner plate is used, this conformance should be checked. Only one thickness of web plate for each group of nominal depth girders is given in the tables. When the ratio of V_{max}/M_{max} is relatively large, overall economy may indicate a thicker web. The listing of S' in the tables enables the designer to rapidly compute the change in section modulus, S, for the girder listed in the table. S', multiplied by the amount of sixteenths-of-an-inch increase in web thickness added to S, gives the increased section modulus value.

Sometimes economy can be achieved by thickening the web plate such that intermediate stiffeners are not required (e.g., when $h/t_w \leq$ the value obtained from Eq. 7-1a and $h/t_w < 260$ and $f_v < F_v$ computed by Eq. 7-6). However, the girder sections listed in the table provide a balanced design relative to bending moment and web shear and provide a reasonable number of intermediate stiffeners. The important point to make is that the lightest design need not necessarily be the most economical design. A thin web with many stiffeners necessitates piece handling, welding, and fabrication, which may result in added costs.

In general, economy is achieved when the girder is made as deep as the governing limitations permit. Putting as much area in the flanges of the plate girder generally gives the best economy unless, as a result.

the increase in stiffener requirements outweighs the savings in main material. Sometimes aesthetics dictates limitations on dimensions.

In the case of girders that have a large span length, reduction of the flange plate size at one or more points close to the girder supports (where the bending moment is reduced) may be economical. Again, this necessitates an overall analysis of economy: does the weight savings achieved in flange plate reduction warrant the added fabrication costs involved in the splicing of the flanges? For long-span plate girders, the splicing of the flange plates may be a necessity to start with, thereby justifying the reduction of flange plates at the ends of the girders.

The Dimensions and Properties Tables list the maximum end reaction, R, permissible without intermediate stiffeners for the tabulated web plate thicknesses for $F_y = 36$ ksi. The value of R would increase in proportion to the increase in web plate area when thicker web plates are used. R is equal to $A_w F_v$, where F_v is obtained from Eq. 7-6 or from the "Over 3" column of Table 7-1.

When intermediate stiffeners are required, their proper spacing can be found with the aid of the appropriate table. Appendix A of the AISC Specification gives tables for $F_y = 36$ ksi and 50 ksi. These tables, *Allowable Shear Stress (F_v) in Webs of Plate Girders (ksi)*, are used for homogeneous (nonhybrid) plate girders. The maximum permissible longitudinal spacing of stiffeners is dependent on the three parameters, a/h, h/t_w, and f_v, and their relationship with one another is presented in these series of tables.

When the designer is given the design loads and desired depth of a plate girder, he or she can determine the shear and then select a t_w (within the h/t_w limitations) such that $f_v < F_v$. With h/t_w and f_v, the a/h required is obtained from the table. After trying two or three t_w trial values, comparison of web plate and stiffener plate material indicates the overall economy. Within these tables are given corresponding gross areas of intermediate stiffeners as a percentage of web area. These values are a solution to Eq. 7-8 expressed as a percentage of web area. These values are shown in italics in the corresponding a/h column and h/t_w row. Stiffeners that provide this area will suffice. The blank spaces in the tables indicate when the a/h and h/t_w ratios are so small as to permit a $F_v >$ $0.35F_y$ (Eq. 7-6 would then govern). F_v is as computed by Eq. 7-6 (or Eq. 7-7 for other than hybrid girders). For Eq. 7-6, tension field action is not relied upon. The table for $F_y = 36$ ksi is shown in Table 7-1. The corresponding table for hybrid girders (for Eq. 7-7 where tension field action is relied upon), *Allowable Shear Stress in Webs of Plate Girders*, is shown in Table 7-2 for $F_y = 36$ ksi. Tension field action is neglected in this table. For both Tables 7-1 and 7-2, the allowable values for shear stress for the case where intermediate stiffeners are not required are given

Table 7-1
ALLOWABLE SHEAR STRESSES (F$_v$) IN PLATE GIRDERS (KSI) FOR 36 KSI SPECIFIED YIELD STRESS STEEL
(TENSION FIELD ACTION INCLUDED, Eq. 7-7)

(*Italic* values indicate gross area, as percent of web area, required for pairs of intermediate stiffeners of 36 ksi yield stress steel.)*

Slenderness ratios h/t: web depth to web thickness — F$_y$ = 36 ksi

h/t	|← Aspect ratios a/h: stiffener spacing to web depth →|													
	0.5	0.6	0.7	0.8	0.9	1.0	1.2	1.4	1.6	1.8	2.0	2.5	3.0	Over 3
60										14.5	14.5	14.5	14.5	14.5
70							14.5	14.5	14.5	14.4	14.2	13.8	13.6	13.0
80					14.5	14.5	14.0	13.4	13.0	12.6	12.4	12.2 *0.3*	12.0 *0.4*	11.4
90				14.5	14.3	13.4	12.5	12.2 *0.6*	11.9 *0.9*	11.8 *1.1*	11.6 *1.2*	11.3 *1.2*	11.1 *1.2*	10.1
100			14.5	13.9	12.8	12.3 *0.5*	11.9 *1.4*	11.6 *1.8*	11.3 *2.0*	11.1 *2.1*	10.9 *2.2*	10.3 *2.3*	10.0 *2.1*	8.3
110		14.5	13.8	12.6	12.2 *0.9*	11.9 *1.8*	11.5 *2.5*	11.0 *3.1*	10.5 *3.5*	10.1 *3.6*	9.8 *3.6*	9.2 *3.4*	8.8 *3.1*	6.9
120		14.3	12.7	12.2 *1.1*	11.8 *2.1*	11.5 *2.8*	10.8 *4.1*	10.2 *4.7*	9.8 *4.9*	9.4 *4.9*	9.0 *4.7*	8.4 *4.3*	8.0 *3.8*	5.8
130	14.5	13.2	12.2 *0.9*	11.9 *2.2*	11.5 *3.2*	11.0 *4.5*	10.3 *5.6*	9.7 *5.9*	9.2 *6.0*	8.8 *5.8*	8.4 *5.6*	7.8 *5.0*	7.3 *4.4*	4.9
140	14.2	12.4 *0.3*	12.0 *1.9*	11.6 *3.2*	11.0 *4.8*	10.5 *5.9*	9.8 *6.7*	9.2 *6.9*	8.7 *6.8*	8.3 *6.6*	7.9 *6.3*	7.2 *5.5*	6.8 *4.9*	4.2
150	13.2	12.2 *1.2*	11.8 *2.8*	11.2 *4.7*	10.6 *6.1*	10.1 *7.0*	9.4 *7.6*	8.8 *7.7*	8.3 *7.5*	7.9 *7.2*	7.5 *6.8*	6.8 *6.0*	6.3 *5.2*	3.7
160	12.4	12.0 *2.1*	11.5 *4.1*	10.9 *6.0*	10.3 *7.2*	9.8 *8.0*	9.1 *8.4*	8.5 *8.3*	8.0 *8.1*	7.6 *7.7*	7.2 *7.3*	6.5 *6.3*		3.2
170	12.3 *0.9*	11.8 *2.8*	11.2 *5.3*	10.6 *7.0*	10.1 *8.1*	9.6 *8.7*	8.9 *9.0*	8.3 *8.9*	7.7 *8.5*	7.3 *8.1*	6.9 *7.7*			2.9
180	12.1 *1.6*	11.6 *4.0*	10.9 *6.3*	10.4 *7.9*	9.9 *8.8*	9.4 *9.4*	8.7 *9.6*	8.1 *9.3*	7.5 *8.9*	7.1 *8.5*	6.7 *8.0*			2.6
200	11.9 *2.9*	11.2 *6.0*	10.5 *8.0*	10.0 *9.2*	9.5 *10.0*	9.1 *10.4*	8.3 *10.4*	7.7 *10.0*	7.2 *9.5*					2.1
220	11.5 *4.8*	10.8 *7.5*	10.3 *9.2*	9.7 *10.2*	9.3 *10.8*	8.8 *11.1*	8.1 *11.0*	7.5 *10.6*						1.7
240	11.2 *6.2*	10.6 *8.6*	10.0 *10.1*	9.5 *11.0*	9.1 *11.5*	8.6 *11.7*								1.4
260	11.0 *7.3*	10.4 *9.5*	9.9 *10.8*	9.4 *11.6*	8.9 *12.0*	8.5 *12.1*								1.2
280	10.8 *8.2*	10.2 *10.2*	9.7 *11.4*	9.2 *12.1*										
300	10.7 *9.0*	10.1 *10.8*	9.6 *11.8*											
320	10.5 *9.5*	10.0 *11.2*												

Girders so proportioned that the computed shear is less than that given in right-hand column do not require intermediate stiffeners.

*For single angle stiffeners, multiply by 1.8; for single plate stiffeners, multiply by 2.4.

Courtesy of American Institute of Steel Construction.

Table 7-2

$F_y = 36$ ksi

PLATE GIRDERS

Allowable shear stress in webs

Tension field action neglected (Eq. 7–6)

$F_y = 36$ ksi

		Aspect ratios a/h: stiffener spacing to web depth													
		0.5	0.6	0.7	0.8	0.9	1.0	1.2	1.4	1.6	1.8	2.0	2.5	3.0	Over 3
Slenderness ratios h/t: web depth to web thickness	60											14.5	14.5	14.5	14.5
	70									14.5	14.5	14.2	13.8	13.6	13.0
	80						14.5	14.1	13.4	13.0	12.6	12.4	12.0	11.9	11.4
	90				14.5	14.3	13.4	12.5	11.9	11.5	11.2	11.0	10.7	10.5	10.1
	100			14.5	13.9	12.8	12.0	11.2	10.7	10.4	10.1	9.9	9.3	9.0	8.3
	110		14.5	13.8	12.6	11.7	10.9	10.2	9.5	8.9	8.5	8.2	7.7	7.4	6.9
	120		14.3	12.7	11.6	10.7	10.0	8.8	8.0	7.5	7.1	6.9	6.5	6.3	5.8
	130	14.5	13.2	11.7	10.7	9.8	8.6	7.5	6.8	6.4	6.1	5.8	5.5	5.3	4.9
	140	14.2	12.2	10.9	9.8	8.4	7.4	6.5	5.9	5.5	5.2	5.0	4.8	4.6	4.2
	150	13.2	11.4	10.1	8.5	7.3	6.5	5.6	5.1	4.8	4.5	4.4	4.1	4.0	3.7
	160	12.4	10.7	9.1	7.5	6.5	5.7	4.9	4.5	4.2	4.0	3.9	3.6	3.5	3.2
	170	11.7	10.1	8.0	6.7	5.7	5.0	4.4	4.0	3.7	3.5	3.4	3.2	3.1	2.9
	180	11.0	9.1	7.2	5.9	5.1	4.5	3.9	3.5	3.3	3.2	3.0	2.9	2.8	2.6
	200	9.9	7.3	5.8	4.8	4.1	3.6	3.2	2.9	2.7	2.6	2.5	2.3	2.3	2.1
	220	8.2	6.1	4.8	4.0	3.4	3.0	2.6	2.4	2.2	2.1	2.0	1.9	1.9	1.7
	240	6.8	5.1	4.0	3.3	2.9	2.5	2.2	2.0	1.9	1.8	1.7	1.6	1.6	1.4
	260	5.8	4.3	3.4	2.8	2.4	2.2	1.9	1.7	1.6	1.5	1.5	1.4	1.3	1.2
	280	5.0	3.7	3.0	2.5										
	300	4.4	3.3	2.6											
	320	3.9	2.9												

Girders so proportioned that the computed shear stress is less than that given in the right-hand column do not require intermediate stiffeners.

Courtesy of American Institute of Steel Construction.

in the column to the extreme right, "Over 3." No stiffener areas are shown in Table 7-2, in which all values of F_v, whether greater or less than $0.35 F_y$, are governed by Eq. 7-6 (no tension field action). No tension occurs when $0.35 F_y \le F_v \le 0.4 F_y$.

In hybrid girders, the reduced maximum allowable bending stress in the flange, F_b', is computed by Eq. 7-5. Table 7-3, *Coefficients C_h for Maximum Allowable Bending Stress in Hybrid Girders,* is useful to com-

Table 7-3

PLATE GIRDERS
Coefficients C_h for maximum allowable bending stress in hybrid girders
$F_b' \le C_h F_b$

$\dfrac{A_w}{A_f}$	α = Ratio F_y (web) to F_y (flange)						
	0.9	0.8	0.7	0.6	0.5	0.4	0.3
0.3	1.00	1.00	0.99	0.99	0.99	0.98	0.97
0.4	1.00	1.00	0.99	0.99	0.98	0.97	0.96
0.5	1.00	1.00	0.99	0.98	0.98	0.97	0.96
0.6	1.00	0.99	0.99	0.98	0.98	0.96	0.95
0.7	1.00	0.99	0.99	0.98	0.97	0.95	0.94
0.8	1.00	0.99	0.99	0.98	0.97	0.95	0.93
0.9	1.00	0.99	0.98	0.97	0.96	0.94	0.93
1.0	1.00	0.99	0.98	0.97	0.96	0.94	0.92
1.1	1.00	0.99	0.98	0.97	0.96	0.93	0.91
1.2	1.00	0.99	0.98	0.97	0.95	0.93	0.91
1.3	1.00	0.99	0.98	0.96	0.95	0.92	0.90
1.4	1.00	0.99	0.98	0.96	0.95	0.92	0.89
1.5	1.00	0.99	0.98	0.96	0.94	0.91	0.89
1.6	1.00	0.99	0.97	0.96	0.94	0.91	0.88
1.7	1.00	0.99	0.97	0.95	0.94	0.90	0.88
1.8	1.00	0.99	0.97	0.95	0.93	0.90	0.87
1.9	1.00	0.99	0.97	0.95	0.93	0.89	0.86
2.0	1.00	0.99	0.97	0.95	0.93	0.89	0.86
2.2	1.00	0.99	0.97	0.94	0.92	0.88	0.85
2.4	1.00	0.98	0.97	0.94	0.92	0.88	0.84
2.6	1.00	0.98	0.96	0.94	0.91	0.87	0.83
2.8	1.00	0.98	0.96	0.93	0.91	0.86	0.82
3.0	1.00	0.98	0.96	0.93	0.90	0.86	0.81

Coefficient $C_h = \left[12 + (A_w/A_f)(3\alpha - \alpha^3)\right] / \left[12 + 2(A_w/A_f)\right]$ from Eq. 7-5.

Note that the lower value using Eq. 7-4 and 7-5 governs for the flanges of a hybrid girder.

F_b = applicable bending stress for flange material.

Courtesy of American Institute of Steel Construction.

pute F_b'. $F_b' \le F_b C_h$, where C_h is defined as a coefficient equal to the term in the [□] in Eq. 7-5. Knowing A_w/A_f and α, a value for C_h can be obtained and, when multiplied by F_b, the applicable allowable bending stress for the flange material, gives F_b'.

Figure 7-10 shows a *section modulus nomograph* for symmetrical cross-sectional girders to assist the designer in rapidly selecting a preliminary trial section for homogeneous or hybrid plate girders. The design of a plate girder is a straightforward process because of the numerous design parameters involved. All the equations and rules discussed appear to be complex, but many of the values of the equations have been tab-

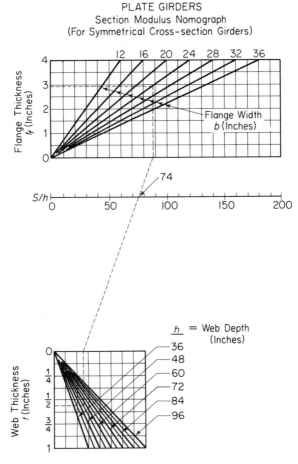

PLATE GIRDERS
Section Modulus Nomograph
(For Symmetrical Cross-section Girders)

FIG. 7-10 *Courtesy of American Institute of Steel Construction*

ulated to minimize computations and to give the designer some insight into the effects on design of many of the parameters involved. The design examples in Section 7-10 illustrate the use of design aids and tables as well as the AISC Specification provisions and Appendix A of the Specification.

7-10 DESIGN EXAMPLES

EXAMPLE 7-2

By means of the various methods of computation, design aids, and tables available, choose the trial profile for a constant-depth plate girder spanning 51 ft. The girder supports a uniformly distributed load of 4 klf and picks up a 50-kip reaction from other girders at its third points (Fig. 7-11). Assume that the girder's compression flange is laterally supported at points of concentrated load only and that all material is A36 steel. Owing to the headroom restrictions, further assume a maximum depth of girder, d, of 66 in.

solution:

Three methods illustrating the selection of a suitable girder trial cross section will be shown. Method A shows how the

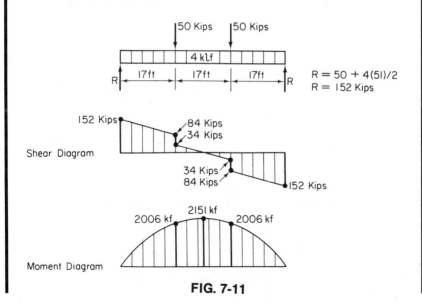

FIG. 7-11

trial section is obtained by using the flange-area method and then checked by the moment-of-inertia method; method B utilizes the Section Modulus Nomograph Selection design aid (Fig. 7-10) to aid in the selection of the trial cross section; method C uses the tables of Dimensions and Properties of Welded Plate Girders (Section 7-9 and AISC Manual, Part 2) for trial section selection.

(A) Flange-Area Selection Method with Moment-of-Inertia Check Method

(1) Preliminary Web Plate Design:

For maximum general economy, the girder is assumed to be as deep as permissible.

Assume that $h = d - 2t_f \approx 64$ in.

To gain an overall view of the options available as to web plate selection, various conditions governed by the design rules are examined prior to selection of web plate size.

(a) For no reduction in flange stress (Section 7-4):

$$h/t_w \leq 760/(F_b)^{1/2} = 760(22)^{1/2} = 162$$
$$t_w \geq h/162 = 64/162 = 0.395 \text{ in.}$$

(b) Utilizing Appendix A of the AISC Specification to solve Eq. 7-1a for the limiting h/t_w ratio:

$$h/t_w \leq 322 \text{ and } t_w \geq h/322 = \tfrac{64}{322} = 0.199 \text{ in.}$$

and, when transverse stiffeners are provided ($a \leq 1.5d$)

$$h/t_w \leq t_w \geq 333 \ h/333 = \tfrac{64}{333} = 0.192 \text{ in.}$$

Select some web plate thickness between that obtained from (a) and (b): *try* web plate $\tfrac{5}{16} \times 64$, $A_w = 20.0$ in.²; $h/t_w = 64/0.313 = 204$.

(2) Preliminary Flange Design:

(a) Assume $t_f = \tfrac{3}{4}$ in. and an allowable bending stress $F_b = 0.60F_y = 22$ ksi (Section 8-2.2; plate girders with thin webs are subject to warping).

$$A_f \approx [12M_{max}/(h + t_f)F_b] - \tfrac{1}{3}(A_w/2)$$
$$\approx [12(2151)/(64 + 0.75)22] - \tfrac{1}{3}(\tfrac{20}{2}) = 14.8 \text{ in.}^2.$$

$b_f = A_f/t_f = 14.8/0.75 = 19.7$. *Try* $\frac{3}{4} \times 22$ plate. $A_f = 16.5$ in.$^2 > 14.8$ in.^{2}.

(b) Local buckling check (Section 6-8 and Section 1.9.1 of the AISC Specification): $b/2t_f = \frac{22}{2}(0.75) = 14.7 < 15.8$.

(3) Trial Girder Section Design:
Web plate $\frac{5}{16} \times 64$ and two flange plates $\frac{3}{4} \times 22$.

(a) Checking by the moment-of-inertia method (Section 7-2 and Fig. 7-12):

Section	A (in.^{2})	y (in.)	Ay^2 (in.^{4})	I_0 (in.^{4})	I_{gr} (in.^{4})
Web $\frac{5}{16} \times 64$	20.00			6,837	6,837
Flange $\frac{3}{4} \times 22$	16.5 ⎫	32.375	34,588	2	34,590
Flange $\frac{3}{4} \times 22$	16.5 ⎭				
			Moment of inertia = 41,427		

$S = 41,427/32.75 = 1265$ in.^{3}. $S_{req} = 12M_{max}/F_b = 12(2151)/22 = 1173$ in.^{3}. $S > S_{req}$.

(b) Lateral buckling check:* $f_{b\,max} = 12M_{max}/S = 12(2151)/1265 = 20.4$ ksi. $I'_{yy} = (I_f + I_{web/6})_{yy} = 0.75(22)^3/12 = 666$ in.^{4}. $A' = A_f + A_w/6 = 16.5 + 20/6 = 19.8$ in.^{2}.

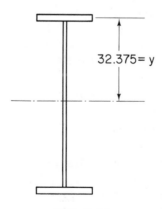

32.375= y

FIG. 7-12

---

*See Section 8-2.2.

r_T = radius of gyration of the section comprising the compression flange + the compression web area/3 taken about the y-y axis. $r_t = (I'/A')^{1/2} = (666/19.8)^{1/2} = 5.80$ in.

(c) Bending stress check at interior 17-ft span: since the interior moment is larger than either of the end moments of the unbraced length (interior 17-ft span), $C_b = 1.0$ (C_b is a bending coefficient dependent upon moment gradient).

From Appendix A of the AISC Specification: $l/r_T = 17(12)/5.80 = 35.2 < 53/(C_b)^{1/2} = 53$. $F_b = 0.60F_y = 22$ ksi.

Since $h/t_w = 204 > 760/(F_b)^{1/2} = 162$, the allowable stress in the compression flange is

$$F_b' = 22.0\left\{1.0 - 0.0005\left(\frac{20.0}{16.5}\right)\left[\frac{64}{0.313} - \frac{760}{(22)^{1/2}}\right]\right\}$$

$$= 21.45 \text{ ksi} > 20.4 \text{ ksi} = f_b \qquad (Eq.7\text{-}4)$$

(d) Bending stress check at exterior 17-ft. span: $C_b = 1.75 + 1.05(M_1/M_2) + 0.3(M_1/M_2)^2 \leq 2.3$; M_1 = smaller moment at ends of unbraced length; M_2 = larger moment at ends of unbraced length.

Then $M_1/M_2 = 0$ and $C_b = 1.75$; $f_{b \text{ max}} = 12(2006)/1265 = 19.0$ ksi; $l/r_T = 12(17)/5.80 = 35.2 < 53(C_b)^{1/2} = 70.1$; $F_b = 0.60F_y = 22$ ksi; $F_b' = 21.45$ ksi $> f_b = 19.0$ ksi.

Use a web plate $\frac{5}{16} \times 64$ and two flange plates $\frac{3}{4} \times 22$.

(B) Section Modulus Nomograph Selection Method (Fig. 7-10)

(1) Preliminary Web Plate Design: same as in method A.

(2) Preliminary Flange Design:

(a) $t_w = \frac{5}{16}$ in., $h = 64$ in. Assume that $F_b' = 0.60F_y = 22$ ksi. $S_{req} = 12M_{max}/F_b' = 12(2151)/22 = 1173$ in.³. $S/h = 1173/64 = 18.3$; assume that $b_f = 22$ in. From Fig. 7-10, *try* $t_f = \frac{3}{4}$ in.

(b) Local buckling check: same as in method A.

(3) Trial Girder Section Design:
Web plate $\frac{5}{16} \times 64$ and two flange plates $\frac{3}{4} \times 22$.

   (a) Check by moment-of-inertia method: same as in method A.

   (b) Lateral buckling check: same as in method A.
   Use the same section as in method A.

(C) Table of Dimensions and Properties of Welded Plate Girders (Section 7-9 and AISC Manual, Part 2)
$F_b = 22$ ksi; $S_{req} = 12 M_{max}/F_b = 12(2151)/22 = 1173$ in.^{3}; $V_{max} = 152$ kips.

Enter the table of Dimensions and Properties of Welded Plate Girders (AISC Manual, Part 2) with assumed $h = 64$ in.: for a section of $\frac{7}{16} \times 60$ web with $\frac{3}{4} \times 20$ flange plates, $S = 1160$ in.^{3} < 1173 in.^{3}; for a section of $\frac{1}{2} \times 66$ web with $\frac{3}{4} \times 22$ flange plates, $S = 1440$ in.^{3} > 1173 in.^{3}.

(1) Preliminary Web Plate Design:
Try web plate $\frac{1}{2} \times 64$; $A_w = 32.0$ in.^{2}.

   (a) Web plate check: $h/t_w = 64/(\frac{1}{2}) = 128$. From Table 7-1: for $h/t_w = 128$, under the column headed "Over 3," the F_v without intermediate stiffeners (Eq. 7-6), by interpolation $= 5.08$ ksi. $f_{v\ max} = 152/32.0 = 4.75$ ksi $< F_v$. Use a $\frac{1}{2} \times 64$ web without intermediate stiffeners. This would be one solution.

   (b) Another solution would be to use a thinner web with intermediate stiffeners as follows: *Try* a $\frac{5}{16} \times 64$ web plate; $A_w = 20.0$ in.^{2} (see method A).

(2) Preliminary Flange Design:

   (a) *Try* $t_f = \frac{3}{4}$ in.; $d = h + 2t_f = 64 + 1.5 = 65.5$ in. $I_{req} = S_{req}(d/2) = 1173(32.75) = 38,416$ in.^{4}. Assuming a web plate $\frac{5}{16} \times 64$, $I_{web} = 21,845(0.313) = 6837$ in.^{4} (AISC Manual, Part 2, Plate Girders Moment of Inertia of One Plate about Axis x-x Tables).*
   I_{req} for two flange plates $= I_{req} - I_{web} = 31,579$ in.^{4}. Distance between flange centroids $= 2y = 64.75$ in. $(2y = h + t_f)$, where $y =$

---

*Seventh Edition.

32.375 in. $2y^2$ = 2096 in.^{2} (AISC Manual, Part 2, Plate Girders Values of $2y^2$ for Computing Moment of Inertia of Areas about x-x Tables).*

Required area, A_f (one flange) = 31,579/$2y^2$ = 31,579/2096 = 15.07 in.^{2}. $\frac{3}{4}$ × 22 = 16.5 in.^{2} > 15.07 in.^{2}.

(b) Local buckling check (Section 6-8 and Section 1.9.1 of AISC Specification): $b_f/2t_f$ = $\frac{22}{2}$ (0.75) = 14.7 < 15.8. *Use* two flange plates $\frac{3}{4}$ × 22.

EXAMPLE 7-3

For the plate girder of Example 7-2 (a trial section of $\frac{5}{16}$ × 64 web plate and two flange plates of $\frac{3}{4}$ × 22), determine all web stiffener requirements and design the stiffener size. All materials to be A36 steel.

solution:

(A) Stiffener Requirements

(1) Intermediate Stiffener Spacing at Ends of Span:
The spacing between intermediate stiffeners at end panels should be such that $f_v \leq F_v$, where F_v is computed by Eq. 7-6 (Section 7-5). $f_v = V_{max}/A_w$ = 152/20.0 = 7.6 ksi. $f_v \leq F_v = C_v(F_y/2.89) \leq 0.4F_y$.

Assume that $f_v = F_v = 7.6$ ksi. 7.6 $\leq$ (36/2.89)C_v. $C_v \geq 0.610$; assume that C_v = 0.610.

When $C_v < 0.8$, $C_v = 45,000k/F_y (h/t_w)^2$ and k = 0.610(36)(204)^2/45,000 = 20.3.

When $a/h < 1.0$ or $a < h$, k = 20.3 = 4.00 + 5.34/$(a/h)^2$ and $(a/h)^2$ = 5.34/16.3 = 0.328, a/h = 0.572 and a = 0.572(64) = 36.6 in.

Use a = 36 in. (from each end of the girder span).

(2) Bearing Stiffener Requirements at Interior Concentrated Loads:
Reference is made to Eq. 6-14. This equation assures that without the use of bearing stiffeners at the interior concentrated loads, the compressive stress at the web toe of the fillets resulting from

---

* Seventh Edition.

these loads is not exceeded. If the stress is exceeded, bearing stiffeners should be used. $R/t_w(N + 2k) \leq 0.75F_y = 27$ ksi.

Assume that N, the length of concentrated load $= k$ (for end reactions), and that $k = t_f + \frac{1}{4} = 1.0$ in. $50/0.313(3.0) = 53.2$ ksi > 27 ksi.

Therefore, *use* bearing stiffeners under interior concentrated loads. The spacing, a, between the first intermediate stiffener at the end of the span and the concentrated interior load is $a = 12(17) - 36.0 = 168$ in.

(3) Intermediate Stiffener Requirements:
$h/t_w = 64/0.313 = 204 < 260$ (Section 7-5). f_v at 36 in. from the end of the girder (end-panel check):
$f_v = \{152 - [4.0(3.0)]\}/20.0 = 7.0$ ksi. $a/h = 168/64 = 2.63$.

From Eq. 7-6: when $a/h > 1.0$, $k = 5.34 + 4.00/(2.63)^2 = 5.92$, $C_v = 45,000(5.92)/36(204)^2 = 0.178 < 0.8$, $F_v = (36/2.89)0.178 = 2.22$ ksi $< 0.4(36) = 14.4$ ksi, and $f_v > F_v$.

Therefore, use an additional intermediate stiffener between the first intermediate stiffener and the bearing stiffener at the first interior concentrated load (at $168/2 = 84$ in. o.c.).

From Section 7-5, $a/h \leq [260/(h/t_w)]^2 = (260/204)^2 = 1.62 \leq 3.0$. Assume that $a/h = 1.62$; therefore, $a \leq 1.62(64) \leq 103.7$ in.; $84 \leq 103.7$.

From Table 7-1, for $a/h = 84/64 = 1.31$ and $h/t_w = 204$: $F_v \approx 8.0$ ksi $> f_v$.

(4) Combined Shear and Tension Stress Check (Section 7-6) at the Interior Concentrated Loads:
$f_v = 84/20.0 = 4.20$ ksi $\leq 0.6F_v = 0.6(8.0)$ (Fig. 7-11).

Therefore, use $F_b : f_b = 2006(12)/1265 = 19.0$ ksi $> 0.75F_b = 0.75(22)$.

Therefore, use Eq. 7-10: $F_b \leq [0.825 - 0.375(4.20/8.0)]36 = 22.6$ ksi > 22.0 ksi.

Therefore, use $F_b = 22.0$ ksi.

Use for intermediate stiffener spacing: from end girder to first intermediate stiffener, 36 in.; from first intermediate stiffener to next intermediate stiffener, 84 in.; from second intermediate stiffener to bearing stiffener at first interior concentrated load, 84 in.

(5) 17-ft. Interior (center) Panel Check: $h/t_w = 204 <$ 260 (Section 7-5); $f_{v\,max} = 34/20.0 = 1.70$ ksi; distance between stiffeners (at concentrated loads) $= 12(17) = 204$ in.; $a/h = 204/64 = 3.19$. *Use* 3.0.

From Eq. 7-6, when $a/h > 1.0$: $k = 5.34 + 4.00/(3.0)^2 = 5.78$; $C_v = 45,000(5.78)/36(204)^2 = 0.174 < 0.80$; $F_v = (36/2.89)0.174 = 2.16$ ksi $< 0.4(36) = 14.4$ ksi; and $f_v < F_v$.

(6) From Section 7-3 (Eqs. 7-2 and 7-3), check web crippling and requirement or nonrequirement for bearing stiffeners due to the sum of the compression stresses resulting from the concentrated and distributed loads bearing directly on the girder flange:

Assuming the compressive flange is restrained by construction against rotation: between intermediate stiffeners $f_1 = 0$ and $f_2 = 4/0.313(12) = 1.07$ ksi. $f_2 < [5.5 + 4/(3.00)^2]10,000/(204)^2 = 1.43$ ksi; 1.07 ksi < 1.43 ksi.

Therefore, no stiffener is required at midspan and the intermediate stiffener spacing is as shown in Fig. 7-13. All intermediate stiffeners (not serving as bearing stiffeners at concentrated interior loads) may be stopped short of the tension flange. A minimum distance of $4t_w$ and a maximum distance of $6t_w$ shall be applied to the termination of the stiffener-to-web weld and web-to-flange weld (Section 7-5). $4t_w = 1.25$ in.; $6t_w = 1.88$ in.

(B) Stiffener Size

(1) Intermediate Stiffeners:

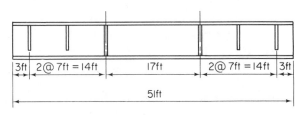

FIG. 7-13

(a) From Section 7-5, Eq. 7-8 can be solved by using Table 7-1 as follows: for $h/t_w = 204$ and $a/h = 1.31$, $A_w = 20.0$ in.^{2}; $A_{st} \approx 0.102A_w = 2.04$ in.^{2}; $F_v \approx 8.0$ ksi $> f_v = 7.0$ ksi (see step A3).

Actual $A_{st \text{ required}} = 2.04(7.0/8.0) = 1.79$ in.^{2}

Try two plates $\frac{1}{4} \times 4$, $A_{st} = 2.0$ in.$^2 > 1.79$ in.^{2}.

(b) Width-to-thickness ratio check (AISC Specification, Appendix A of the Specification, and Section 6-6): $4/0.25 = 16.0 \approx 95/(F_y)^{1/2} = 15.8$.

(c) Moment-of-inertia check (Fig. 7-14; see also Section 7-5): $I_{\text{req}} \geq (h/50)^4 = (64/50)^4 = 2.68$ in.^{4}.

$I_{\text{actual}} = 2bd^3/3 = 2(0.25)(4.1565)^3/3 = 11.97$ in.$^4 > I_{\text{req}}$.

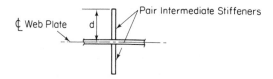

$\mathbb{C}$ Web Plate $\quad$ d $\quad$ Pair Intermediate Stiffeners

FIG. 7-14

(d) Maximum length of stiffener: $h - 4t_w = 64 - 1.25 = 62.75$ in. *Use* 5 ft, 2 in. *Use* intermediate stiffeners: two plates $\frac{1}{4} \times 4 \times 62$ bearing on the compression flange of the girder.

The means of connecting the stiffeners and their connection design will not be included in this example, although the governing specifications were discussed in Section 7-5.

(2) Bearing Stiffeners (at end bearing and at interior concentrated loads): From Section 7-5, with $\frac{3}{4} \times 22$ flange plates. *Try* two plates $\frac{5}{8} \times 10$.

(a) Width-to-thickness ratio check (AISC Specification, Appendix A of AISC Specification, and Section 6-8): $10/0.625 = 16 \approx 95/(F_y)^{1/2} = 15.8$.

(b) Check for compressive stress (end bearing) (Section 7-5, Figs. 7-7 and 7-8):

(i) For end-bearing stiffeners (Fig. 7-7a): $I = 2bd^3/3 = 2(0.625)(10.1565)^3/3 = 437$ in.^{4}. $A_{effective} = 2A_{st} + 12t_w = 2(10)(0.625) + 12(0.313) = 16.26$ in.^{2}; $r = (I/A_{eff})^{1/2} = (437/16.26)^{1/2} = 5.18$ in.; $I_{eff} = 0.75I_{actual} = 0.75(64) = 48.0$ in.; $I/r = 48.0/5.18 = 9.27$.

From Sections 5-3 and 5-6, Eq. 5-3 is solved with the aid of Fig. 5-2 as follows: Assuming that $K = 1.0$, for $KI/r = 9.27$ and $F_a = 21.20$ ksi.

From Fig. 7-11, $f_a = V_{max}/A_{eff\,st} = 152/16.26 = 9.35$ ksi $< F_a$. *Use* for end-bearing stiffeners two plates $\frac{5}{8} \times 10$ with close bearing on compression flange receiving reaction.

(ii) For bearing stiffeners at concentrated interior loads (Fig. 7-7b): $I = 437$ in.^{4}; $A_{eff} = 2A_{st} + 25t_w = 2(10)(0.625) + 25(0.313) = 20.3$ in.; $r = (437/20.3)^{1/2} = 4.64$ in.; $I_{eff} = 48.0$ in.; $I/r = 48.0/4.64 = 10.34$.

Assuming that $K = 1.0$, $KI/r = 10.34$, and, from Fig. 5-2, $F_a = 21.14$ ksi.

From Fig. 7-11, $f_a = V_{max}/A_{eff\,st} = 84/20.3 = 4.14$ ksi $< F_a$.

Use for intermediate (bearing stiffeners) at interior concentrated loads two plates $\frac{5}{8} \times 10$ with close bearing on compression flange receiving concentrated load.

Once again, the means of connecting the stiffeners and their connection design will not be included in this example, although the governing specifications were discussed in Section 7-5.

EXAMPLE 7-4

Choose a trial section for the girder in Example 7-2. For this example assume the girder to be hybrid (for the flanges, $F_y = 50$ ksi; for the web, $F_y = 36$ ksi) and that the compression flange is fully laterally supported throughout the span. Determine whether intermediate stiffeners are required.

solution (see Fig. 7-11):

(1) Preliminary Web Plate Design:
Assume that $h = 64$ in.

(a) For no reduction in flange stress (Section 7-4): $h/t_w \leq 760/(30)^{1/2} = 139$; $t_w \geq 64/139 = 0.460$ in.

(b) From Appendix A of the AISC Specification, solving Eq. 7-1a for the limiting h/t_w ratio: $h/t_w \leq 243$; $t_w \geq 64/243 = 0.263$ in.

(c) From Section 4-3, Eq. 4-3, the maximum allowable shear stress in the web is $F_v = 0.40F_y = 14.4$ ksi. Therefore, $t_w \geq V_{max}/hF_v = 152/64(14.4) = 0.165$ in. *Try a web plate $\frac{5}{16} \times 64$: $A_w = 20.0$ in.^{2}; $h/t_w = 204$.*

(2) Preliminary Flange Design:

(a) $t_w = \frac{5}{16}$ in., $h = 64$ in. Assume that $F_b' = 0.60F_y = 30$ ksi. $S_{req} = 12(2151)/30 = 860$ in.^{3}; $S/h = 860/64 = 13.4$; assume that $b_f = 20$ in. From Fig. 7-10, *try $t_f = \frac{1}{2}$ in.*

(b) Local buckling check (Section 6-8 and Section 1.9.1 of the AISC Specification): $b/2t_f = 20/2(0.5) = 20 > 13.4$ won't work. *Try $t_f = \frac{3}{4}$ in.; $b/t_f = 20/2(0.75) = 13.3 < 13.4$.*

(3) Trial Girder Section Design:
Web plate $\frac{5}{16} \times 64$ and two flange plates $\frac{3}{4} \times 20$.

(a) Checking by the moment-of-inertia method (Section 7-2):

Section	A (in.^{2})	y (in.)	Ay^2 (in.^{4})	I_0 (in.^{4})	I_{gr} (in.^{4})
Web $\frac{5}{16} \times 64$	20.0			6,837[a]	6,837
Flange $\frac{3}{4} \times 20$	15.0	32.375	31,455[b]	1	31,456
Flange $\frac{3}{4} \times 20$	15.0				
			Moment of inertia		= 38,293

[a] From the table Plate Girders, Moment of Inertia of One Plate About Axis x-x, AISC Manual, Part 2, Seventh Edition, with $d = h = 64$ in. and $t = t_w = 1$ in. $= 21,845(0.313) = 6837$ in.^{4}. $S = 38,293/32.75 = 1169$ in.^{3}.

[b] Enter the table Plate Girders, Values of $2y^2$ for Computing Moment of Inertia of Areas About Axis x-x, AISC Manual, Part 2, Seventh Edition, with $2y = 64.75$ and read 2097 by interpolation. $2097A_f = 2097(15.0) = 31,455$ in.^{4}.

(b) Check allowable flange stress: Since the compression flange is fully laterally supported for the entire span, $F_b = 0.6F_y = 30$ ksi, $h/t_w \leq 760/(F_b)^{1/2}$ $= 139 < 204$. Therefore, the flange stress must be reduced (Eqs. 7-4 and 7-5): $f_{b\ max} = 12(2151)/1169$ $= 22.1$ ksi.

From Eq. 7-4, $F_b' \leq 30.0 \{1.0 - 0.0005(20/15.0) [64/0.313 - 760/(30)^{1/2}]\} = 28.7$ ksi > 22.1 ksi. $\alpha = F_{y\ web}/F_{y\ flanges} = 36/50 = 0.72$.

From Table 7-3, with $\alpha = 0.72$ and $A_w/A_f = 20/15 = 1.33$, $C_h = 0.982$, and in Eq. 7-5, $F_b' \leq C_h F_b$ $= 0.982(30) = 29.5$ ksi > 22.1 ksi.

Use a web plate $\frac{5}{16} \times 64$ ($F_y = 36$ ksi) and two flange plates $\frac{3}{4} \times 20$ ($F_y = 50$ ksi).

(4) Intermediate Stiffener Requirements.

From Section 7-5, tension field action is not permitted in hybrid girder design. The need for intermediate stiffeners is determined by Eq. 7-6 or by Table 7-2, where tension field is neglected. For $a/h =$ "Over 3," $F_{y\ web}$ $= 36$ ksi, $h/t = 204 < 260$.

From Table 7-2 (Eq. 7-6), $F_v = 2.02$ ksi. From Fig. 7-11, $V_{max} = 152$ kips. $f_v = V_{max}/A_w = 152/20.0 = 7.6$ ksi $> F_v$. Therefore, intermediate stiffeners are required.

Example 7-2 illustrated various methods that can be used to choose a trial plate girder section. The flange-area method, checked by the moment-of-inertia method (method A), the plate-girder-section modulus nomograph method (Fig. 7-10), and tables, Welded Plate Girders Dimensions and Properties (AISC Manual, Part 2), were each used to determine a trial section.

Example 7-3 illustrated the design of bearing and intermediate web stiffeners. The use of Table 7-1 (allowable F_v in plate girders, tension field action assumed) was shown.

Example 7-4 illustrated the design of a hybrid plate girder. The use of Tables 7-2 and 7-3 (allowable F_v for tension field neglected) and the solution of Eq. 7-5 were shown, respectively. In addition, the use of the tables found in the AISC Manual, Part 2, Seventh Edition, to facilitate the computation of Ay^2 for flanges and I_0 for web (I_{gr} of the plate girder section) was illustrated.

Use of the design aids (tables, charts, nomographs, etc.), with experience, enables the engineer to design more easily.

PROBLEMS

(7-1) Determine the section modulus of a welded plate girder consisting of a $\frac{3}{4}$ in. × 48 in. web plate welded to 1 in. × 18 in. flange plates.

(7-2) Determine the weak axis section modulus for the plate girder described in Problem 7-1.

(7-3) The plate girder shown in cross section on the sketch carries a maximum shear of 130 kips. Compute the shear stress at the neutral axis, in addition to the shear flow (kips/in.) that must be carried by each of the welds shown (web to flange plate, and flange plate to cover plate). (Refer to Chapter 4.)

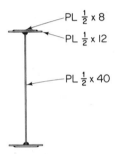

(7-4) Compute the maximum bending stress on the section shown in Problem 7-3, assuming a maximum bending moment of 800 kip-ft.

(7-5) The plate girder (A36 steel) shown has been partially designed as indicated:
   (a) If the compression flange is fully braced laterally, is the cross section adequate? Check web adequacy after completing parts (b) and (c).
   (b) Determine spacing and *area* needed for intermediate stiffeners in the length *ab*.
   (c) Are intermediate stiffeners needed in length *bc*?
   Section: Web, 70 × $\frac{3}{8}$
             Flange, 17 × $\frac{7}{8}$
   Load: 3 kips/ft including girder weight

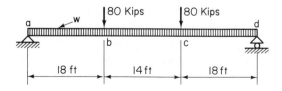

(7-6) A plate girder has been designed of A36 steel. It is laterally braced at the ends and at each load.

    (a) Using maximum value of C_b, what is the maximum allowable bending moment, M_2?

    (b) What is the shear capacity based on the stiffener spacing in the end 15-ft section (assume adequate stiffeners)? Assume tension field action where applicable.

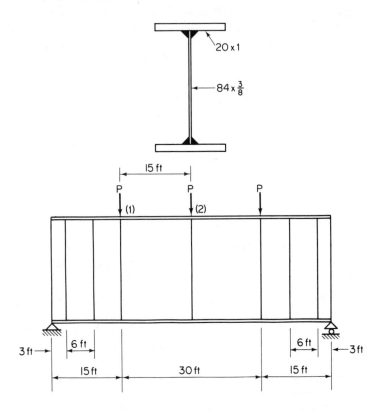

(7-7) Design a welded plate girder (choose the appropriate profile), using the section modulus nomograph design aid, to carry a uniform load of 2.5 kips/ft in addition to a concentrated load of 60 kips (assume point bearing) at the center of a 48-ft span. Assume A36 steel and limit the depth to 60 in. Compression flange has continuous lateral support.

(7-8) Determine all web stiffener requirements for the girder section chosen in Problem 7-7. Assume stiffener material to be A36 steel.

(7-9) Design a "hybrid girder" to support a uniform load of 3 kips/ft in addition to concentrated loads of 150 kips at the third points (assume point bearing) of its 60-ft span. Assume lateral support is provided to the compression flange only at the concentrated loads. Clearance limitations permit a maximum depth of $5\frac{1}{2}$ ft. F_y = 50 ksi for flanges, and F_y = 36 ksi for web.

(7-10) Determine all stiffener requirements for the girder section chosen in Problem 7-9. Assume stiffener material to be A36 steel.

8-1   INTRODUCTION

This chapter will discuss the use of AISC Specification formulas for beams not continuously braced. The phenomenon of beam buckling will be reviewed. The beam formulas will be explained, suggesting which is proper to use. The use of the AISC Manual design aids and tables will be shown, enabling the designer to design with speed and confidence. Several design examples will be presented to illustrate the design methodology and use of the design aids. The last subject to be discussed will be bracing. The types of beam bracing and column bracing design will be reviewed.

8-2   BUCKLING

In review, curve 5 of Fig. 6-5 is typical of members with large unbraced lengths that are not torsionally stiff, where L_b, the unbraced length, is larger than L_c and L_u. The yield-point stress at any place on the section is not achievable before some buckling phenomenon takes place. L_c is the maximum unbraced interval of the compression flange at which $F_b = 0.66F_y$ or as determined by Eq. 6-11, whichever is applicable. L_u is the maximum unbraced interval of the compression flange at which $F_b = 0.60F_y$. Lateral bracing and buckling were discussed in Section 6-10.

To more properly understand the design formulas for beams not continuously braced, a review of the buckling phenomenon is in order. An unbraced segment of a beam is shown in Fig. 8-1. The segment is considered to be braced at its ends and to be subject to moments M_{cr} at its ends as indicated. M_{cr} is the moment at which the beam buckles, as shown in Section 1-1 of Fig. 8-1. The solid cross section indicates the unloaded position. The uniform moment condition shown is considered to be the basic case. Other loading and support conditions will be handled by modifying factors as discussed in Section 8-2.5.

Upon the application of load, the beam cross section deflects vertically until the increased loading reaches M_{cr}, at which time the cross

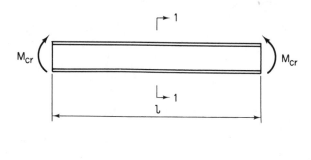

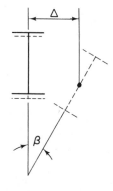

Section 1-1

FIG. 8-1

section twists, moves laterally Δ, and rotates through the angle β. It is assumed that the cross section maintains its shape.

When a W shape is in a state of lateral buckling, it is subject to torque. The torque or twist, T, in a beam subjects it to torsion and warping as shown in Fig. 8-2a and b. If there is no restraint, a plane surface before bending becomes a warped surface after bending, as shown in (b).

The equation $T = GJ\beta'$ presents a relationship between applied torque, T, and angle of twist, β. β' is the first derivative of the angle of twist with respect to length. Buckling is resisted by T. G is the shearing modulus of elasticity or shear modulus (ksi) and J is a torsional constant of the cross section (in.⁴). For the case of a doubly symmetrical W shape,

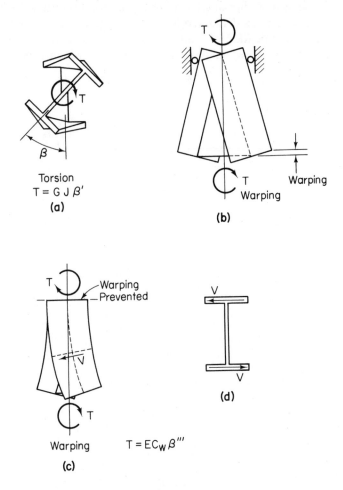

Torsion
$T = G J \beta'$
(a)

Warping
T
Warping
(b)

Warping $T = EC_W \beta'''$
(c)

(d)

FIG. 8-2

$J = 2b_f t_f^3/3 + ht_w^3/3$. Or, more simply, $J = \frac{1}{3}\Sigma$ width multiplied by (thickness)^3 for all the plate elements of the section. Values for J for the various cross sections are listed in the AISC Manual, Part 1, Properties for Designing Tables. GJ is analogous to EI (stiffness) for bending. It is the torsional stiffness. For structural steel, G is usually 11.5 to 12.0 ksi. G is equal to $E/2(1 + \mu)$, where μ is Poisson's ratio (equal to 0.25 to 0.33 in the elastic range). It is the ratio of transverse strain to longitudinal strain under axial load.

If warping is prevented at a particular location (Fig. 8-2c) such as a support, the cross section provides increased resistance to twist. Resistance to warping results in flange bending. The bending produces shear forces, V (Fig. 8-2d). V multiplied by the depth of section is the twist resistance. The equation $T = EC_w\beta'''$ represents the warping resistance to twist (β''' is the third derivative of the angle of twist with respect to length). E is the modulus of elasticity (ksi) and C_w is a warping constant for a section (in.^{6}). As with J, C_w is listed in the AISC Manual, Part 1, Properties for Designing Tables. In addition, equations for C_w for various cross sections are given elsewhere.* C_w is related to the moment of inertia of the flange and section depth in the case of a W shape.

The total resistance to twist, or torsional strength, is the sum of pure torsional strength and warping strength. Pure torsional strength governs when the section is shallow with thick plate elements and large flange area. Warping strength controls for deep thin sections. Strength or resistance to twist is the vector sum of the two aforementioned resistances:

$$\text{strength} = [(\text{torsional resistance})^2 + (\text{warping resistance})^2]^{1/2} \qquad (8\text{-}1)$$

Resistance to lateral movement, Δ, as shown in Fig. 8-1, may be expressed as $EI_{yy}\Delta''$. EI_{yy} is the resistance to this movement in the weak-axis direction. Δ is the movement in the yy direction at any point along the span of the beam, and Δ'' is the second derivative of this movement in the yy direction. Combining the twist and lateral resistance equations to satisfy equilibrium yields the critical moment when lateral-torsional buckling occurs:

$$M_{cr} = \frac{\pi}{L_b}[EI_y(GJ + EC_w\pi^2/l^2)]^{1/2} \qquad (8\text{-}2)$$

In summary, the factors that affect buckling are as follows:

(1) Lateral end restraint: the basic case assumes no lateral restraint with twist prevented. Lateral movement is assumed. If the beam

* F. Bleich, *Buckling Strength of Metal Structures* (New York: McGraw-Hill Book Company, 1952).

ends were fully restrained laterally (basic case assumes pinned ends), M_{cr} would be increased two to four times the M_{cr} for pinned ends (increase is twice if torsional strength dominates and four times if lateral bending strength dominates). It is safe to assume pinned ends, and this assumption is recommended for design. Of additional interest is that for continuous beams there is no lateral bending restraint if the spans buckle simultaneously.

(2)   Loading condition: the basic case assumes uniform moment, which is the most critical condition. For other loading conditions, a modifying factor, C_b, is used to correct for this. This is discussed in Section 8-2.5.

(3)   Loading location: a load applied above the centroid of the cross section, such as in Fig. 8-3a, causes an additional overturning and is more critical than if the load is applied at the centroid (Fig. 8-3b). A load applied above or below the centroid (Fig. 8-3c) can change

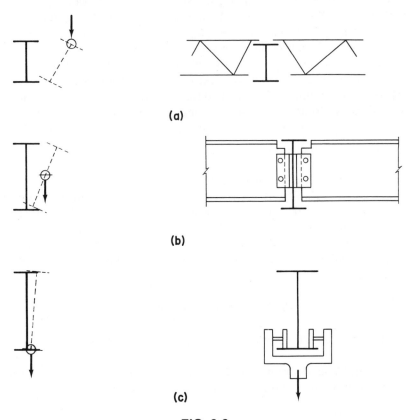

(a)

(b)

(c)

FIG. 8-3

the buckling load by $\pm 40\%$. If the load is applied below the centroid, it produces a stabilizing effect. If the load point is a braced point, as is the usual case, the location of load application has no effect.

8-2.1 Beam Formulas

M_{cr}, divided by the section modulus, S_x, gives the value for the critical buckling stress, F_{cr}. For an I-shape cross section, this formula for F_{cr} can be expressed as in Eq. 8-3 for the basic case discussed in Section 8-2 after substituting the cross-sectional properties in Eq. 8-2 divided by S_x:

$$F_{cr} = \left\{ \left[\frac{20,000}{(ld/A_f)} \right]^2 + \left[\frac{\pi^2 E I_y}{l^2 2(A_f + A_w/6)} \right]^2 \right\}^{1/2} \qquad (8\text{-}3)$$

F_{cr} is the elastic critical buckling stress. The first term, the torsion term, involves the approximation of neglecting the web. The error is small. There are no approximations in the second term, the warping term. The warping term can be simplified by introducing a new parameter, r_T. r_T is defined as the radius of gyration of a section comprising the compression flange plus one third of the compression web area, taken about the yy axis (in.). $(I_y/2)/(A_f + A_w/6)$ is exactly the same as r_y^2 of a cross section with two thirds of the web removed and is equal to r_T^2. Values for r_T for various shapes are listed in the AISC Manual, Part 1, Properties for Designing Tables. Equation 8-3 may now be written as

$$F_{cr} = \left\{ \left[\frac{20,000}{(ld/A_f)} \right]^2 + \left[\frac{\pi^2 E}{(l/r_T)^2} \right]^2 \right\}^{1/2} \qquad (8\text{-}4)$$

This formula is for the basic case defined in Section 8-2. The assumption of no lateral movement at the ends of the unbraced length (pinned ends) is a safe one. Lateral restraint can increase buckling strength but is difficult to evaluate and is therefore neglected in design. Although top flange loading can be more critical than that in Eq. 8-4, the usual framing members, transferring load to the beam, provide some twisting restraint and lateral bracing, and so this factor is generally neglected in design (although any unusual cases should be checked).

C_b is a moment modifying factor that permits the use of the basic case of uniform moment buckling. A value of C_b equal to unity is used if the moment is uniform or if maximum between brace points. Other values of C_b are used if the maximum moment is in close proximity to a braced point. For other values and a more detailed discussion of C_b, see Section 8-2.5. Use of $C_b = 1.0$ is always conservative. In order to cover, more accurately, all cases of loading conditions, Eq. 8-5 is applicable.

$$F_{cr} = \left(\frac{M_{cr}}{S_x}\right) \approx C_b \text{ (Eq. 8-4)} \qquad (8\text{-}5)$$

The use of the usual factor of safety of 1.67 gives

$$F_b = \frac{F_{cr}}{1.67} = C_b \left\{ \left[\frac{12{,}000}{(ld/A_f)}\right]^2 + \left[\frac{0.6\,\pi^2 E}{(l/r_T)^2}\right]^2 \right\}^{1/2} \qquad (8\text{-}6)$$

The total strength of a W shape is always greater than either the torsional strength or the lateral bending strength (warping strength), and since the total strength is a vector sum of the two terms (Eq. 8-1), it is conservative to use the higher value as an estimate of strength. The AISC Specification approach is to use the larger of the two terms. This is a conservative approach and involves neglecting the lesser of the two terms.

The AISC Specification states that F_b for compression for members having an axis of symmetry about, and loaded in the direction of, the yy-axis and bent about the xx-axis shall be the larger of the following equations, as applicable, but not more than $0.60F_y$. When

$$\left[\frac{102(10)^3 C_b}{F_y}\right]^{1/2} \le \frac{l}{r_T} \le \left[\frac{510(10)^3 C_b}{F_y}\right]^{1/2}$$

$$F_b = \left[\left(\frac{2}{3}\right) - \frac{F_y (l/r_T)^2}{1530(10^3)C_b}\right] F_y \qquad (8\text{-}7a)$$

and when $l/r_T > [510(10)^3 C_b/F_y]^{1/2}$,

$$F_b = \frac{170(10)^3 C_b}{(l/r_T)^2} \qquad (8\text{-}7b)$$

or, when the compression flange is solid and approximately rectangular in cross section with its area not less than that of the tension flange,

$$F_b = \frac{12(10)^3 C_b}{(ld/A_f)} \qquad (8\text{-}7c)$$

where l = unbraced length of compression flange, in.
   r_T = radius of gyration of a section comprising one third of the compression web area plus the compression flange area taken about the yy axis, in.
   A_f = area of the compression flange, in.^{2}
   C_b = $1.75 + 1.05(M_1/M_2) + 0.3(M_1/M_2)^2 \le 2.3$

M_1 is the smaller and M_2 the larger bending moment at the ends of the unbraced length, l, taken about the xx axis. M_1/M_2, the end-moment ratio, is positive for reverse curvature bending and negative for single curvature bending. When the bending moment at any point within l is larger than that at both ends of this length, C_b will be unity. C_b can, conservatively, always be taken as unity.

F_y is the yield stress of the compression flange. Equation 8-7c is not applicable to hybrid girders. Only Eq. 8-7c is applicable to channel sections (allowable F_b compression).

Equations 8-7b and 8-7c represent the elastic determination of warping resistance and torsional resistance, respectively. Equation 8-7a represents the inelastic determination of warping resistance. There is no inelastic counterpart to the torsional resistance, since tests have shown little effect on inelastic behavior for this case.

Equations 8-7a, b, and c provide an allowable compressive bending stress for members bent about their major axis having an axis of symmetry and loading in the plane of the minor axis, which are braced at greater intervals than $76.0b_f/(F_y)^{1/2}$ [when $l \leq 76.0b_f/(F_y)^{1/2}$, $F_b = 0.60F_y$]. These formulas provide reasonable design criteria to somewhat account for the members' torsional rigidity about their longitudinal axis (St. Venant torsion), as well as the bending stiffness of their compression flange between points of lateral support (warping torsion).

Figure 8-4a and b shows the curves for F_b plotted against l/r_T and ld/A_f for warping and torsion resistance, respectively. A value of C_b of unity and $F_y = 36$ ksi is assumed. For any specific beam section, with F_b (ksi) as an ordinate and L_b(ft) as an abscissa, the interrelationships among all the allowable bending stress formulas can readily be seen. Figure 8-5 illustrates this.

In summary, for W 21 × 44, A36 steel, and $C_b = 1.0$ assumed (conservative):

F_b	Explanation
$0.66F_y$	Limited by L_c (unbraced length)
$0.60F_y$	Limited by L_u (unbraced length)
Eq. 8-7a	Warping controls
Eq. 8-7b	Warping controls (warping stress $< F_y/3$)
Eq. 8-7c	Torsion controls

The AISC Allowable Beam Moment Curves, which were discussed in Section 6-14.2, are constructed in much the same way as Fig. 8-5. If the stress ordinates of Fig. 8-5 were multiplied by the section modulus of the shape, bending moment would be the ordinate, and a beam-moment curve exists.

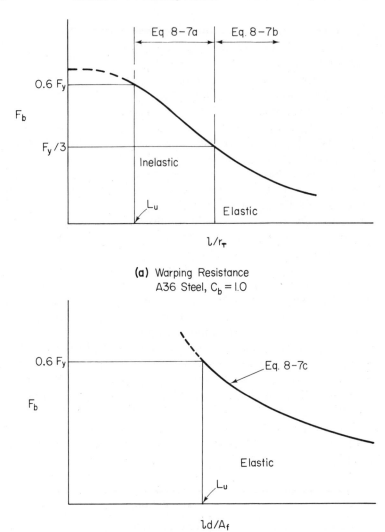

(a) Warping Resistance
A36 Steel, $C_b = 1.0$

(b) Torsion
A36 Steel, $C_b = 1.0$

FIG. 8-4

8-2.2   *Plate Girders /Channels / Crane Girders*

Plate girders usually have thin flexible webs that give little torsional strength. The d/A_f ratio is usually much higher than the corresponding ratio for rolled W shapes. Equation 8-7c for plate girders may therefore err on the conservative side. As a general rule, the larger value permitted by Eq. 8-7a or b governs. Large flanges give substantial warping strength.

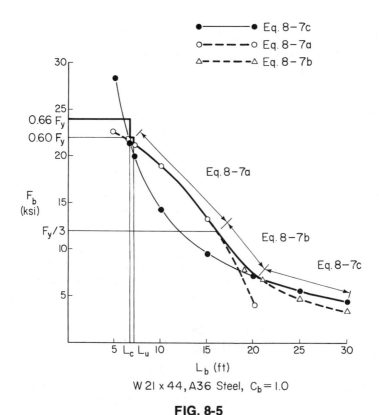

FIG. 8-5

Equation 8-7c, only, is applicable to channels. One need not be overly concerned about warping in the case of a channel because its shape and the means of loading (not through its shear center) subject it to torsion.

Crane girders often have an increased compression flange area to resist bending due to lateral loading and vertical loading (Section 7-8, Fig. 7-9). Equation 8-7a or b is used to determine the permissible bending stress in this case. Equation 8-7c may yield unconservative results where the tension flange is very much larger than the compression flange. Hence, as a general rule, Eq. 8-7c is not permitted for use where the compression flange area is less than the tension flange area.

8-2.3   Cantilevers

In general, the torsional equation, Eq. 8-7c, governs for cantilever beams. Cantilever beams are usually framed to a column flange as shown in Fig. 8-6 and the column usually provides a weak restraint for the beam. The column usually is able to rotate unless its flanges are very husky, in which

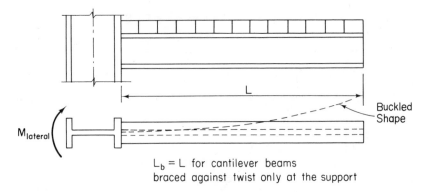

$L_b = L$ for cantilever beams
braced against twist only at the support

FIG. 8-6

case the designer may want to investigate the warping equation also. Unless the column can resist a lateral moment or the beam-to-column connection can restrain the beam laterally, as a conservative approach, the torsion term (Eq. 8-7c) should be used to investigate buckling.

8-2.4   Box Girders/W Shape (Weak-Axis Bending)

Section 6-13 discussed box sections and stated that box girders are very stable laterally and torsionally are very stiff. It also stated that lateral torsional buckling was not a problem when the depth $\leq$ 6 times its width. $F_b = 0.60F_y$ when $190/(F_y)^{1/2} < b_f/t_f < 238/(F_y)^{1/2}$. Similarly, compact box-shaped members have no lateral torsional buckling problem as long as $d < 6b_f$. $F_b = 0.66F_y$ provided that l_b is limited following Eq. 6-13, and $t_f \leq 2t_w$.

As a general rule, for box-shaped sections, if $I_{xx} \approx I_{yy}$ or if the torsional stiffness is greater than the in-plane bending stiffness, there is no buckling problem (i.e., $GJ + EC_w\pi^2/L_b^2 > EI_{xx}$). Weak-axis bending of W shapes and box sections was discussed in Section 6-5. For this case there is no buckling problem and $F_b = 0.75F_y$ (Eq. 6-5) and $F_b = 0.66F_y$, respectively.

8-2.5 C_b Factor

C_b is a modifying factor used to account for loading conditions other than the basic case of uniform moment and is discussed and defined in Sections 8-2 and 8-2.1. All the AISC formulas (Eqs. 8-7a, b, and c) are based on this uniform moment condition. To repeat, it is always conservative to use a value of unity for C_b (Figs. 8-4 and 8-5), and the AISC uses this assumption.

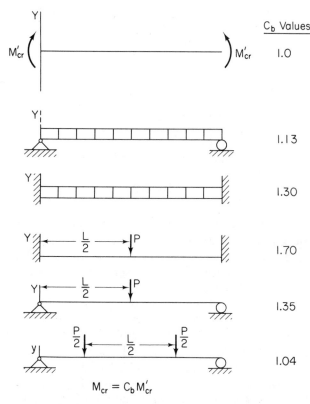

$$M_{cr} = C_b M'_{cr}$$

Note : All Beams are Assumed with No End Restraint
About the y−y Axis at Each Support

FIG. 8-7 *Column Research Council, Guide to Stability Design Criteria for Metal Structures, 3rd Ed., p. 133.*

Appendix A of the AISC Specification has a design aid that helps to determine C_b values for various end-moment ratios (moments at ends of unbraced length). C_b values for various other loading and end-support conditions are shown in Fig. 8-7. All beams are braced at support points only and all loads are applied through the centroids. The equation for C_b is repeated here for convenience and given an equation number:

$$C_b = 1.75 + 1.05\left(\frac{M_1}{M_2}\right) + 0.3\left(\frac{M_1}{M_2}\right)^2 \le 2.3 \qquad (8\text{-}8)$$

where $M_1 < M_2$, each at the ends of the unbraced length, L_b. M_1 and M_2 are taken about the strong axis of the beam. For single curvature, M_1/M_2 is −; for double (reverse) curvature,

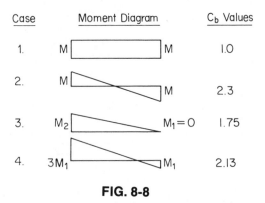

Case	Moment Diagram	C_b Values
1.	M [ ] M	1.0
2.	M [ ] M	2.3
3.	M_2 [ ] $M_1 = 0$	1.75
4.	$3M_1$ [ ] M_1	2.13

FIG. 8-8

M_1/M_2 is $+$. It follows, then, that, for the moment conditions shown in Fig. 8-8, the C_b values vary and the moment diagram has a significant effect on buckling strength. Case 2 (reverse curvature) is 2.3 times stronger than case 1 (uniform moment, single curvature). The buckling of a cantilever beam is different from that of a beam with two end supports. A value of $C_b = 1.0$ should be conservatively used for cantilever beams. A value of $C_b = 1.75$ is incorrect for cantilever beams (for example, case 3 in Fig. 8-8, where the cantilever span is L_b, $M_1 = 0$ and M_2 is as shown). The Column Research Council, *Guide to Stability Design Critera for Metal Structures,* Third Edition, p. 135, states that other values larger than unity may be used (see Fig. 8-9).

Finally, the supports of the beam should offer restraint against rotation about the longitudinal (zz) axis of the beam, as well as restraint for the vertical end reaction of the beam, so C_b values of 1.0 or 1.3 and 2.05 for the cases in Fig. 8-9 can be used.

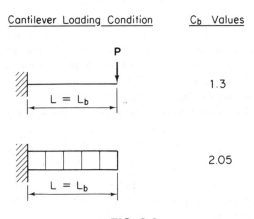

Cantilever Loading Condition	C_b Values
P, $L = L_b$	1.3
$L = L_b$	2.05

FIG. 8-9

8-3 DESIGN AIDS

To expedite design and to make life a little more pleasant for the designer, the AISC has developed several useful design aids. Three of the more popular ones, which have been included in the AISC Manual, will be discussed: the Allowable Stress Design Selection Tables and the Plastic Design Selection Tables (Beam Selection Tables); the Allowable Moments in Beam Charts (Beam Moment Curves); and the Uniform Load Constants for Beams Laterally Supported Tables (Beam Load Tables). The Beam Selection Tables are the fastest to use and are used most often. The Beam Moment Curves and Beam Load Tables follow in that order.

8-3.1 Allowable Stress Design and Plastic Design Selection Tables

The Beam Selection Tables were discussed in some detail in Section 6-14.1. The parameters found in the table were explained and the process of selection utilizing the tables was discussed. Of importance relative to the subject matter of this chapter were the parameters L_c and L_u as related to the unbraced length, L_b. These tables are most useful when $L_b \leq L_c$. For values of $L_b > L_u$, Beam Moment Curves are recommended for use.

The use of this design aid can best be illustrated by a design example.

EXAMPLE 8-1

Select the most economical rolled beam of $F_y = 50$ ksi steel which spans 24 ft, supports a uniform load of 4 klf, and is braced at the compression flange at 5 ft on center. Assume no depth restriction and ASD.

solution:

$V_{max} = 4(24)/2 = 48$ kips; $M_{max} = wL^2/8 = 4(24)^2/8 = 288$ kip-ft.

(a) Enter the ASD Selection Table looking for an M_R value $\geq M_{max}$ under the $F_y = 50$-ksi column.

For example, $M_R = 294$ kip-ft (for W 12 × 79 section) is closest to M_{max}. However, a W 24 × 55 (above) is a more economical section ($M_R = 314$ kip-ft).

The corresponding $L_c = 5.0$ ft $\geq L_b$ (bracing adequate), which indicates that the listed M_R is valid (based on $0.66F_y = F_b$). In addition, $F_y < F'_y$. This latter observation indicates a compact section. Finally, V_{allow} = $F_v dt_w$ = 0.4(50)(23.57)(0.395) = 186 kips $\gg V_{\text{max}}$. Use W 24 × 55.

(b) Assume that F_b = $0.66F_y$ = 33 ksi; $S_{x\,\text{req}}$ = $12M_{\text{max}}/F_b$ = 12(288)/33 = 100.5 in.^{3}.

Enter the ASD Selection Table looking for an S_x $\geq S_{x\,\text{req}}$. For example, S_x = 103 in.^{3} for a W 14 × 68, which is closest to $S_{x\,\text{req}}$.

However, a W 24 × 55 (above and in boldface type) is a more economical section (S_x = 114 in.^{3}). For a W 24 × 55, L_c = 5.0 $\geq L_b$ (bracing is adequate).

Therefore, the section is compact and the use of F_b = $0.66F_y$ to find $S_{x\,\text{req}}$ is valid.

In addition, observe that $F_y < F'_y$ and F''_y (in this case both are indicated by a dash). $V_{\text{allow}} \gg V_{\text{max}}$. Use W 24 × 55.

8-3.2   Allowable Moment Charts

This design aid was discussed in Section 6-14.2. To repeat, these charts give the total allowable moments in kip-feet for all unbraced lengths (L_b < L_c, $L_b > L_c$, $L_c < L_b < L_u$, and $L_b > L_u$). The curves drawn automatically account for the proper bending stresses as a function of $b_f/2t_f$. The charts are most useful for beams where $L_b > L_u$. Without these charts, a beam could not be selected knowing solely its S_x, since depth and flange proportions influence bending strength. Charts are presented for W and M shapes of F_y = 36- and 50-ksi steels.

The unbraced lengths L_c (represented by solid dots) and L_u (represented by open dots) are shown in the charts. The charts were plotted for values of C_b = 1.0. The lightest available beam is indicated by the solid portion of the curve. The dashed portions of the curves indicate that a lighter section, which is able to support the load, exists.

The use of the charts is illustrated for the selection of a beam where $L_b > L_u$.

EXAMPLE 8-2

Select the most economical beam section of A36 steel when the largest L_b is 13 ft and M_{max} = 200 ft-kips. Use allowable stress design and assume that C_b = 1.0.

solution:

(a) From the ASD Selection Table, for M_R = 206 kip-ft (W 14 × 68),

L_c = 10.6 ft (most economical section, W 24 × 55, L_c = 7.0 ft.).

$L_b > L_c$, therefore, go to the Beam Moment Charts for F_y = 36 ksi.

For an ordinate, M_{allow} = 200 ft-kips, and an abscissa, L_b = 13 ft: any beam above and to the right of the point satisfies ASD. For the most economical shape, select a W 21 × 62 (M_{allow} = 206 ft-kips).

Note: If depth is limited, the following beams may be used: W 14 × 74, W 12 × 87, or W 18 × 65.

(b) As a matter of interest, the ASD Selection Table could have been used. For a M_{max} = 200 kip-ft and L_b = 13 ft, *try* W 14 × 68, M_R = 206 kip-ft, and $L_c < L_b < L_u$.

Therefore, F_b = 0.60F_y = 22 ksi. M_R corrected = 206(0.60F_y/0.66F_y) = 188 kip-ft.

This is no good because it is less than 200 kip-ft, so *try* W 12 × 79: M_R = 214 kip-ft. M_R corrected = 214($\frac{22}{24}$) < 200 kip-ft.

This is also no good, so *try* W 14 × 74; M_R = 224($\frac{22}{24}$) = 206 kip-ft > 200 kip-ft.

Use W 14 × 74. (Going up the table, the following shapes can also be used: W 18 × 65, W 16 × 67, and W 18 × 71.)

Note that although method (b) gave a solution using the ASD Selection Table, it did not necessarily give the lightest solution. Therefore, when $L_c \le L_b < L_u$ or $L_b > L_u$, use the Beam Charts.

There is an advantage to using C_b values greater than unity. This is apparent when examining Eqs. 8-7a, b, and c, which indicate that F_b increases with increasing values of C_b. For values of C_b greater than 1, the Allowable Moment Charts, which are based upon C_b = 1.0, can be used. Equation 8-8 is used to determine the value of C_b for any unbraced segment of beam span (exclusive of cantilever beams or segments of the span that are cantilevers). After computing C_b, enter the proper chart, using C_b = 1.0, with M and move horizontally until a solid curve is reached with L_u (indicated by the open circle symbol) above the line. By eye, multiply L_u by the computed C_b value and extend the curve (L_u and beyond) by $C_b L_u$. If this modified curve extends beyond the L_b involved in the design, the section is satisfactory. This procedure is valid because L_u is directly proportional to C_b for almost all beam sections. L_u is usually derived from Eq. 8-7c and at times from the lower-bound limit on l/r_T from Eq. 8-7a, in which case the curve can only be extended to $C_b^{1/2}L_u$.

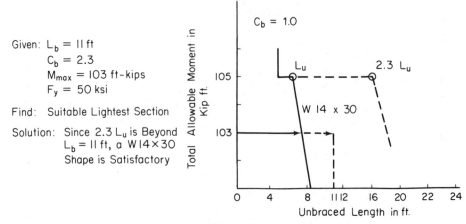

Given: $L_b = 11$ ft
$\quad\quad C_b = 2.3$
$\quad\quad M_{max} = 103$ ft-kips
$\quad\quad F_y = 50$ ksi

Find: Suitable Lightest Section

Solution: Since 2.3 L_u is Beyond
$\quad\quad\quad L_b = 11$ ft, a $W14 \times 30$
$\quad\quad\quad$ Shape is Satisfactory

FIG. 8-10

The graphical procedure is shown in Fig. 8-10, which is a sketch representing the Uniform Moment Chart as found in the AISC Manual.

8-3.3   Allowable Load Tables

Tables for uniform load constants for beams laterally supported were discussed in Section 6-14.1. Tables are given for $F_y = 36$ ksi steel for M, S, MC, and L shapes and for $F_y = 50$ ksi steel for W and M shapes.

For compact shapes, the listed constant is based upon $F_b = 0.66F_y$ and for noncompact shapes on an allowable stress of $0.60F_y$ or a value between $0.60F_y$ and $0.66F_y$, depending on $b_f/2t_f$ (Eq. 6-11). The F_b value from Eq. 6-11 is flagged by reference to a footnote to the table referring to AISC Specifications Section 1.5.1.4.2. Loading is assumed applied normal to the beam's x-x axis. To repeat, for symmetrical shapes (W, M, and S shapes), the load constants are based on $L_b \le L_c$. For compact shapes, when $L_c < L_b < L_u$, multiply the constant by the ratio of $0.60F_y$ divided by the allowable stress used to compute its capacity. The tables are not applicable for $L_c > L_u$.

These tables give a load constant, $W_c = 2S_xF_b/3$ in kip-feet, and when W_c/L, the total allowable uniformly distributed load on a simply supported member is arrived at, where $L_b \le L_u$ (the aforementioned adjustments are applicable when $L_c < L_b \le L_u$). These tables are also applicable to laterally supported, simply supported beams with equal concentrated loads, P spaced at equal intervals, as shown in Table 8-1. This table gives an equivalent uniform load equal to ΣP on the beam span.

L_c (the smaller value of $76.0b_f/[12(F_y)^{\frac{1}{2}}]$ or $20,000/[12(d/A_f)F_y]$) and L_u (the larger value of $20,000/[12(d/A_f)F_y]$ or $(102,000/F_y)^{\frac{1}{2}}(r_T/12)$) are also tabulated. When the value of L_c does not appear in the uniform load

Table 8-1

Loading Type	Equivalent Uniform Load (kips)	Deflection Coefficient
P at midspan	$wL = 2.00P$	0.80
P at third points	$wL = 2.67P$	1.02
P at quarter points	$wL = 4.00P$	0.95
P at fifth points	$wL = 4.80P$	1.01

constant tables, L_u is the maximum unbraced length for which the load and deflection constants are valid.

The tables can be entered with an equivalent uniform load constant equal to the equivalent uniform load times the span except where shear controls the design (short spans). The equivalent uniform load values are computed by setting $wL^2/8$, the maximum moment for a uniformly loaded simply supported beam, equal to the maximum moment caused by the concentrated load case and solving for $wL = W$. For relatively short spans, shearing stress in the web may control rather than maximum bending stress in the flanges for beams and channels. Limiting lengths to avoid this are listed in the table as L_v. For spans $< L_v$, the maximum total uniform load the beam can support is twice F_v (where $F_v = V/dt_w$), the allowable beam shear. When span lengths $< L_v$ and concentrated loads are involved, $\Sigma P \leq 2V$. For L shapes listed, F_v generally is greater than F_b by far, and shear only has to be considered for very short spans.

For C and MC shapes used as beams, the tabulated constants are based on $F_b = 22$ ksi (Section 8-2.1). It is assumed that the compression flange is supported at intervals $\leq L_u$.

Tabulated values for L shapes assume $F_b = 0.60F_y$, and assume adequate support of the leg subject to compression. The b/t ratios are in accordance with the AISC Specification with the width, b, defined as the width from neutral axis of angle to the free edge of the compression leg.* For W, M, S, C, and MC shapes, a deflection constant, D_c, is given for each beam when supporting its maximum uniformly distributed load. Deflections, Δ, for the symmetrical uniformly loaded simple beam $= 5Wl^3/384EI$ in inches. $W = wL$ (kips) and includes the beam weight. l

---

* For a more detailed discussion of angles used as beams with no support of compression flange, reference is made to the Australian Institute of Steel Construction, *Safe Load Tables for Laterally Unsupported Angles;* and B. F. Thomas, J. M. Leigh, and M. G. Lay, *The Behavior of Laterally Unsupported Angles,* The Civil Engineering Transactions of the Institution of Engineers, Australia, 1973.

is the span length (in.) and $E = 29,000$ ksi. Substituting for E, the equation becomes $D_c L^2/1000$, where L is in feet; $D_c = 30F_b/29d$ (in./ft²); d = depth of section (in.); F_b = allowable bending stress (ksi). The deflection constants listed are computed on the basis of $F_b = 0.66F_y$ for compact sections, or for a reduced value if the member is noncompact (Eq. 6-11). Therefore, the tabulated D_c values must be reduced for cases where L_b is between L_c and L_u. These reduction factors are as follows: $L_c < L_b \leq L_u$, multiply D_c by $0.60F_y/0.66F_y$ for compact sections and by $0.60F_y$ divided by $0.66F_y$ or the value computed by Eq. 6-11 for noncompact sections. The deflection values need not be reduced when $L_b \leq L_c$ (compact sections). For the usual span lengths used, when $M_{LL}/M_{DL} \approx 1.0$, ΔLL approaches $L/360$, the generally accepted maximum live-load deflection for beams and girders supporting plastered ceilings (Section 1.13.1, AISC Specifications). This stipulation is made to prevent excessive cracking of the plaster.

D_c, as listed in the tables, must be multiplied by the appropriate deflection coefficient (Table 8-1) to determine the concentrated load deflection.

D_c values for channels are computed on the basis of 22 ksi.

Finally, the uniform load constants for beams laterally supported tables give various property and reaction values: values for S (in.³); V, the maximum web shear ($0.40F_y dt_w$ in kips); R, the maximum end reaction for $3\frac{1}{2}$ in. of bearing = $0.75F_y t_w$ $(3.5 + k)$ (Eq. 6-14); R_i, the increase in R for each additional inch of beam bearing beyond $3\frac{1}{2}$ in. (kips) = $0.75F_y t_w$; N_e, the length of bearing to develop V (in.), $N_e = (V/R_i) - k$.

EXAMPLE 8-3

Select a 12-in. W-shape to span 28 ft. The beam is loaded with concentrated loads of 25 kips located at the third points of the span. Use $F_y = 50$ ksi. Assume that proper lateral support of the compression flanges occurs at support points and at only one concentrated load point.

solution:

(a) From Table 8-1, $W_{equiv} = wL = 2.67P = 66.75$ kips and deflection coefficient = 1.02.

(b) Entering the Uniform Load Constants Tables for W shapes in the AISC Manual, Part 2: for $F_y = 50$ ksi, W12, $L = 28$ ft. and $W_c = 1896$ kip ft.
Select W 12 × 65. F_b (Eq. 6-11) = 32.5 ksi.

$L_{b\ max} = L - L/3 = 28 - 9.33 = 18.67$ ft and therefore $L_c < L_b \le L_u$ and the tabulated load constant $W_c = 1896$ kip ft must be reduced as follows: $1896(30.0/32.5) = 1750(1/28) = 62.51$ kips $\not> 66.75$ kips. *Use* W 12×72.

(c) Deflection Computation

The table indicates $D_c = 2.8$ in/ft². Therefore, the deflection must be computed as follows: $I = 597$ in.⁴; $\Delta = D_c(28)^2/1000 = 2.20$ in.

This deflection is based upon a uniformly loaded, simply supported beam where $L_b \le L_c$.

This value must be corrected by multiplying by two factors: by the reduction factor, $30.0/33 = 0.909$, due to the section being compact with $L_c < L_b \le L_u$, and by the deflection coefficient, 1.02, due to the loading being concentrated loads. Therefore, $\Delta = 2.20(0.909)1.02 = 2.04$ in.

8-4   BRACING

The AISC Specification, as well as good engineering design practice, dicates that all beams, girders, and trusses should be restrained against rotation about their longitudinal axis at support points or connection points. Web connections with adequate depth restrain the beam against rotation. They tend to resist torsion (torque about y-y axis) as well as warping (torque about longitudinal axis). Connection of the beam flange to a metal deck (i.e., flange bracing) is adequate to restrain the beam against rotation only when warping controls.

An often-overlooked case is where a single simple-span steel beam rests on a block wall. To provide adequate rotational restraint about its longitudinal axis, diaphragms (beams) running normal to the beam webs at the block wall supports and connected to the beam webs or blocking in the webs of the beam at the block wall support should be employed.

Another case requiring the attention of the designer is that shown in Fig. 7-4. A similar case would be where a W-shape beam is continuous over the top of a W-shape column. This is common to the industrial-type building, where bay after bay of this type of construction occurs in both directions. Rotational restraint is needed where beam rests atop each column. One solution would be to use web stiffeners on the beam, plus providing adequate connection of beams to columns. Figure 8-11 shows this case.

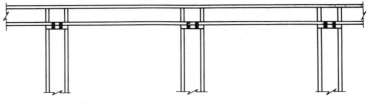

FIG. 8-11

The practical design of sign supports is illustrated in Fig. 8-12. The unbraced length of the cantilever leg supports need not be the full height, h. If diaphragm beams are provided, connecting leg to leg, preventing rotation, the use of L_b as the unbraced length of column can be used.

For cantilever beams that are loaded through the top (tension) flange, the most effective bracing is provided at the tension flange at the free end of the beam (position 1). This beam brace is more effective than bracing the lower (compression) flange at the end. Bracing is always more effective when it is located so as to counteract the twist caused by the load. Placing the brace at this end position on the tension flange, the buckling pattern changes to that more like a simply supported beam. Any additional braces should be placed at the compression flange level (position 2). Figure 8-13 illustrates this for a cantilever beam. At position 1, the tension flange tends to kick out (rotate). At position 2, the compression flange tends to kick out (rotate). Lateral bracing at these two positions tends to right the tendency to rotate.

Another area of question: Is an inflection point (a point of zero moment where moment reverses sign interior to support points, at times termed a point of contraflexure) a braced point? (See Fig. 8-14.)

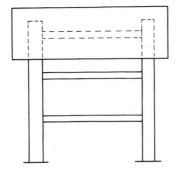
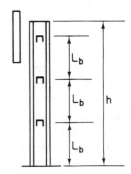

FIG. 8-12

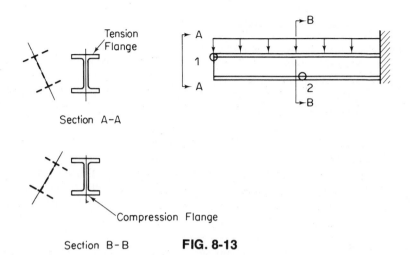

Section A-A

Section B-B   **FIG. 8-13**

Tests have shown that lateral movement and twist occur at the inflection points. However, tests also show that when no brace is placed at the inflection point, the buckling load greatly exceeds the theoretical prediction. Better correlation occurs when the distance to the inflection point is used as the unbraced length.

Current theories and design approaches are conservative, utilizing conservative approximations. Until theory improves, reflecting higher allowable stresses than those now employed in the AISC Specification, it is probably acceptable to consider an inflection point as a braced point.

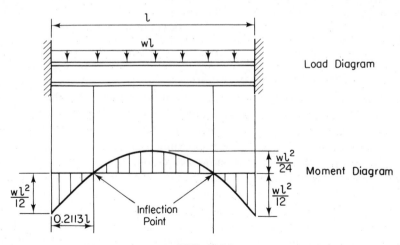

FIG. 8-14

8-5   BRACING DESIGN

Bracing has long been designed based upon strength considerations.* The necessary area of a diagonal brace in a simple or multibay frame to resist the horizontal deflection (sidesway) caused by gravity loading has long been computed by the conservative rule of thumb that the area is equal to that necessary to resist a horizontal force equal to 2% of the sum of the vertical loads on the frame. Formulas based upon strength have also been developed to determine the thickness of a masonry wall within a bay of a multibay frame to resist sidesway due to vertical loads applied.

If the steel diagonal-brace member or masonry wall-brace member is required to resist lateral (wind, seismic, etc.) load, deflection (drift), in addition to sidesway, must be taken into account. $F_{horiz} = 0.02 \Sigma P + F$ due to lateral load.

The discussion to follow was developed by Joseph A. Yura at the University of Texas at Austin and was subsequently presented at the AISC National Engineering Conference in Cleveland, 1971. Yura developed a quantitative method for designing and/or evaluating bracing as opposed to the rule of thumb, 2% of summation vertical load approach. The approach in the following two sections is to refer to bracing as stability bracing, as opposed to bracing used to resist horizontal lateral load. The approach results in the determination of the amount of bracing required for stability.

Bracing is utilized for three general purposes: to reduce the effective length factor in column design; to reduce the lateral unbraced length, L_b, of a beam's compression flange; and to provide overall structural stability for a structure under gravity loads only. A brace must provide stiffness in addition to strength. The formulas to be used in analyzing and designing braces will use the approach of adequate stiffness and adequate strength.

8-5.1   Stiffness and Strength

Figure 8-15a shows the case where a column is pinned at the top and bottom. The column shown is a Euler column. The Euler equation for this ideal pinned-end, centrally loaded column is as shown in the figure by P_e, and it represents the critical buckling load (see Eq. 5-1).

In place of the top pin, assume the column to be held in position with a brace as in Fig. 8-15b. P can be increased to P_e only if the brace

---

* For details, see T. V. Galambos, Lateral Support for Tier Building Frames, *AISC Engineering Journal*, January 1964; and I. Hooper, discussion of Galambos, *AISC Engineering Journal*, October 1964.

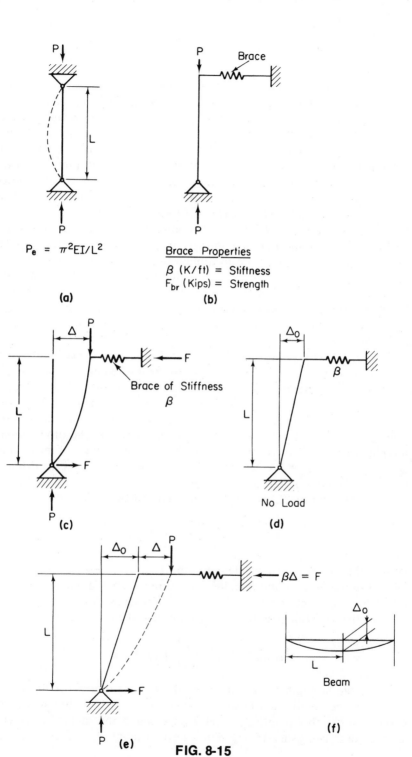

$P_e = \pi^2 EI/L^2$

(a)

Brace Properties

β (K/ft) = Stiffness
F_{br} (Kips) = Strength

(b)

Brace of Stiffness β

(c)

No Load

(d)

$\beta\Delta = F$

Beam

(f)

(e)

FIG. 8-15

does not fail. For the brace not to fail, it must have adequate stiffness and strength. Stiffness (in kips per foot) is designated as β and strength (in kips) is designated as F_{br}. Figure 8-15c illustrates the significance of β, the stiffness. As the column is loaded, it sways to the side a distance Δ. The brace, which has a stiffness or spring constant, β, is compressed and, due to compression, develops a force F. For equilibrium in addition to this force F being developed at the brace, the same force is developed at the base of the column and $F = \beta\Delta$. When $\beta\Delta L > P\Delta$, there is no sidesway. When $\beta\Delta L < P\Delta$, there is sidesway. When $\beta\Delta L = P\Delta$, equilibrium exists and $\beta L = P_e$. Up to this point we have been assuming the existence of a column that was initially perfectly straight, an unreal assumption. Every column, due to human inaccuracies during the manufacturing process, is not perfectly straight. Real columns have an initial out-of-plumbness, and its value will be designated as Δ_0, as shown in Fig. 8-15d. The brace or spring constant at this point does not react. The column leans on the brace and is stable. There is no column load and no brace reaction.

After a load is applied as in Fig. 8-15e, the top of the column moves from the solid position toward the right to the dashed position as a result of additional moments and stresses exerted within the column. This is similar to a spring. When a spring is compressed a distance Δ, it develops a resistance, $\beta\Delta$, where β is the spring constant. In this case, β is the brace stiffness. The brace force, $\beta\Delta$, is developed as the brace comes into play to resist this new deformation. ΣM about the base of the column gives $\beta\Delta L = P(\Delta + \Delta_0)$, where P is the axial compressive load that a particular member or part of a member is required to carry. β is the brace stiffness. Solving for β, we get the required bracing stiffness:

$$\beta_{\text{req}} = \frac{P}{L}\left(1 + \frac{\Delta_0}{\Delta}\right) \qquad (8\text{-}9)$$

This is the stiffness requirement for the brace to keep the top of the column in the equilibrium position shown in Fig. 8-15e when loaded by a load P. The brace must also be of sufficient strength.

Multiplying β_{req} (Eq. 8-9) by Δ, we obtain the force in the brace, F:

$$F_{br\ \text{req}} = \beta_{\text{req}}\Delta = \left(\frac{P}{L}\right)(\Delta + \Delta_0) \qquad (8\text{-}10)$$

$F_{br\ \text{req}}$ is the strength requirement for the brace to keep the top of the column in the equilibrium position shown in Fig. 8-15e when loaded by a load, P. At this point, β_{req} and $F_{br\ \text{req}}$ define the minimum bracing requirements necessary to say that a certain point is, in fact, a braced point, or that a column is a braced column, or that the beam has an unbraced length, L_b, or the frame is a braced frame.

The AISC Code of Standard Practice for Steel Buildings and Bridges, which is found in the AISC Manual, is a document that provides certain standards considered to be of good practice relative to structural steel design, fabrication, and erection. The section on erection tolerances says that, in the erection process, individual steel pieces are considered plumb if the out-of-plumbness is equal to or less than 1:500 (e.g., 0.288 in., or about $\frac{9}{32}$ in., in a height of 12 ft). Using this criterion, assume Δ_0 = $0.002L$ as an initial out-of-plumbness of a column (Fig. 8-15e) or beam (Fig. 8-15f). Assume Δ, the distance of sway due to load P, equal to Δ_0. These assumptions seem to be reasonable.

Substituting these values of Δ and Δ_0 into Eqs. 8-9 and 8-10, stiffness $\beta = 2P/L$, and assuming a factor of safety of 2,

$$\beta = \frac{4P}{L} \qquad (8\text{-}11)$$

Strength is

$$F_{br} = 0.004P \qquad (8\text{-}12)$$

F_{br} does not necessitate a factor of safety because of the use of allowable stresses, which already have a factor of safety accounted for. Compared to the rule of thumb, $0.02P = F$, it is seen that the 2% rule is very conservative. It is five times more conservative than Eq. 8-12. However, the 2% rule did not consider stiffness, which can govern.

8-5.2   Horizontal Brace

The schematic representation of a brace (Fig. 8-15) can be replaced by an actual horizontal brace (normal to the axially loaded column) as in Fig. 8-16. For the column to move Δ, a force, F, is developed in the brace and for the axially loaded tension brace, by Hooke's law, $\Delta = FL_b/A_bE$. Solving for A_b: $A_b = FL_b/\Delta E$. Since $F = \beta\Delta$, then $F/\Delta = \beta$ and $\beta = 4P/L$ (Eq. 8-11), where $L = L_c$. Therefore, substituting (stiffness requirement: A_b = gross area),

$$A_b = \frac{4PL_b}{L_cE} \qquad (8\text{-}13)$$

which is the required gross area of the horizontal brace of ample stiffness to stabilize the system.

Dividing Eq. 8-12 by the allowable stress in tension, $F_t = 0.60F_y$,

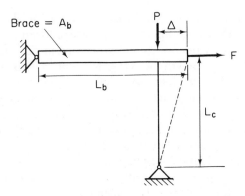

FIG. 8-16

the required net area of the horizontal brace of ample strength to develop the bracing resistance, F_{br}, is obtained.

$$\frac{F_{br}}{F_t} = A_b$$

Then (strength requirement: A_b = net area),

$$A_b = \frac{0.004P}{0.60F_y} \qquad (8\text{-}14)$$

In general, strength governs. Exceptions to this may be situations where the brace is very long, or the column length is very short, or a high-strength steel is used, or a combination of these. In these cases, stiffness may govern, and design by the old 2% rule might be unconservative.

8-5.3   Inclined Brace

In Fig. 8-17 is shown a column braced by a diagonal member, an inclined brace. For the horizontal brace discussed in Section 8-5.2, the brace force, F, and the deflection, Δ, were perpendicular to the column. In the case of the inclined brace, the F and Δ components along the incline are determined and the procedure is then similar to the case for the horizontal brace.

The brace force is $F_{br} = F / \cos \theta$ and, dividing by F_t, we obtain the required net area for the inclined brace to have ample strength to develop the bracing force, $F_{br} = 0.004P$ (Eq. 8-12). Then (strength requirement: A_b = net area),

$$A_b = \frac{0.004P}{0.60F_y \cos \theta} \qquad \cdot \qquad (8\text{-}15)$$

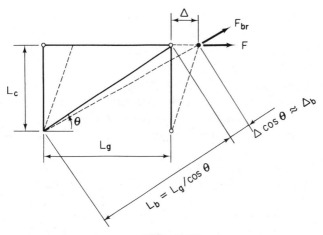

FIG. 8-17

For the column to move Δ, a force F_{br} is developed in the brace, and for the axially loaded inclined tension brace, by Hooke's law, $\Delta_b = \Delta \cos \theta = F_{br}L_b/A_bE$ and, solving for A_b, $A_b = F_{br}L_b/\Delta \cos \theta E$. Since $F_{br} = F/\cos \theta$ and $A_b = FL_b/\Delta \cos^2 \theta E$, $F/\Delta = \beta$, and $\beta = 4P/L$ (Eq. 8-11), where $L = L_c$. Therefore, substituting (stiffness requirement: A_b = gross area),

$$A_b = \frac{4PL_b}{L_c \cos^2 \theta E} \qquad (8\text{-}16)$$

This is the required *gross* area of the inclined brace of ample *stiffness* to stabilize the system.

In general, stiffness will govern for this case and, although this is the case, the application of the 2% rule would be conservative.

8-5.4   Additional Information and References

Several references for determining both stiffness and strength of brick masonry and light-gage-steel diaphragms used as bracing are as follows:

(1) *Recommended Practice for Engineered Brick Masonry,* Brick Institute of America, McLean, Virginia.

(2) *Design of Light Gage Steel Diaphragms,* American Iron and Steel Institute, Washington, D.C., 1967.

(3) J. R. Benjamin and H. A. Williams, The Behavior of One-Story Brick Shear Walls, *Proceedings of the American Society of Civil Engineers,* vol. 84, July 1958.

(4)   G. Winter, Lateral Bracing of Columns and Beams, *Proceedings of the American Society of Civil Engineers,* vol. 84, March 1958.

(5)   W. McGuire, *Steel Structures* (Englewood Cliffs, N.J.: Prentice-Hall, Inc., 1968), pp. 562–573.

(6)   M. A. Larson, Discussion of (4), *Proceedings of the American Society of Civil Engineers,* vol. 84, September 1958.

(7)   L. D. Luttrell, Strength and Behavior of Light-Gage Steel Shear Diaphragms, *Cornell Engineering Research Bulletin 67-1,* July 1967.

(8)   L. D. Luttrell, *Tentative Recommendations for the Design of Steel Deck Diaphragms,* Steel Deck Institute, Oct. 12, 1972.

The discussions have been limited to single-story and single-bay frames. In reality, if simple connections are used, the formulas are equally applicable to multistory and multibay frames. Conceivably, only one bay can be braced, thus attempting to stabilize the entire story frame. The formulas for strength and stiffness for a single frame need only be altered by substituting ΣP for P. The brace in one bay is stabilizing all the columns of the frame in that story.

8-5.5   Relative and Single-Point Braces

There are two distinctly different types of braces. One is a relative brace (everything discussed thus far has been about relative braces), where relative movement between two points is controlled. The second type of brace is a single-point brace, where absolute control of deformation or movement at a single point is desired. Requirements for a single-point brace are more severe than the relative brace requirements.

The stiffness, β, for a single-point brace is two to four times higher than for a relative brace. For conservatism, a multiple of 4 is used. The brace force or strength, F_{br}, for a single-point brace is twice the requirement for a relative brace because the single-point brace is acted upon by a level above and a level below, and in a sense there are two column loads trying to move the single point. With a relative brace, we are concerned with only one level at a time.

For a single-point brace, the stiffness and strength may conservatively be taken to be

$$\beta = \frac{16P}{L} \tag{8-17}$$

$$F_{br} = 0.008P \tag{8-18}$$

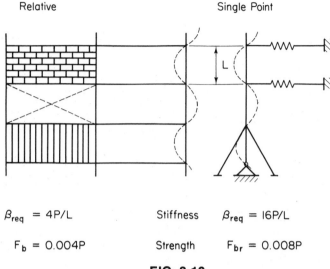

Relative   Single Point

$\beta_{req} = 4P/L$ Stiffness $\beta_{req} = 16P/L$

$F_b = 0.004P$ Strength $F_{br} = 0.008P$

FIG. 8-18

The derivation of formulas for strength and stiffness for a single-point brace are complex and, in lieu of these, Eqs. 8-17 and 18 are suggested for use.

Figure 8-18 shows a relative and a single-point brace. In addition, it summarizes the stiffness and strength requirements for each. For multistory or multibay frames, $P = \Sigma P$ for a single brace stabilizing many bays, or ΣP is the entire vertical load being stabilized by that particular brace.

8-5.6   Design Examples

To gain a better understanding of the bracing design procedures discussed and to see more clearly where they can be applied, several practical design examples are presented in this section.

EXAMPLE 8-4

Design a horizontal brace to adequately provide lateral support to a W 8 × 21 pinned-end column 22 ft long. Assume the brace to be at midspan, that it acts to prevent buckling about the column's weak axis, it is 8 ft long, and that all material is A36. The column is laterally supported at its ends in both planes.

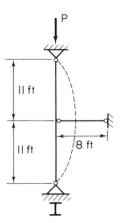

FIG. 8-19

solution (Fig. 8-19):

(a) Finding the capacity of the column (see Chapter 5): for the y-y axis: $K = 1.0$; $KL = 1.0(11.0) = 11.0$ ft, $P_{allow} = 76$ kips; for the x-x axis: $K = 1.0$; $KL = 1.0(22) = 22$ ft; $r_x/r_y = 2.77$.

The corresponding KL relative to the y-y axis: $22.0/2.77 = 7.94$ ft < 11.0 ft. Therefore, the effective length for the y-y axis is critical. *Use* $P_{allow} = 76$ kips.

(b) Design of horizontal brace (see Section 8-5.2): from Eqs. 8-13 and 8-14, considering the brace to be a relative point brace.

Stiffness requirement: $A_b = 4PL_b/L_cE = 4(77)(96)/132(29,000) = 0.008$ in.^{2}; strength requirement: $A_b = 0.004P/0.60F_y = 0.004(77)/22 = 0.014$ in.^{2}. The strength requirement controls.

Try WT 3 × 4.5. $A = 1.34$ in.$^2 \gg 0.014$ in.^{2}; $l/r_y = 96/0.905 = 106 < 300$; $d/t_w = 17.4 < 127/(F_y)^{1/2} = 21.1$. *Use* W T3 × 4.5 as the horizontal brace.

EXAMPLE 8-5

Figure 8-20 shows the typical framing and loading for a building. Find the size of the tension bracing system necessary to stabilize the framing and to support the loads shown. Assume that the roof acts as a rigid diaphragm, that all material is A36 steel, and that every third bent is braced.

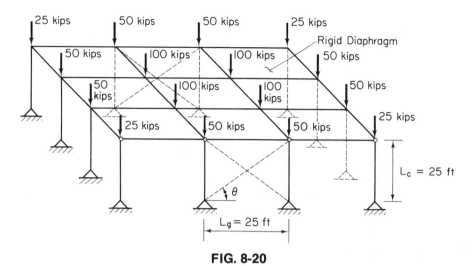

FIG. 8-20

solution (see Section 8-5.3):

P/brace = 450 kips; relative brace. Stiffness requirement (A_b = gross area) (Eq. 8-16): $L_b = L_g/\cos\theta = 25/0.707 = 35.36$ ft; $A_b = 4PL_b/L_c \cos^2\theta E$.

$A_b = 4(450)35.36/25(0.707)^2 29{,}000 = 0.176$ in.². A bar $\frac{1}{2}\phi$ would be okay (AISC Manual, Part 1, Square and Round Bar Weight and Area Table).

Strength requirement (A_b = net area) (Eq. 8-15): $A_b = 0.004P/0.60F_y \cos\theta = 0.004(450)/22(0.707) = 0.116$ in.².

A bar $\frac{1}{2}\phi$ threaded would be okay ($A_{root} = 0.126$ in.²) (Threaded Fastener Table, AISC Manual, Part 4). Stiffness governs, so *use* bar $\frac{1}{2}\phi$.

EXAMPLE 8-6

How much weld is required so that the open web steel joists adequately brace the W 14 × 26? Assume that the W shape is A36 steel.

solution (Fig. 8-21):

From the ASD Selection Table, for $F_y = 36$ ksi, $M_r = 71$ kip-ft ($L_b < L_c$). As in Fig. 8-21b, treating the compression region of the flange as a column, $P = 12M_R/d = 12(71)/13.91 = 61.3$ kips. For a point brace (at each joist, it is desired to have no movement of the beam flange):

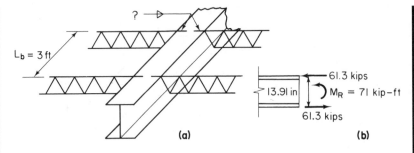

FIG. 8-21

Strength requirement: $F_{br\ req} = 0.008P$ (Eq. 8-18); F_{br} = 0.008 (61.3) = 0.49 kips. This brace force is transferred by shear through the welds. For a $\frac{1}{4}$-in. fillet weld, $\frac{1}{4}$ in. long (E70 electrode), 0.3(70)(0.707)(0.25)(2)(0.25) = 1.86 kips > 0.49 kip. This is okay.

Stiffness requirement: the floor diaphragm will tend to stabilize the beam's compression flange rather than the weld stiffness. This is a relative brace that prevents the relative movement of adjacent joints.

Stiffness requirement: $\beta_{req} = 4P/L_b = 4(61.3)/3 = 81.7$ kips/ft. A typical metal floor deck provides 10 times this required stiffness. A typical metal floor deck with $2\frac{1}{2}$ in. of concrete fill provides about 30 times more. So the stiffness is there, as is the force to brace the top flange of this beam.

Use a $\frac{1}{4}$-in. fillet tack weld, $\frac{1}{4}$ in. long.

EXAMPLE 8-7

In Fig. 8-22, will the column at the stairwell be stable for the loads it carries and can it be designed for a value of K = 1.0? (In other words, are the beams adjacent to the column of sufficient strength and stiffness to brace the column at the floor level under consideration?) Assume that the column height is 12 ft and that all spandrel beams are braced laterally where beams normal to their span frame.

solution:

The bracing (the two spandrel beams) must prevent the kicking out of the column, as shown dashed in Fig. 8-22b. Since it is the absolute movement at this one point that is

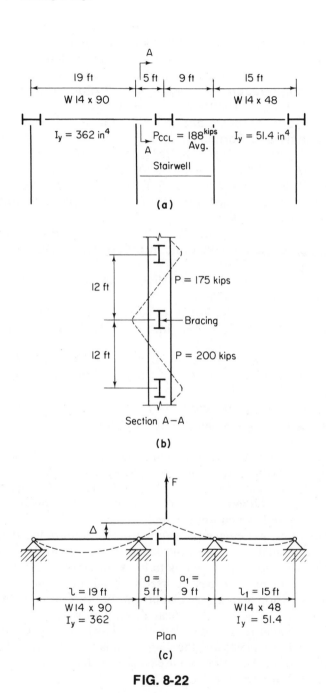

FIG. 8-22

desired to be prevented, the point is a single brace point.

Therefore, to prevent this kicking out at the particular level, the spandrel beams must provide: stiffness requirement $\beta_{req} = 16P/L$ (Eq. 8-17); $\beta_{req} = 16(188)/12 = 251$ kips/ft; strength requirement: $F_{br} = 0.008P$ (Eq. 8-18); strength supplied: $F_{br} = 0.008(188) = 1.50$ kips.

From Fig. 8-22c and the AISC Manual, Part 2, Beam Diagrams and Formulas: $\Delta = Fa^2(l + a)3EI$ or $Fa_1(l_1 + a_1)/3EI$ and $\beta = F/\Delta = 3EI/a^2 (a + l)$ or $3EI/a_1^2(a_1 + l_1)$.

W 14 × 90: $\beta = 3(29,000)(362)/(5)^2(5 + 19)144 = 365$ kips/ft; W 14 × 48: $\beta = 3(29,000)(51.4)/(9)^2(9 + 15)144 = 16$ kips/ft; $\Sigma\beta = 381$ kips/ft; $\Sigma\beta > 251$ kips/ft, therefore, the stiffness is okay.

Strength supplied: $F_{br\ req} = 1.50$ kips, and this will be distributed to the spandrels in proportion to their relative stiffnesses.

W 14 × 90: 1.50(362/381) = 1.43 kips; W 14 × 48: 1.50(16/381) = 0.06 kips.

See Fig. 8-22c: For W 14 × 90, 1.43a = 1.43(5) = 7.15 kip-ft $= M_{yy\ max}$. $S_y = 49.9$ in.^{3}; $f_{by} = 12M_{yy}/S_y = 1.72$ ksi; (assume that $f_{by} + f_{bx} < F_b$.)

For W 14 × 48: 0.06a = 0.06(9) = 0.54 kip-ft $= M_{yy\ max}$. $S_y = 12.8$ in.^{3}; $f_{by} = 12M_{yy}/S_y = 0.51$ ksi; (assume that $f_{by} + f_{bx} < F_b$).

Use W 14 × 90 and W 14 × 48 as spandrels to prevent the column from kicking out. $K = 1.0$ is valid.

Although the number of design examples presented was great and the type of problems varied, they were each presented to show how these stiffness and strength criteria as applied to the design of adequate bracing can be used. In addition, the problem was chosen to show real situations of relative as well as point-brace applications.

Although this text concentrates on structural (hot-rolled shapes) steel design, an attempt is made to introduce the reader to diaphragm (cold form/thin gage) design, as well as to brick masonry (shear wall) design. These means of bracing may be used in much the same way. A more detailed discussion of thin-gage diaphragms or brick masonry bracing will not be made. The action of a shear diaphragm with perimeter framing members is similar to that of a plate girder in which two perimeter members are the flange plates and the diaphragm is the web plate. For a more in-depth study of the latter two areas, the reader is referred to the references listed in Section 8-5.4.

PROBLEMS

(8-1)   Using the curves of Allowable Moments in Beams, choose the most economical section to satisfy the following requirements. Assume no lateral support unless otherwise indicated. Using the AISC equations, verify the load-carrying capacity of the beam chosen in part (a).

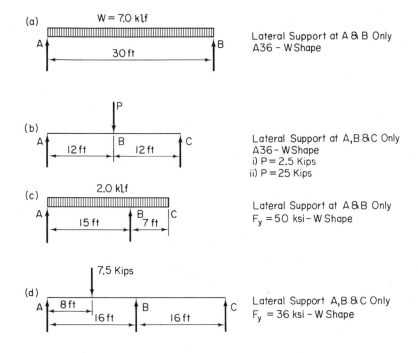

(a) $W = 7.0$ klf
A — 30 ft — B
Lateral Support at A & B Only
A36 – W Shape

(b)   P
A — 12 ft — B — 12 ft — C
Lateral Support at A,B & C Only
A36 – W Shape
i) $P = 2.5$ Kips
ii) $P = 25$ Kips

(c)   2.0 klf
A — 15 ft — B — 7 ft — C
Lateral Support at A & B Only
$F_y = 50$ ksi – W Shape

(d)   7.5 Kips
A — 8 ft — 16 ft — B — 16 ft — C
Lateral Support A,B & C Only
$F_y = 36$ ksi – W Shape

(8-2)   Using ASD, what is the allowable concentrated midspan load (neglect beam weight) for a beam span of 30 ft? The beam is W 14 × 53 (A36 steel). The distance between lateral support points is (a) 7.5 ft, (b) 15 ft, and (c) 30 ft. (Assume $C_b = 1.0$.)

(8-3)   Rework Problem 8-2b except that bending is about the minor axis.

(8-4)   Determine the net permissible load (above and beyond the beam's dead weight) for each of the beams shown. Lateral support conditions are indicated for each.

(a)

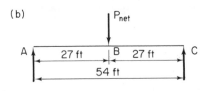

C12 x 25 (A36 Steel)
i) Full Lateral Support
ii) Lateral Support (Provided at
Maximum Interval of 6 ft)
$(C_b = 1)$

(b)

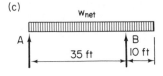

W 36 x 194 $(F_y = 50$ ksi$)$
i) Full Lateral Support
ii) No Lateral Support (Except
at A,B,C)$(C_b = 1)$
iii) No Lateral Support (Except at
A,B,C)$(C_b = $ max. Allowable Value)

(c)

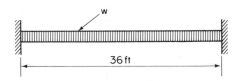

W 24 x 117 $(F_y = 50$ ksi$)$
i) Full Lateral Support
ii) No Lateral Support (Except
at A & B)$(C_b = 1)$

(8-5) Using ASD, design the least-weight beam for each condition shown. The uniform load, including the beam weight, is 2.75 kips/ ft. Assume A36 steel and $C_b = 1$.
(a) $L_b = 0$.
(b) $L_b = 18$ ft.

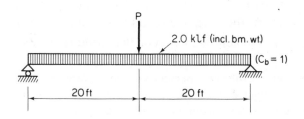

(8-6) The beam shown is a W 27 × 178 of A588 steel $(F_y = 50)$. What is the allowable load P (using ASD) that can be carried if the compression flange is supported laterally at (a) the quarter points, (b) the ends and at the center of span, and (c) the ends only.

(8-7) Design the most economical W shape of A36 steel for the loading
       shown for the following conditions. Use ASD and a modifying
       factor (C_b) of 1.
       (a) $L_b = 0$.
       (b) $L_b = 10$ ft.
       (c) $L_b = 20$ ft.

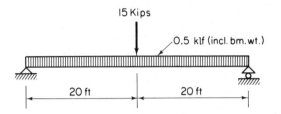

(8-8) If the maximum value of C_b is used, how much *additional* con-
       centrated load could the beam selected in Problem 8-7c carry?

(8-9) Determine the allowable load P if the beam is a W 24 × 55. Use
       ASD. Show the effect of the modifying factor (C_b). Assume A36
       steel. What is P neglecting the beam weight, if (a) $C_b = 1$; (b) C_b
       = calculated value (Eq. 8-8).

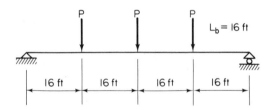

(8-10) A W 27 × 146 of A36 steel is to be used to span a length of 48
       ft. If the floor slab supported by the beam weighs 1 kip/ft, what
       concentrated midspan live load P (using ASD and neglecting
       beam weight) can the beam carry, when
       (a) $L_b = 0$.
       (b) $L_b = 24$ ft ($C_b = 1$).
       (c) $L_b = 24$ ft (C_b = maximum allowable value).

(8-11) Design the most economical W shape using ASD for each of the
       conditions shown. A36 steel. $C_b = 1$. Neglect beam weight.
       (a)   Full lateral support, no depth limit.
       (b)   Full lateral support, depth not greater than 18 in.

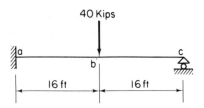

(c)   Laterally braced every 8 ft.
(d)   Laterally braced every 16 ft.

(8-12)   In Problem 8-11, what is the maximum value of C_b for the lengths of ab and bc?

(8-13)   Determine the maximum deflection for each of the following beams. Assume A36 steel. Use the laterally supported beam loading tables:

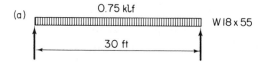

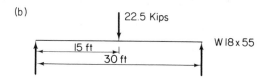

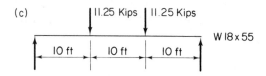

(8-14)   Using the appropriate deflection formulas, compute the maximum deflection and its location for the following beams. Use the Beam Diagrams and Formulas as found in Part 2 of the AISC Manual.

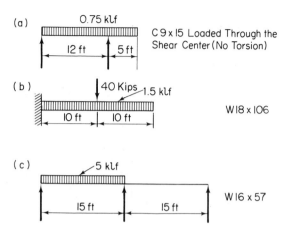

(8-15) Explain briefly how a change in direction (sign) of the bending moment between points of lateral support can influence the buckling strength of a beam.

(8-16) Briefly explain how the location of the applied load (relative to the centroid of the cross section) can influence the critical buckling load on a laterally unsupported beam.

(8-17) Of the three AISC formulas for allowable bending compressive stress on flexural members (Eqs. 8-7a, b, and c):
    (a)   Which formula applies primarily to sections with high torsional (St. Venant) resistance?
    (b)   Which formulas apply to sections with high warping resistance? Differentiate between the two formulas that apply in this case.

(8-18) Explain briefly how a steel section can be considered *compact* for one yield strength of steel, and *noncompact* for others.

(8-19) Determine the adequacy of bracing provided by a ST 1.5 × 2.85 member (A36) 6 ft in length used to horizontally brace a column, W 10 × 45, 24 ft long, at its midlength. The A36 column is pinned and laterally restrained at both ends. The brace resists buckling about the column's weak axis. Check strength and stiffness.

beam
columns

9-1   INTRODUCTION

As the name suggests, *beam columns* are members subjected to axial compressive loads (column action) and bending moments (beam action). Beam columns combine beam action, which involves bending and lateral torsional buckling, with column action, which involves compression buckling. As a result, all the factors that affect beams and columns will also influence the behavior, strength, and design of beam columns. These factors include the magnitude and distribution of loads and moments, the use or nonuse of lateral bracing, sidesway permitted or not permitted, and effective length kl.

Solutions to these problems, arrived at over a period of many years, have been reasonably exact for most of the problems. Beam-column behavior presents a complex problem for which rather simple design procedures have been made available.

This chapter will review the AISC Specification interaction equations that are used for design. The terms of these equations will be defined. A discussion of column end support conditions, loading, and slenderness ratios will illustrate the proper purpose and application of the various parameters in the interaction conditions.

The use of the Column Load Tables in the AISC Manual, Part 3, will be discussed in some detail (they were briefly mentioned in Chapter 5). These discussions will be followed by a detailed description of the beam-column selection and design process. The chapter will conclude with design examples that illustrate the trial section selection and design processes.

9-2 INTERACTION FORMULAS

A beam column may be an eccentrically (nonaxially) loaded column, the legs or beams of rigid frames that may be subject to moment and axial load, or a frame leg subject to lateral (e.g., wind) load and axial compressive load.

The AISC Specification interaction formulas for proportioning members subjected to axial compression and bending are

$$\frac{f_a}{F_a} + \frac{C_m f_b}{(1 - f_a/F_e')F_b} \leq 1.0 \qquad (9\text{-}1)$$

$$\frac{f_a}{0.60F_y} + \frac{f_b}{F_b} \leq 1.0 \qquad (9\text{-}2)$$

When $f_a/F_a \leq 0.15$, Eq. 9-3 is used instead of Eqs. 9-1 and 9-2:

$$\frac{f_a}{F_a} + \frac{f_b}{F_b} \leq 1.0 \qquad (9\text{-}3)$$

where

$f_a = P/A$ is the computed axial stress, ksi

$f_b = M/S$ is the computed compressive bending stress at a particular point, ksi

$F_a =$ axial stress permitted if only axial load acts (ksi) (Chapter 5)

$F_b =$ compressive bending stress permitted if only bending exists (ksi) (Chapter 6)

$F_e' = 12\pi^2 E/23(Kl_b/rb)^2 =$ Euler stress (see Eq. 5-2) divided by a factor of safety $\frac{23}{12}$ and K

$l_b =$ unbraced length in the plane of bending

$r_b =$ radius of gyration corresponding to the plane of bending

K = effective length factor (Section 5-5) in the plane of bending

C_m = coefficient applied to the bending term of the interaction formula, which varies with column curvature as caused by applied moments (values of C_m will be discussed in more detail; C_m, often called a *reduction* or *moment factor,* is a modifying factor used to account for different shapes of moment diagram)

$1/(1 - f_a/F_e')$ = *amplification factor*

$C_m f_b/(1 - f_a/F_e')$ = modified actual bending stress

Equations 9-1, 9-2, and 9-3 are ASD equations and provide solutions to design problems that are safe and accurate, giving satisfactory predictions of column strength. PD methods of design are presented in Part 2 of the AISC Specification. The interaction formulas are in a somewhat different form, using loads and moments instead of stresses. Prior to 1963, only Eq. 9-3 was used for design. The current specification only uses this equation when $f_a/F_a \leq 0.15$. The influence of the term $C_m/(1 - f_a/F_e')$ is generally small for this case and can be neglected (Eq. 9-3). If $f_a/F_a >$ 0.15, Eqs. 9-1 and 9-2 must be checked. The second term of each equation, the bending term, can actually be two terms, f_{bx}/F_{bx} and f_{by}/F_{by} or $C_{mx}f_{bx}/(1 - f_a/F_{ex}')F_{bx}$ and $C_{my}f_{by}/(1 - f_a/F_{ey}')F_{by}$, depending on whether bending occurs about both the major and minor axes or either the major or the minor axis.

The term *interaction* can be explained by graphical representation. If for a column subject to compressive axial load and bending moment a plot is made where the ordinate is P/P_{max} and the abscissa is M/M_{max}, P and M are of varying magnitude with a maximum value of P_{max} and M_{max}, respectively. Figure 9-1 shows this. The maximum ordinate and the maximum abscissa values are 1.0 and define the limiting conditions of P/P_{max} and M/M_{max}. These limiting points are joined by some type of curve (a straight line, a convex line, a concave line, etc.). This figure graphically represents interaction. The ratios are used to have dimensionless units. As a matter of interest, Eq. 9-3 would be represented by the straight line in Fig. 9-1.

Figure 9-2a shows a column subjected to axial load, P, and end moments, M_0. The column is one that is bent into single curvature, as shown by the dashed lines. Figure 9-2b shows the deflection, Δ_0, produced by the bending moments, M_0. The addition of axial load, as shown in Fig. 9-2c, produces additional bending and deflection. The total deflection due to axial load and bending moment is Δ. Figure 9-2d shows

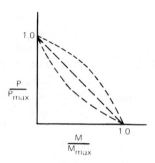

FIG. 9-1

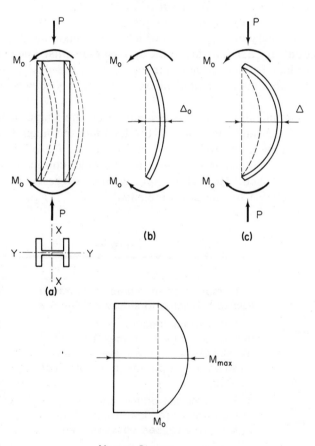

Moment Diagram

(d)

FIG. 9-2

the moment diagram produced by the initial end moments, M_0, and the moment due to the subsequent addition of axial load, $P\Delta$. For most cases of column interaction of moment and axial load, complex calculations are required to determine M_{max}. For the case under discussion, $M_{max} \approx M_0/(1 - P/P_e)$ and bending stress $= f_b/(1 - f_a/F_e')$, where f_b is the calculated stress due to M_0. M_0 is magnified by the amplification factor, $1/(1 - P/P_e)$, where P_e (the Euler load, Eq. 5-1) is computed about the bending axis only. The amplification factor, $1/(1 - f_a/F_e')$, magnifies f_b in the bending stress computation. The amplification factor is particularly significant for slender columns. For these cases, the factor is large because P_e or F_e' is small.

Summarizing, the application of axial load in addition to moment creates an axial displacement in the plane of bending. This displacement causes a secondary moment ($P\Delta$), which is not accounted for in the term f_b; f_b accounts for bending due to the applied moments. The amplification factor provides for this added moment, $P\Delta$, in the design of members subjected to combined bending and axial stress. It is always greater than unity.

The amplification factor is dependent on the shape of the applied moment diagram and therefore on the location and magnitude of Δ. It may under some circumstances overestimate the $P\Delta$ effect. To adjust this overestimate of secondary moment, the reduction factor or moment factor, C_m, is used. The AISC Specification suggests the C_m values shown in Table 9-1 for the cases indicated ($f_a > 0.15F_a$).

Table 9-1

Case	C_m
1. Compression members in *unbraced* frames subjected to *sideway* (Fig. 5-3)	0.85
2. For restrained compression members in *braced* frames (no joint translation) with *no transverse load* between their supports in the plane of bending	Use Eq. 9-4
3. For compression members in *braced* frames (no joint translation) with *transverse load* between their supports:	
(a) members with *restrained ends*	0.85
(b) members with *unrestrained ends*	1.0

0.4   0.6   1.0
(min)   (min)

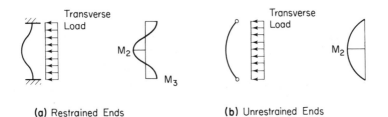

(a) Restrained Ends   (b) Unrestrained Ends

$$C_m = 0.6 - 0.4 \left(\frac{M_1}{M_2} \right) \geq 0.4 \qquad (9\text{-}4)$$

where M_1 is the lesser of the end moments and M_2 the larger of the end
moments. M_1/M_2 is + for a compression member bent in double (reverse)
curvature (Fig. 9-3a) and − for a compression member bent in single
curvature (Fig. 9-3b). C_m is always less than or equal to unity. For case
3, the maximum moment between the supports, M_2, should be reduced
by C_m. C_m factors alter the various moment diagrams to the equivalent
uniform moment diagram condition (Fig. 9-2a). More exact values of C_m
are provided in the Commentary to the AISC Specification for various
end and load conditions.

  In the interaction equations, F_a is governed by the critical (larger)
slenderness ratio (Kl/r). The critical Kl/r depends upon effective length
factors and whether bracing exists or does not exist in addition to column
slenderness. $(Kl/r)_x$ or $(Kl/r)_y$ can control, whichever is larger.

  To repeat, F_e' always depends upon the Kl/r in the plane of bending.
The effective length factor, K, as used in both the F_a and F_e' allowable
stresses, is of significant value only when the term f_a/F_a is large relative
to the bending term. For example, in a single-story rigid frame, a $K >$
2 conceivably may not change the column size from that design utilizing
$K = 1.0$. Further, as l/r is reduced, the influence of K reduces.

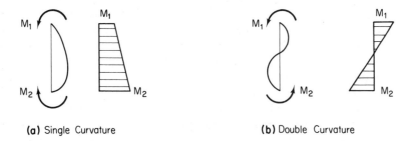

(a) Single Curvature   (b) Double Curvature

FIG. 9-3

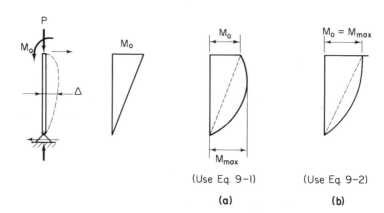

FIG. 9-4

For braced frames (sidesway prevented) the maximum $P\Delta$ (secondary moment) is between the ends of the compression member. There is no deflection (no $P\Delta$) at the two ends of the member. This is illustrated in Fig. 9-4a for the case indicated.

The actual distribution of the total moments, $M_0 + P\Delta$, along the length of the beam column may be either as shown in Fig. 9-4a (M_{max} between ends) or Fig. 9-4b, where the maximum moment is at the end of the member. The controlling case depends upon the slenderness ratio in the plane of bending. For the case shown in Fig. 9-4b, Eq. 9-1 may overestimate the strength [underestimate the modified actual bending stress, $C_m f_b/(1 - F_a/F_e')$]. Failure may actually be at the ends if this is the location of the maximum moment. Equation 9-2 should be used for this case. This equation is used only at the ends of the column (e.g., using M_2 at the ends in the bending term f_b). Equation 9-1 uses M_2 intermediate to the ends in the bending term.

Equation 9-1 provides a check on the overall stability of the member (stability from interaction of M_0 and $P\Delta$). Equation 9-2 provides a check on the stress conditions at the ends of the member. If a member is bent in reverse curvature (Fig. 9-3b), Eq. 9-2 usually governs for braced frames because the C_m values will be 0.4 to 0.6, $C_m/(1 - f_a/F_e')$ will be very much less than 1.0, and the modified bending stress will be low.

In unbraced frames, secondary moments are maximum at the ends and are usually important unless the axial stress is low, as would be the case for low buildings. The interaction equations, although originally derived for the sidesway-prevented (no joint translation) case ($K = 1.0$), can be used to evaluate the strength of unbraced frames if $C_m = 0.85$ and Kl is used in both the axial load and bending terms. By so doing, the $P\Delta$ moments are included in the interaction equations in an indirect manner for unbraced frames.

Equations 9-1 and 9-2 should always be checked unless $f_a/F_a \leq$ 0.15, in which case *only* Eq. 9-3 need be checked. For wind and seismic stresses, the one-third increase in allowable stresses applies to F_a, F_b, $0.60F_y$, and F_e'. Should there be a member subjected to both axial *tension* and bending, Eq. 9-2 should be used with f_b taken as the calculated bending *tensile* stress. f_b, the computed bending compressive stress taken alone, however, shall not be greater than the applicable F_b value. Axial tension tends to reduce the bending stress between points of lateral support ($P\Delta$ is of opposite sense to the applied bending moment).

9-3 COLUMN LOAD TABLES

A review of the Column Load Tables is in order. Although discussed in Section 5-6, the tables will be discussed again in this section because they will be used to select beam-column sections. The procedure of selection of beam-column sections will be discussed after this review.

The Column Load Tables present the allowable loads for axially loaded members having effective unsupported lengths (KL) in feet listed on the leftmost column of each table. All the tabulated loads are in kips and are for main members. The heavy horizontal lines in the tables indicate a Kl/r value of 200. Tabulated values are omitted for values of Kl/r exceeding 200. Values of $Kl/r > 200$ require more than a routine use of the Specification in the areas of accuracy of the analyses and load evaluation, construction tolerances, accuracy of KL evaluation, attention to small eccentricities, and so on.

For secondary members or bracing, when $Kl/r > 120$ (K may be taken as 1.0), the increased load as per AISC Specification is obtained by dividing the value taken from the table by $(1.6 - l/200r)$ (from Eq. 5-6).

To obtain an effective length with respect to the minor-axis equivalent in load-carrying capacity to the actual effective length about the major axis, Kl_x/r_x should be divided by r_x/r_y. If the resulting equivalent length is larger than the actual minor-axis effective length, this length is the controlling one. In other words, the larger of the two effective lengths (about the minor axis) is used to enter the Column Load Tables to obtain the allowable load. r_x/r_y is listed at the bottom of the Column Load Tables together with other properties and factors useful for checking the strength of the column about the strong axis, and, as will be seen later, for checking combined loading conditions. The tables include shape listings for two strength levels of material, as listed in Table 9-2.

Once again, as a review and summary, the significant unbraced lengths listed in the Column Load Tables are explained. KL is used to obtain the allowable axial loading. L_c, listed at the bottom of the tables

Table 9-2

Shapes	Materials Listed in Column Load Tables
W, M, and S	$F_y = 36$ ksi (A36)
	$^aF_y = 50$ ksi (A441, A242, A572, and A588)
Pipe (round)	$F_y = 36$ ksi (A53[b] and A501)
Tubing (square	$F_y = 36$ ksi (A501)
and rectangular)	$F_y = 46$ ksi (A500)
Double angles	$F_y = 36$ ksi (A36)
	$F_y = 50$ ksi (A441, A242, A572, and A588)

[a] All columns listed meet the width-to-thickness requirements of the AISC Specification except W 14 × 43 ($F_y = 50$ ksi) for KL values less than 2 ft (Appendix C design controls for KL up to 2 ft, d/t_w ratio violated).

[b] Actually $F_y = 35$ ksi, but 36 ksi may be used and is tabulated.

for each shape, indicates the maximum unbraced length for bending beyond which the section is no longer compact. L_u, listed along with L_c, indicates the maximum unbraced length for bending beyond which F_b is less than $0.60F_y$ because of lateral buckling. For the maximum use of section, the designer should try to stay within L_u. Beyond L_u, F_b must be reduced (see Eq. 8-7a, b, or c, as applicable).

Figure 9-5 shows a page from the AISC Manual. The Column Load Table shown will be used in the following design example to illustrate its application.

EXAMPLE 9-1

Choose a W 14 column section for a member axially loaded with a 685-kip load. The column is A36 steel, 17 ft long, and braced in both directions at its pinned-end supports only.

solution:

For pinned-end supports, $K = 1.0$ and $KL_y = KL_x = 17$ ft; therefore, KL_y controls ($r_y < r_x$). From Fig. 9-5, for $KL = 17$ ft, select W 14 × 132. $P_{allow} = 697$ kips > 685 kips.

EXAMPLE 9-2

All data the same as in Example 9-1, except that $KL_x = 30$ ft and $KL_y = 17$ ft (braced at supports, pinned end, $K = 1.0$). Is the W 14 × 132 satisfactory?

F_y = 36 ksi

F_y = 50 ksi

COLUMNS
W Shapes

Allowable axial loads in kips

Designation		W14							
Wt./ft.		193		176		159		145	
F_y		36	50	36	50	36	50	36	50
Effective length in ft. KL with respect to least radius of gyration r_y	0	1227	1704	1119	1554	1009	1401	922	1281
	6	1178	1620	1074	1477	968	1331	884	1217
	7	1167	1603	1064	1461	959	1317	877	1203
	8	1157	1584	1054	1444	950	1301	869	1189
	9	1146	1565	1044	1426	941	1285	860	1174
	10	1134	1545	1034	1407	931	1268	851	1159
	11	1122	1524	1022	1388	921	1250	842	1142
	12	1110	1502	1011	1368	911	1232	832	1125
	13	1097	1479	999	1347	900	1213	822	1108
	14	1083	1455	987	1325	889	1193	812	1090
	15	1069	1431	974	1302	877	1173	801	1071
	16	1055	1406	961	1279	865	1152	790	1051
	17	1040	1380	947	1255	853	1130	779	1031
	18	1025	1353	933	1231	840	1107	767	1011
	19	1010	1326	919	1205	827	1085	755	990
	20	994	1298	904	1179	814	1061	743	968
	22	961	1239	874	1125	786	1012	718	923
	24	927	1178	842	1069	758	960	691	875
	26	891	1113	809	1009	727	906	663	825
	28	853	1046	775	947	696	850	634	773
	30	814	976	739	882	663	791	604	719
	32	774	902	701	815	629	729	573	662
	34	732	826	662	744	594	665	540	603
	36	688	745	622	670	558	598	507	541
	38	643	669	580	601	520	537	472	486
	40	596	604	537	543	480	484	435	438
Properties									
U		2.29	2.29	2.31	2.31	2.32	2.32	2.34	2.34
P_{wo} (kips)		340	473	299	415	251	349	214	298
P_{wi} (kips)		32	45	30	42	27	37	24	34
P_{wb} (kips)		1542	1817	1250	1474	904	1066	688	810
P_{fb} (kips)		467	648	386	536	319	443	267	371
L_c (ft.)		16.6	14.1	16.5	14.0	16.4	13.9	16.4	13.9
L_u (ft.)		68.1	49.0	62.6	45.0	57.2	41.2	52.6	37.9
A (in.^{2})		56.8		51.8		46.7		42.7	
I_x (in.^{4})		2400		2140		1900		1710	
I_y (in.^{4})		931		838		748		677	
r_y (in.)		4.05		4.02		4.00		3.98	
Ratio r_x/r_y		1.60		1.60		1.60		1.59	
B_x } Bending		0.183		0.184		0.184		0.184	
B_y } factors		0.477		0.484		0.485		0.489	
a_x } *		358		319		283		255	
a_y }		139		125		111		101	

*Tabulated values of a_x and a_y must be multiplied by 10^6

FIG. 9-5 *Courtesy of American Institute of Steel Construction*

				F_y = 36 ksi
	COLUMNS			F_y = 50 ksi

COLUMNS
W Shapes
Allowable axial loads in kips

Designation		W14								
Wt./ft.		132		120		109		99		90
F_y		36	50	36	50	36	50	36	50†	36	50†

Effective length in ft. KL with respect to least radius of gyration r_y

	36	50	36	50	36	50	36	50†	36	50†
0	838	1164	762	1059	691	960	629	873	572	795
6	801	1101	729	1002	661	908	600	825	547	751
7	794	1088	722	990	654	897	595	815	541	742
8	786	1074	714	977	647	885	589	805	536	732
9	777	1060	707	963	640	873	582	793	530	722
10	768	1044	699	949	633	860	575	782	524	711
11	759	1028	690	935	626	847	568	769	517	700
12	750	1011	682	919	618	833	561	757	511	689
13	740	994	673	903	609	818	554	743	504	676
14	730	976	663	887	601	803	546	730	497	664
15	719	958	654	870	592	788	538	715	489	651
16	708	938	644	852	583	772	529	701	482	637
17	697	919	633	834	574	755	521	685	474	624
18	686	898	623	815	564	738	512	670	466	609
19	674	877	612	796	554	721	503	654	458	595
20	662	856	601	776	544	703	494	637	449	580
22	637	811	578	735	523	665	475	603	432	548
24	610	764	554	692	501	626	454	567	413	515
26	583	714	528	647	478	585	433	529	394	481
28	554	663	502	599	454	541	411	489	374	444
30	524	608	475	549	429	496	388	448	353	406
32	493	551	446	497	403	449	365	404	331	366
34	461	492	416	443	376	399	340	359	308	325
36	427	439	385	395	348	356	314	320	285	290
38	392	394	353	355	319	320	287	288	260	261

Properties										
U	2.47	2.47	2.48	2.48	2.49	2.49	2.50	2.28	2.52	2.29
P_{wo} (kips)	196	272	173	240	148	205	125	174	109	151
P_{wi} (kips)	23	32	21	30	19	26	17	24	16	22
P_{wb} (kips)	587	692	449	529	316	373	249	294	186	220
P_{fb} (kips)	239	332	199	276	166	231	137	190	113	158
L_c (ft.)	15.5	13.2	15.5	13.1	15.4	13.1	15.4	13.0	15.3	13.0
L_u (ft.)	47.7	34.4	44.1	31.7	40.6	29.2	37.0	26.7	34.0	24.5

A (in.^{2})	38.8		35.3		32.0		29.1		26.5	
I_x (in.^{4})	1530		1380		1240		1110		999	
I_y (in.^{4})	548		495		447		402		362	
r_y (in.)	3.76		3.74		3.73		3.71		3.70	
Ratio r_x/r_y	1.67		1.67		1.67		1.66		1.66	
B_x } Bending	0.186		0.186		0.185		0.185		0.185	
B_y } factors	0.521		0.523		0.523		0.527		0.531	
a_x }	228.0		204.8		184.5		165.1		148.9	
a_y } *	81.7		73.6		66.3		59.7		54.1	

*Tabulated values of a_x and a_y must be multiplied by 10^6.
†Flange is non-compact.

FIG. 9-5 (*Continued*)

solution:

$r_x/r_y = 1.67$; therefore, x-x axis equivalent $KL_y = 30/1.67$ $= 18$ ft; y-y axis, $KL_y = 17$ ft. $18 > 17$; therefore, x-x axis controls. Entering Fig. 9-5 with $KL = 18$ ft, for a W 14 × 132, $P_{allow} = 686$ kips; $686 > 685$ kips. *Use* W 14 × 132.

9-4   BEAM-COLUMN SELECTION USING THE COLUMN LOAD TABLES

To design beam columns utilizing the Column Load Tables, the inter-action formulas can be manipulated in the way illustrated. The interaction equations are repeated here to begin the manipulation. The manipulation is necessary to convert stress (interaction formulas) to load (Column Load Tables).

$$\frac{f_a}{F_a} + \frac{C_{mx}f_{bx}}{(1 - f_a/F'_{ex})F_{bx}} \leq 1.0 \qquad (9\text{-}1)$$

(This is Eq. 9-1 with the x-x axis bending term. The y-y axis bending term would be the same except with y subscripts substituted for the x subscripts.)

Since $f_a = P/A$, $f_b = M/S$ and $(1 - f_a/F'_e)$ is equivalent to $(F'_e - f_a)/F'_e$:

$$\left(\frac{P}{A}\right)\left(\frac{1}{F_a}\right) + \left(\frac{C_{mx}M_x}{S_x}\right)\left[\frac{1}{(F'_{ex} - f_a)/F'_{ex}}\right]\left(\frac{1}{F_{bx}}\right) \leq 1.0$$

Multiplying all the terms by AF_a,

$$P + \left(\frac{C_{mx}M_xA}{S_x}\right)\left[\frac{F'_{ex}}{(F'_{ex} - f_a)}\right]\left(\frac{F_a}{F_{bx}}\right) \leq AF_a \qquad (A)$$

In the second term, C_{mx} is dimensionless and therefore $(C_{mx}M_xA/S_x)$ is equivalent to a load in kips. The remaining part of the second term is dimensionless. Therefore, say that P', an equivalent axial load due to the bending component in the interaction formula, equals $(C_{mx}M_xA/S_x)$ $[F'_{ex}/(F'_{ex} - f_a)](F_a/F_{bx})$. Therefore, $P + P'_x = $ axial load capacity $= AF_a$, where P'_x is the second term of Eq. (A). Assume that $B_x = A/S_x$, $F'_{ex} = 12\pi^2E/23(Kl/r_x)^2$, and $f_a = P/A$. Then the second part of the second term in Eq. (A) becomes

$$\frac{F'_{ex}}{(F'_{ex} - f_a)} = \frac{12\pi^2E/23(Kl/r_x)^2}{12\pi^2E/23(Kl/r_x)^2 - P/A}$$

Multiplying the numerator and denominator by $A(Kl)^2$,

$$\frac{12\pi^2 E r_x^2 A/23}{(12\pi^2 E r_x^2 A/23) - P(Kl)^2}$$

Assume a constant, $a_x = 12\pi^2 E r_x^2 A/23 = 0.149(10)^6 A r_x^2$. By substitution,

$$\frac{F'_{ex}}{(F'_{ex} - f_a)} = \frac{a_x}{a_x - P(Kl)^2}$$

Substituting back in Eq. (A) gives

$$P + P'_x = P + B_x M_x C_{mx} \left(\frac{F_a}{F_{bx}}\right) \left[\frac{a_x}{a_x - P(Kl)^2}\right] \leq AF_a$$

$$= \text{tabular load in Column Load Tables}$$

Note that for the term $(Kl)^2$ in this modification of Eq. 9-1, K is the effective length factor and l is the actual unbraced length in the plane of bending. Values for $a_x = 0.149(10)^6 A r_x^2$ and $a_y = 0.149(10)^6 A r_y^2$ are listed at the bottom of the Column Load Tables.

In a similar manner, Eqs. 9-2 and 9-3 are modified respectively to become

$$P + P'_x = \frac{PF_a}{0.6F_y} + B_x M_x \left(\frac{F_a}{F_{bx}}\right) \leq AF_a$$

$$= \text{tabular load in Column Load Tables}$$

When $f_a/F_a \leq 0.15$,

$$P + P'_x = P + B_x M_x \left(\frac{F_a}{F_{bx}}\right) \leq AF_a$$

$$= \text{tabular load in Column Load Tables}$$

Not shown in these modifications of Eqs. 9-2 and 9-3 (or in the modified Eq. 9-1) is the y-y-axis bending term. It would be similar to the x-x term shown except with y-y subscripts. Values for $B_x = A/S_x$ and $B_y = A/S_y$ are listed at the bottom of the Column Load Tables.

The three modified interaction equations shown here have converted stress to load. By evaluating $P = P'_x + P'_y$, the designer can enter the Column Load Table to choose a column shape having equal or larger capacity than $P + P'_x + P'_y$. The following example illustrates the design procedure utilizing these equations.

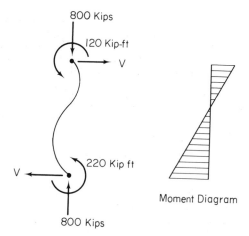

FIG. 9-6

EXAMPLE 9-3

Choose a W 14 column section in a multistory building for a 12-ft story height. The section is to support an 800-kip axial gravity load. The wind moments, each in the same direction, are 120 kip-ft top and 220 kip-ft bottom. Assume that bending is about the major axis, $K = 1.0$, relative to both axes, and A36 steel.

solution (Fig. 9-6):

(a)   For gravity loads only, from the Column Load Table for $KL = 12$ ft: *try* W 14 × 145: $P_{allow} = 832$ kips > 800 kips.

(b)   For gravity loads plus wind: the Column Load Table does not reflect the permissible $33\frac{1}{3}\%$ increase in allowable stress for wind (see Section 2-2).

Therefore, as an equivalent to this, reduce the existing load condition by 25% [i.e., $1.33F_{allow}$ = function of P; F_{allow} = (function of P)/1.33, F_{allow} = 0.75 (function of P)].

Therefore, $P = 600$ kips and $M_{max} = 165$ kip-ft.

(c)   Trial section: from the Column Load Table, the average $B_x = 0.184$ and for wind plus gravity loads: $P + P'_x \approx P + B_x M_x = 600 + 0.184(165)12 = 965$ kips.

From the Column Load Table, choose a column section with a capacity slightly less than 965 kips for KL = 12 ft.

Try W 14 × 159: P_{allow} = 911 kips.

(d) Final check by modified interaction formulas from the Column Load Table for a W 14 × 159: L_c = 16.4 ft > L_b = 12 ft; therefore, the section is compact.

Use F_b = 0.66F_y = 24 ksi. r_y = 4.00 in.; B_x = 0.184; a_x = 283(10)^6.

Assuming a braced frame (K = 1.0), with transverse loads (wind load), with restrained ends (wind moments): therefore, C_m = 0.85. Kl/r_y = 12(12)/4.00 = 36.0

From Section 5-3.1, Fig. 5-2 (Allowable Stress Table): for Kl/r_y = 36.0, F_a = 19.49 ksi. f_a/F_a is proportional to P_{actual}/P_{allow} = 600/911 = 0.659 > 0.15.

Therefore, our modified Eq. 9-3 need not be checked. From the modified Eq. 9-1:

$$P + P_x' = 600 + \left\{ 0.184(165)12(0.85)\left(\frac{19.49}{24}\right) \right.$$

$$\left. \left[\frac{283(10)^6}{283(10)^6 - 600(144)^2} \right] \right\}$$

$$= 600 + 263 = 863 \text{ kips}$$

From the modified Eq. 9-2:

$$P + P_x' = 600\left(\frac{19.49}{22}\right) + 0.184(165)12\left(\frac{19.49}{24}\right)$$

$$= 532 + 296 = 828 \text{ kips}$$

(e) Final selection: modified Eq. 9-1 requires an equivalent axial load capacity, $(P + P_x')$, of 863 kips, which is larger than 828 kips (modified Eq. 9-2) and larger than 800 kips (gravity loads only).

Therefore, from the Column Load Tables with KL = 12 ft and $P + P_x'$ = 863 kips, choose W 14 × 159, P_{allow} = 911 kips > 863 kips. *Use* W 14 × 159.

EXAMPLE 9-4

Rework Example 9-3 using the actual AISC Specification interaction formulas (refer to Section 9-2).

solution (Fig. 9-6):

(a) For gravity loads plus wind, Example 9-3, step (b):

$$P = 600 \text{ kips and } M_{max} = 165 \text{ kip-ft}$$

(b) Trial section; for gravity loads only, from the column load table for $KL = 12$ ft: trial section would be W 14 $\times$ 145 ($P_{allow} = 832$ kips > 800 kips). For gravity load plus wind, choose the next stronger column section, W 14 $\times$ 159, $P_{allow} = 911$ kips and $911(0.75) = 683$ kips > 600 kips.

(c) Cross section properties for W 14 $\times$ 159:

$$A = 46.7 \text{ in.}^2 \qquad S_x = 254 \text{ in.}^3 \qquad r_x = 6.38 \text{ in.}$$
$$L_c = 16.4 \text{ ft} \qquad S_y = 96.2 \text{ in.}^3 \qquad r_y = 4.00 \text{ in.}$$

Effective slenderness ratios:

$$Kl/r_x = \frac{1.0(12)12}{6.38} = 22.6$$

$$Kl/r_y = \frac{1.0(12)12}{4.00} = 36.0 \quad \text{(governs for } F_a\text{)}$$

From Example 9-3, step (d), $c_{m_x} = 0.85$ (only x-axis bending).

For $Kl/r_y = 36.0$, $F_a = 19.50$ ksi   (Fig. 5-2)

$$f_a = \frac{P}{A} = \frac{600}{46.7} = 12.85 \text{ ksi} > 0.15F_a$$

For $Kl/r_x = 22.6$, $F'_{e_x} = 300.67$ ksi (Table 9, AISC Specification, Appendix A).

Since $L_c = 16.4$ ft $> L_b = 12$ ft, the section warrants use of $F_{bx} = 0.66F_y = 24$ ksi (section is compact).

$$f_{bx} = \frac{12M_x}{S_x} = \frac{12(165)}{254} = 7.80 \text{ ksi}$$

(d) Check interaction formulas:
Since $f_a > 0.15F_a$, only Eqs. 9-1 and 9-2 need be checked.

Eq. 9-1: $\quad \dfrac{f_a}{F_a} + \dfrac{C_{mx}f_{bx}}{(1 - f_a/F'_{ex}) F_{bx}} \leq 1.0$

$$\frac{12.85}{19.50} + \frac{0.85(7.80)}{[1 - (12.85/300.67)]24} = 0.948 \leq 1.0$$

Eq. 9-2: $\dfrac{f_a}{0.60F_y} + \dfrac{f_{bx}}{F_{bx}}$ ≤ 1.0

$$\frac{12.85}{22} + \frac{7.80}{24} = 0.909 \qquad\qquad \leq 1.0$$

Use W 14 × 159.

A few additional comments relating to the combined axial and bending (interaction) formulas follow. The design of beam-columns involves a trial and error process; a trial section is chosen and checked for compliance with the interaction formulas. A quick method for selecting an economical trial W, M, or S shape uses the aforementioned equivalent axial load method combined with the AISC Manual's Table B (p. 3–10) and the U values that are listed in the Column Load Tables. U is a factor used to convert M_{yy} to an equivalent bending moment with respect to the major, x-x axis. $U = (F_{bx}S_x)/(F_{by}S_y)$. The quick method steps are as follows:

(1) With given KL, select a first approximation value of m from Table B (m is defined as a factor used to convert bending to an approximate equivalent axial load in columns subject to combined loading).

(2) Using $U = 3$, solve for $P_{eff} = P_{actual} + M_x m + M_y mU$, where P_{eff} is an effective value for P accounting for concentric load, P_{actual}, M_x, and M_y.

(3) From the Column Load Tables, select a tentative column section to support P_{eff}.

(4) Using a U value for the tentative column section selected in step (3) (from Column Load Table) and a "subsequent approximate" value of m from Table B, solve for P_{eff} using these new U and m values.

(5) Repeat steps (3) and (4) until the values of m, U, and P_{eff} cease to change results significantly.

Values for P_{wo}, P_{wi}, P_{wb}, and P_{fb} are listed in the Column Load Tables for W, M, and S shapes. These terms are useful in determining whether a column web requires stiffeners due to forces transmitted into it from a rigid beam connected to the column flange. Column stiffener design as well as these terms are discussed and illustrated in Chapter 11.

9-5   ADDITIONAL BEAM-COLUMN DESIGN AIDS

(1)   Lewis B. Burgett, Selection of a "Trial" Column Section, *AISC Engineering Journal,* April 1973.

(2)   Moe A. Rubinsky, Rapid Selection of Beam Columns, *AISC Engineering Journal,* July 1968.

(3)   Michael K. S. Phang and Carey R. Babyak, Simplified Solution to Interaction Equation, *AISC Engineering Journal,* January 1975.

(4)   William Y. Liu, Steel Column Bending Amplification Factor, *AISC Engineering Journal,* April 1965; William Y. Liu, Here are Charts to Rapidly Determine the Bending Amplification Factor for Steel Compression Member Both with or without the Effect of Wind or Seismic, *Consulting Engineer,* October 1966, R. W. Roe and Partners, St. Joseph, Michigan.

(5)   Ea-Lu Ting, Calculator for Beam-Column Design, *AISC Engineering Journal,* April 1970.

(6)   Steel Design File, Beam-Column Tables, *Handbooks 2192 and 2193,* Bethlehem Steel Corporation, Bethlehem, Pa.

(7)   Column Design Curves, *Publication ADUSS 27-3065-01,* U.S. Steel Corporation, Pittsburgh, Pa., 1969.

(8)   Suresh T. Dalal, Some Non-conventional Cases of Column Design, *AISC Engineering Journal,* January 1969.

(9)   Balbir S. Sandhu, Effective Length of Columns with Intermediate Axial Load, *AISC Engineering Journal,* October 1972.

(10) John P. Anderson and James H. Woodward, Calculation of Effective Lengths and Effective Slenderness Ratios of Stepped Columns, *AISC Engineering Journal,* October 1972.

(11) Specifications for the Design and Construction of Mill Buildings, *AISE Standard 13,* Association of Iron and Steel Engineers, Pittsburgh, Pa., 1969, Appendix E.

(12) Richard W. Furlong, Concrete Encased Steel Columns—Design Tables, *Journal of the Structural Division, ASCE,* vol. 100, no. ST 9, Proc. Paper 10807, September 1974, pp. 1865–1882.

The method listed in (1) is one based upon an effective axial load resulting from axial load combined with bending moments. The selection method is rapid and direct and the section will always be a practical one (approaching a value of unity in the interaction equations). Often the section selected is the most economical. This method can be used for preliminary design as well as final design. In the latter case it is suggested that a formal check be made using the appropriate interaction equations. The method uses tabulated values of parameters along with the Column Load Tables, leading to the solution of a single equivalent axial load equation and subsequent selection of section size. A quick convergence leading to the desired section is apparent.

Method (2) is one that was developed to enable an understaffed design group to employ novice engineering technicians to rapidly select an efficient beam-column section. Solution can be performed almost mechanically through the aid of "format-type tables." The methods employed convert direct load and/or minor-axis moment into an equivalent major-axis moment, reducing the problem to one involving bending about the x-x axis only. Another method presented converts the moment about the major and/or minor axis into direct loads, reducing the problem to the selection of a section subject to solely axial load. The methods are simple and direct.

Method (3) makes it easy to find the modified bending amplification factor, $C_m/(1 - f_a/F_e')F_b$ through the use of charts for $F_y = 36$ ksi and 50 ksi, because the chart values for C_m and F_b are taken as 1.0 and $0.60F_y$, respectively. If this is not the case under actual design computations, a simple calculation is all that is needed to correct this assumption. Entering the chart with known values of f_a and Kl/r in the plane of bending, values of $1.0/(1 - f_a/F_e')0.6F_y$ are read.

Method (4) is similar, graphically speaking, to (3). The method graphically obtains the modified amplification factor, $C_m/(1 - f_a/F_e')$ by assuming that $C_m = 0.85$. Knowing f_a and Kl/r in the plane of bending, the modified amplification factor, $0.85/(1 - f_a/F_e')$, is read.

Method (5) describes the design of a special manual slide-rule-type calculator for arriving at a solution to the interaction equations. The calculator eliminates long-hand computations. It can be a time-saving tool during the trial-and-error stage in the selection of a proper beam-column section.

References (6) and (7) present design tables and charts, respectively, for the design of beam columns. Both references use a modified interaction formula. The latter reference was the source for method (1).

Reference (8) presents a design approach and tables for "nonconventional" cases of columns subjected to various types of axial loading. The subjects of stepped columns and prismatic columns with intermediate axial loads are discussed. Useful design examples are presented. As a supplement to reference (8), references (9) and (10) are given relative to the effective length of columns with intermediate axial load and effective lengths and effective slenderness ratios of stepped columns. A stepped column and a prismatic column with intermediate load are

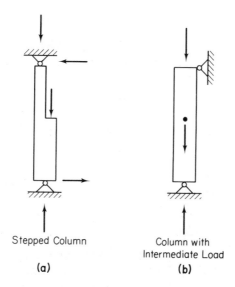

Stepped Column

(a)

Column with
Intermediate Load

(b)

FIG. 9-7

shown in Fig. 9-7a and b, respectively. Stepped column design is also discussed in reference (11).

Tables are presented in reference (12) for the design of structural-steel-shape beam columns encased in concrete. The design tables include A36 rolled shapes from depths of 6 to 14 in. and weights less than 114 lb/linear ft embedded in lightweight concrete.

The author has attempted to present several approaches to the design of beam columns. References have been made to approaches to design problems often encountered in everyday practice. These references are intended for the designer when these problems are encountered. The alternative approaches to conventional beam-column design problems have been selected to offer the designer the option of another approach to design.

The late Hardy Cross* said that it was a good idea to familiarize oneself with all the design and analysis methods available. However, he cautioned one not to be like the centipede with one hundred legs, each representing a method of design, each moving in nonunison, resulting in the centipede falling off the path onto its back, thus making no progress at all. Master one method and use it to facilitate a solution.

---

* Former Chairman, Civil Engineering Department, Yale University; originator of the moment distribution method.

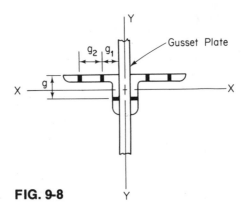

FIG. 9-8

9-6   ANGLE COLUMNS

Double-angle columns, fabricated from two angles connected to opposite gusset plate sides, may be designed, neglecting the eccentricity between the gravity axis and the gage lines (Fig. 9-8). Tabulated allowable concentric load tables are given in the AISC Manual, Part 3, for double angles referred to the y-y axis assuming a gusset plate thickness of $\frac{3}{8}$ in. The tabulated loads are conservative for thicker gusset plates. KL values for either axis are listed for various angle sections (wthin allowable b/t ratios) for both $F_y = 36$ ksi and 50 ksi, and the corresponding allowable concentric loads are listed. For thicker gusset plates, the table is entered with an adjusted $KL = (r$ for $\frac{3}{8}$-in. gusset$/r$ for t-in. gusset$)$ (KL).

Allowable concentric loads on *single-angle struts* are not tabulated in the AISC Manual because physically it is not possible to load such columns concentrically. In these cases, since the actual eccentricity caused by loading single angles is large, the eccentricity in design should *not* be neglected.

A suggested approximate approach to design is illustrated in the following example. To avoid eccentricity about both the x-x and y-y axes, the applied load delivered to the angle through its fasteners should if possible be placed along the x-x axis as shown in Fig. 9-9.

EXAMPLE 9-4

Determine the approximate allowable compressive load on a single-angle strut L8 × 4 × 1. Assume A36 steel, $KL = 10$ ft, and the connection (as indicated in Fig. 9-9) made by means of one row of bolts.

solution:

In Fig. 9-9 (from Properties for Designing Tables for Unequal Angles), $x = 1.05$ in.; $y = 3.05$ in.; $e_y = x - 0.5 =$

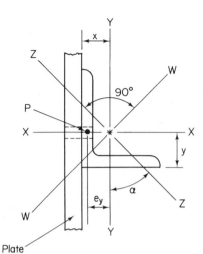

FIG. 9-9   Plate

0.55 in.; $A = 11.0$ in.^{2}; $S_x = 14.1$ in.^{3}; $S_y = 3.94$ in.^{3}; $r_z = 0.846$ in.; $Kl/r_z = (10) 12/0.846 = 142$.

From Fig. 5-2, $F_a = 7.41$ ksi (main member); $P_{allow} = F_aA = 7.41(11.0) = 81.5$ kips (for the z-z axis).

With reference to the x-x and y-y axes, eccentricity exists for bending about the y-y axis only: $M_y/S_y = f_y$; $f_yA = P_{actual}$ due to M_y; $M_y = 0.55P$.

Therefore, $P + (M_y/S_y)A \leq P_{allow} = 81.5$ kips; $P_{actual} = (0.55P/3.94) 11.0 = 81.5$; $P + 1.54P = 81.5$ kips; $P = 32.1$ kips = approximate allowable compressive load.

Sometimes it is not possible to apply the load as shown in Fig. 9-9 so as to avoid eccentricity about the x-x axis. In this case the approximate axial load can be calculated by solving the following equation for P:

$$P + \frac{Pe_xA}{S_x} + \frac{Pe_yA}{S_y} \leq P_{allow}$$

where e_x is the eccentricity about the x-x axis.

The AISC Manual presents another approximate procedure for determining the allowable load on a single-angle column strut. This procedure is to calculate the bending stress from the scaled eccentricity using a to-scale sketch, such as Fig. 9-9, which shows the z-z and w-w principal axes. Part 6 of the AISC Manual gives the necessary information for computing the properties with respect to these axes. α and r_z are given in the Properties for Designing Tables for angles.

The design of single angles as columns must comply with the beam-column interaction formulas. When biaxial (about z-z and w-w axes) bending occurs, Eq. 9-1 is critical. For the case on hand this equation may be expressed as

$$\frac{f_a}{F_a} + \frac{M_W}{S_W F_b \left(1 - \dfrac{f_a}{F'_{eW}}\right)} + \frac{M_Z}{S_Z F_b \left(1 - \dfrac{f_a}{F'_{eZ}}\right)} \leq 1.0 \qquad (9\text{-}5)$$

where F_a is computed with respect to r_Z (z-z axis), the F'_e terms with respect to r_W and r_Z (the principal w-w and z-z axes), respectively, and the F_b terms with respect to S_W and S_Z (the bending stress with respect to the principal w-w and z-z axes), respectively. C_m is taken as unity in Eq. 9-5. The procedures used in Example 9-4 and those relating to the principal axes as discussed usually provide results that are conservative.

PROBLEMS

(9-1)   Choose a W 12 column (A36) to carry a 350-kip axial load in addition to equal and opposite bending moments at each end (minor axis) of 50 kip-ft. Column length is 10 ft. Assume $K = 1$ and $C_m = 0.85$. Use ASD.

(9-2)   Choose a W 14 column ($F_y = 36$ ksi) to support an eccentric load of 1200 kips. Assume a 4-in. eccentricity produces bending about the strong axis. Column length is 20 ft. Assume $K = 1$ and $C_m = 0.85$. Use ASD.

(9-3)   Choose a W 14 column to carry the loads shown as directed in (a) to (c) below. The member is braced in both directions at the ends and braced in the weak axis at mid-length. Design a member for each combination.

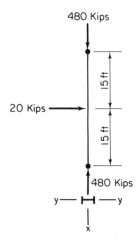

      (a)   Axial load only.

      (b)   Lateral load only ($C_b = 1$).

      (c)   Axial and lateral loads ($C_m = 1$).

(9-4)   A W 14 × 109 has been selected for a beam column to carry an axial load of 450 kips and a uniform load of 1.8 kips/ft (strong axis bending). The member is 20 ft long and is braced at both ends. Is the section satisfactory? (Assume ends are pinned, A36 steel and ASD.)

(9-5)   Design a W 14 column to withstand an axial load of 380 kips and a moment of 75 kip-ft. The column is 30 ft long. The moment is applied about the strong axis at the top only. In the strong direction (i.e., rigid frame), assume member is fixed at the top end and hinged at the bottom (sidesway possible). In the weak direction (between frames), assume column hinged at both ends and no sidesway. Assume A36 steel and ASD.

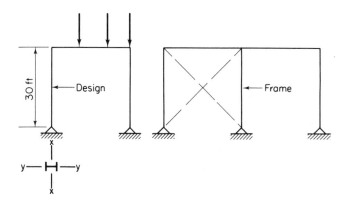

(9-6)   A double angle column (5 × 5 × ¾, A36) is fabricated using a ⅜ in. gusset plate. Determine the permissible axial column load acting in conjunction with equal and opposite end moments of 1.5 kip-ft (x-x axis). Assume a column length of 14 ft. Assume K = 1.0 and ASD.

(9-7)   Determine the permissible compressive load on a single angle strut, 8 × 6 × ¾, A36 steel. Assume simple supports and an effective length of 15 ft. A single row of bolts is used as in Fig. 9-9. (Assume that the applied load is delivered to the angle through the fasteners as shown).

(9-8) The eccentrically loaded column shown in the cross section re-
       ceives bending moment about both axes. By using the appropriate
       interaction formula, determine whether conditions satisfy AISC
       criteria for interaction. Assume $C_m = 1$ and ASD.

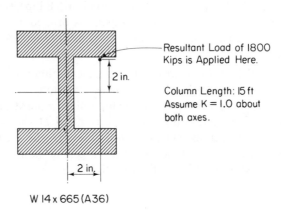

Resultant Load of 1800
Kips is Applied Here.

Column Length: 15 ft
Assume K = 1.0 about
both axes.

2 in.

2 in.

W 14 x 665 (A 36)

10

composite design

10-1   INTRODUCTION

Thus far we have been studying the design of the structural elements of a steel frame—tension members, members subject to shear, axially loaded columns, beam bending members, plate girders, and beam-column members. This chapter will discuss a technique of design in which two different structural materials work together integrally to give an element that acts more efficiently and, under the correct conditions, gives a more economic design.

Composite design, as discussed in this chapter, will apply to the marriage of the two major structural materials, steel and concrete: more specifically, the action of a steel beam together with a concrete slab. Instead of the concrete slab merely resting on the steel beam, the slab, through a mechanical device that will be discussed, is made to act together with the steel beam to carry bending loads. The result is a stiffer and more efficient structural unit, which employs lighter steel beams.

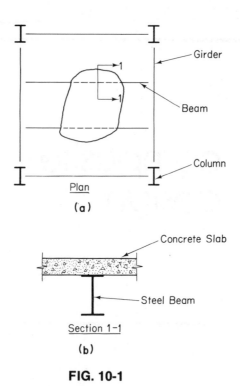

Plan

(a)

Section 1-1

(b)

FIG. 10-1

10-2   BASIC PRINCIPLES: NONCOMPOSITE, FULL, AND PARTIAL COMPOSITE ACTION

Figure 10-1a shows a plan view of a typical bay of a building. The steel framing members are shown together with the concrete floor slab. Section 1-1 in Fig. 10-1b shows a typical cross section of the structural system. Shown is a concrete slab sitting on a steel beam. A common assumption is that the slab, when loaded, carries a load to the beams, which, in turn, carries the load to the girders, and the girders transfer the load to the columns. In reality, the slab deflects along with the steel beam and helps to carry some of the load to the supports. However, when there is no physical connection between slab and beam, the slab resists a small portion of the load. A section in which there is no connection between slab and steel section is termed a *noncomposite section*. When this section deforms under vertical loading, the bottom surface of the slab, which is in tension, elongates, while the top surface of the slab, which is in compression, shortens. As a result of this type of action, there is a separation

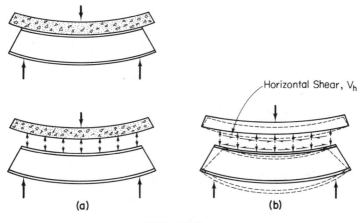

FIG. 10-2

at the ends of the span of concrete and steel at the contact plane between the two. Figure 10-2a shows this, together with the internal forces at the contact plane of slab and beam. Only vertical internal forces are transmitted from slab to beam.

If horizontal forces were to be applied along the top surface of the steel beam as shown in Fig. 10-2b, an elongation of the top surface and an upward deflection would occur as shown. In a similar manner, the horizontal forces shown on the bottom of the slab surface would cause a shortening of this surface and an upward deflection of the slab. If these horizontal forces on the bottom of the slab and on the top of the steel are equal, the forces at the contact surfaces are in equilibrium, and if the magnitude of these forces is large enough, discontinuity in the contact plane is eliminated and a complete interaction of slab and beam is achieved. This would be *full composite action*. The deflection of the composite beam is less than the deflection of a similar noncomposite beam, and the former is significantly stronger than the latter. This can result in a savings in steel, a decrease of beam depth, and greater spans with the more economical rolled shapes.

The noncomposite beam and the composite beam with complete interaction of concrete and steel represent the two extremes or limits of composite action. The actual condition may be somewhere between the two extremes. If the connection between slab and beam is designed such that no slip between slab and beam is assumed to occur, the section is fully composite. If this connection is designed assuming limited slip to occur, the section is termed a *partial composite section*. Finally, if full slip is assumed to occur, the section is *noncomposite*. On the one hand, 100% composite action, in reality, is impossible to achieve, owing to the elasticity of the connection of slab to steel. This elasticity causes a negligible slip. On the other hand, it is not possible to obtain absolute non-composite action, due to the friction and adhesion between slab and steel section.

For practical cases, composite design can increase the ultimate strength of the section by greater than 50%. Although composite construction is appropriate for any condition of loading or geometry, it is most advantageous with very heavy loading, relatively long spans, and beams that are spaced as far apart as possible.

10-3   ENCASED AND MECHANICALLY CONNECTED COMPOSITE BEAMS

Composite construction and design methods may be applied to steel beams when they are totally encased in concrete with a concrete cover of 2 in. or more on their sides and soffit when the concrete is cast together with the concrete slab. This integral action between concrete and steel is by natural bond, provided this is assured by the top of the beam being a minimum of $1\frac{1}{2}$ in. below the top of and 2 in. above the bottom of the concrete slab. In addition, it is required that adequate mesh or reinforcing steel be used throughout the depth and across the soffit of the beam. The mesh prevents spalling of the concrete and assures the two materials working together. Adequate mesh or reinforcing steel is considered to be quite nominal in amount or size. According to the requirements of a large urban center's building code, adequate mesh is considered to be that which weighs at least $1\frac{1}{2}$ lb/yd^2. In the author's opinion, any mesh or reinforcement this side of "chicken wire" should be adequate. The AISC Specification requirements for encased composite beams are shown in Fig. 10-3a.

Composite construction and design methods may also be applied to steel beams that are connected to the concrete slab by means of mechanical shear connectors, as shown in Fig. 10-3b. The types of mechanical shear connectors are discussed in Section 10-9.

10-4   EFFECTIVE WIDTH

The AISC Specification presents rules to determine how large a width of slab is effective to work with the steel section. These limitations on the effective slab width are expressed in terms of the span length of the beam, the clear distance to the adjacent beam, and the slab thickness. These requirements have long been in use in both composite and reinforced concrete tee-beam design for buildings. They are not very restrictive, since any further increase in effective slab width usually has no

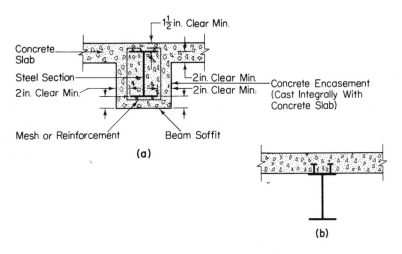

FIG. 10-3

appreciative effect on the cross-sectional design properties. The effective width criteria are expressed in Fig. 10-4a and b for interior beams and exterior (with slab on one side only) beams, respectively. This effective width criteria applies for mechanically connected composite sections. Normally, the third rule is the one that governs because composite construction usually involves relatively long spans with wide beam spacing. Naturally, the smallest value thus obtained should be used as the effective width.

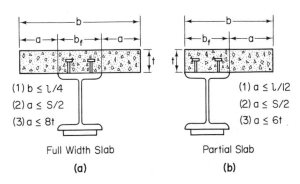

FIG. 10-4

10-5   *TRANSFORMED SECTION*

As will be seen, the design provisions for composite sections are written in terms of working loads and allowable stresses. The computation of working load stresses, ASD, uses the theory of transformed section. This principle assumes an equivalent steel plate by dividing the effective slab width, b, by the modular ratio, n. b/n is then the width of an equivalent steel plate t inches thick (t = slab thickness). n is the modular ratio and is equal to E_s/E_c, where E_s is the modulus of elasticity of steel (29,000 ksi) and E_c is the modulus of elasticity of the concrete. The transformed section principle is shown in Fig. 10-5. All concrete tension stresses should be neglected. Only the compression area on the compression side of the neutral axis is transformed into an equivalent area of steel.

The moment of inertia, the position of the neutral axis, and the section modulus of the transformed section can now be computed as one would do for an ordinary built-up all-steel section. The modular ratio, n, for normal-weight concrete of strength specified is used to determine the section properties (for stress computations) for both normal (stone) and lightweight concrete. However, for deflection computations, the transformed section properties will be based on the actual modular ratio, n, for the specified strength and weight of concrete used. The modular ratio will be clarified in the next section.

10-6   *MODULAR RATIO OF STONE AND LIGHTWEIGHT CONCRETE*

Both stone concrete (normal-weight concrete made with aggregates conforming to ASTM C33) and lightweight concrete (concrete made with rotary kiln-produced aggregates conforming to ASTM C330) of unit weight not less than 90 pcf can be used for mechanically connected composite design.

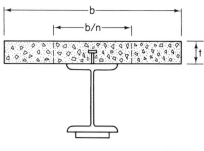

FIG. 10-5

According to the ACI (American Concrete Institute) governing specification, the modulus of elasticity of concrete is

$$E_c = w^{1.5}(33)(f_c')^{1/2} \qquad (10\text{-}1)$$

where w is the weight of concrete (pcf) and f_c' is the specified ultimate compressive strength of the concrete. Repeating, the modular ratio, $n = E_s/E_c$, is computed by using $E_s = 29,000$ ksi and E_c equal to the aforementioned formula value for deflection computations. For stress computations for both lightweight and stone concrete, a value of n for normal (stone) weight concrete of a specified strength can be used.

The flexural capacity of beams designed for full composite action is the same for lightweight or normal-weight concrete when the same area of concrete slab and concrete strength is involved. The number of shear connectors used must be appropriate to the type of concrete. The same concrete design stress level for both types of concrete may be used. For the typical conditions in buildings, values of $f_c' = 3.0$ ksi, $w = 145$ pcf, and therefore $n = 9$ ($E = 29,000$ ksi for steel) are commonly used for normal-weight (stone) concrete.

10-7 HORIZONTAL SHEAR

For mechanically connected composite sections, the entire horizontal shear at the plane where the slab and steel beam join is assumed to be transferred by shear connectors welded to the top flange of the beam and embedded in the concrete slab. For full composite action the total horizontal shear to be resisted by the shear connectors is based upon the ultimate compressive or tensile force at this interface of beam and slab, divided by 2.

Depending upon the geometry of slab relative to steel beam, the neutral axis may fall either in the concrete slab or within the steel section. To cover both possibilities, two formulas for the horizontal shear, V_h, to be resisted by the shear connectors are derived, and the lesser of the two values is used to compute the number of shear connectors necessary. The horizontal shear distribution and material capacity for both conditions are shown in Fig. 10-6a and b for the case where the neutral axis is in the concrete slab and the case where it is in the steel section, respectively. The area of the ultimate stress blocks gives the ultimate V values and, when divided by a factor of 2, the design V_h values are obtained.

For the steel ultimate capacity (Fig. 10-6a), $V = A_s F_y$,

$$V_h = \frac{A_s F_y}{2} \qquad (10\text{-}2)$$

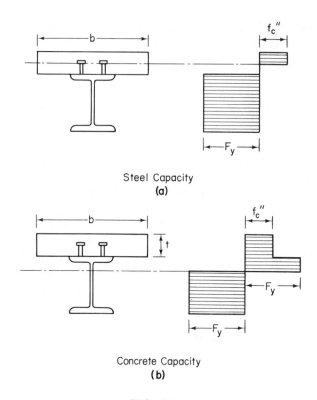

Steel Capacity
(a)

Concrete Capacity
(b)

FIG. 10-6

For the concrete ultimate capacity (Fig. 10-6b), $V = A_c f_c''$,

$$V_h = \frac{0.85 f_c' A_c}{2} \qquad (10\text{-}3)$$

where f_c'' = ultimate strength of the concrete.

ϕ = 0.85 = capacity reduction factor to allow for fluctuation in material, dimensions, the relative importance of the members, and the behavior reliability of the member under shear-type loading (ACI Specification)

f_c' = specified compressive strength of the concrete

A_c = bt = area of concrete effective cross section (see Section 10-4)

A_s = area of steel cross section

The case where the neutral axis is located in the steel section is rare in practical situations. Therefore, Eq. 10-2 usually controls.

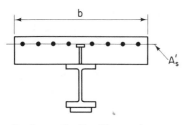

Section at Positive Moment Area

(a)

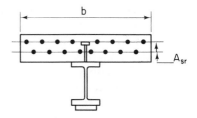

Section at Negative Moment Area

(b)

FIG. 10-7

The smaller V_h value is the total horizontal shear to be resisted between the point of maximum positive moment and points of zero moment. Using the lower of these two values (Eq. 10-2 or 10-3) assures nonfailure in the other (concrete or steel as the case may be) area. If the higher value were to be used, the other element would fail (yielding of the steel or rupture of the concrete).

When longitudinal reinforcing steel with area A'_s is located within the effective width of slab, b, Eq. 10-3 becomes

$$V_h = \frac{0.85f'_c A_c}{2} + \frac{A'_s F_{yr}}{2} \qquad (10\text{-}4)$$

where A'_s = area of the longitudinal compression reinforcing steel in the positive moment area located within the concrete slab effective width

F_{yr} = specified minimum yield stress of the longitudinal reinforcing steel

Equation 10-4 recognizes the possibility of compression steel in the positive moment area of the concrete slab and allows for this influence on the design horizontal shear compressive force of the slab. This is of practical importance where shallow beam depths are desirable (see Fig. 10-7a).

In continuous composite beams in the negative-moment areas, where longitudinal slab reinforcing steel is considered to act compositely with the steel section, the total design horizontal shear, V_h, to be resisted by shear connectors between an interior support and each adjacent point of zero moment (inflection point) is

$$V_h = \frac{A_{sr} F_{yr}}{2} \qquad (10\text{-}5)$$

where A_{sr} is the total area of longitudinal reinforcing steel in the negative moment area located within the concrete slab effective width (Fig. 10-7b) and F_{yr} is as defined before. Equations 10-2 through 10-5 are for full composite action.

For partial composite action with concrete under flexural compression, the design horizontal shear, V'_h, to be resisted is limited by the smaller value of Eq. 10-6 or 10-7.

$$V'_h \geq \text{Eq. 10-2}(0.25) \qquad\qquad\qquad (10\text{-}6)$$

$$V'_h \geq \text{Eq. 10-3}(0.25) \quad \text{or} \quad \text{Eq. 10-4}(0.25) \text{ as applicable} \qquad (10\text{-}7)$$

This means that the minimum partial composite action permitted is controlled by $V'_h/V_h > 0.25$. V'_h is equal to q, the allowable horizontal shear value for one connector, multiplied by the number of shear connectors used.

Essentially, partial composite action permits a savings in the amount of mechanical shear connectors to be used at the expense of using a deeper and/or heavier beam. Sometimes the geometry of the cross section does not permit the amount of connectors to be used that would be necessary for full composite action.

10-8   SHEAR CONNECTORS

Section 10-7 discussed the design horizontal shear to be resisted by mechanical shear connectors. This section will describe the type of connectors in use, the allowable shear load for each type, and how to space and place them throughout the beam span. Figure 10-8 shows the type of shear connectors in common use, although the spiral shear connector has become less popular. By far, the headed-stud connector is the one most frequently used.

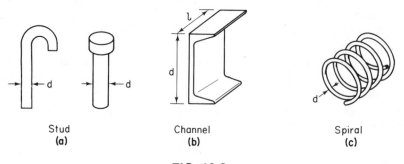

Stud   Channel   Spiral
(a)   (b)   (c)

FIG. 10-8

Based upon tests, allowable horizontal shear loads (kips) for various types and sizes of shear connectors are given in the AISC Specification. These are repeated in Table 10-1 for various f_c' (ksi) concrete strengths for stone (ASTM C33 aggregates) type concrete. Dividing the appropriate total horizontal design shear, V_h or V_h', by the allowable horizontal shear load for the type of connector used gives the total number of shear connectors necessary for flat soffit slabs between the point of maximum moment and each point of zero moment. The spacing of connectors can be equal.

When lightweight concrete is used (conforming to ASTM C330 aggregates), the allowable horizontal shear load for one connector may be obtained by multiplying the values obtained from Table 10-1 by the reduction coefficient obtained from Table 10-2. In a manner similar to that used for stone concrete, dividing the appropriate total horizontal design shear, V_h or V_h', by the allowable horizontal shear for the type of connector used gives the total number of shear connectors necessary for flat soffit lightweight concrete slabs between the point of maximum moment and each point of zero moment. The spacing of connectors can be equal.

Table 10-1

	q = Allowable Horizontal Shear Load(Kips) for Stone Concrete (ASTM C33 Aggregates)[b]		
	$f'c$	(ksi)	
Connector[a]	3.0	3.5	$\geq$ 4.0
$\frac{1}{2}$ in. diam. $\times$ 2 in. hooked or headed stud	5.1	5.5	5.9
$\frac{5}{8}$ in. diam. $\times$ 2$\frac{1}{2}$ in. hooked or headed stud	8.0	8.6	9.2
$\frac{3}{4}$ in. diam. $\times$ 3 in. hooked or headed stud	11.5	12.5	13.3
$\frac{7}{8}$ in. diam. $\times$ 3$\frac{1}{2}$ in. hooked or headed stud	15.6	16.8	18.0
3 in. channel, 4.1 p/f	4.3l	4.7l	5.0l
4 in. channel, 5.4 p/f	4.6l	5.0l	5.3l
5 in. channel, 6.7 p/f	4.9l	5.3l	5.6l

l = length of channel in inches

[a] The allowable horizontal tabulated loads may also be used for studs longer than shown. Research has shown that for a height of stud-to-stud diameter ratio of 4 or greater, the failure occurs in the stud as a shear failure.

[b] For flat soffit slabs. Values are based upon an approximate factor of safety of 2.50 against ultimate strength.

Courtesy of American Institute of Steel Construction.

Table 10-2

Air Dry Unit Weight, pcf[a]	90[b]	95	100	105	110	115	120
Reduction coefficient for $f'c$, $\leq$ 4.0 ksi	0.73	0.76	0.78	0.81	0.83	0.86	0.88
Reduction coefficient for $f'c$, $\geq$ 5.0 ksi	0.82	0.85	0.87	0.91	0.93	0.96	0.99

Interpolation is permitted for values of $f'c$ between 4.0 and 5.0 ksi.

[a] For flat soffit slabs made with rotary kiln-produced aggregates conforming to ASTM C330.

[b] Minimum permissible for composite assumptions.

Courtesy of American Institute of Steel Construction.

Figure 10-9 shows a beam span subjected to a concentrated load. The span shown is subject to positive moment and bending. The amount of connectors required from the point of maximum moment to the points of zero moment (hinges or points of contraflexure) either side may be uniformly distributed. However, the number of shear connectors required between any concentrated load in the positive-moment area and the nearest point of zero moment, N_2, shall not be less than that computed by Eq. 10-8. This equation is not applicable in negative-moment areas of continuous beams where connectors may be spaced uniformly. The balance of the shear connectors, $(N_1 - N_2)$, may be spaced uniformly between the concentrated load point and the point of maximum moment.

$$N_2 = \frac{N_1[(M\beta/M_{\max}) - 1]}{\beta - 1} \qquad (10\text{-}8)$$

where M = moment ($< M_{\max}$) at the concentrated load point

N_1 = number of connectors required between the point of maximum moment and the point of zero moment (V_h/q, V_h'/q, as applicable, where q is the allowable horizontal shear load per connector as determined from Tables 10-1 and/or 10-2)

β = S_{tr}/S_s or S_{eff}/S_s, as applicable

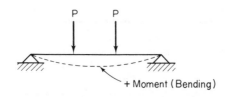

FIG. 10-9

This equation enables the designer to quickly and easily check to see if a uniform spacing of studs along the beam span will, at the same time, provide the required number of studs between concentrated loads and the end supports.

S_{tr}, S_s, and S_{eff} are the section modulus of the transformed composite cross section referred to the bottom flange, the steel beam used in the composite section referred to the bottom flange, and the effective section modulus corresponding to partial composite action, in that order. These terms will be clarified below.

For continuous beams, in the negative-moment (bending) area, connectors may be uniformly spaced between the point of maximum moment and each adjacent point of zero moment. V_h for this case is determined by Eq. 10-5.

The uniform spacing provisions of mechanical shear connectors are based upon the idea that the more heavily (high statical shear areas) stressed shear connectors yield and redistribute further load to the less heavily (lower statical shear areas) stressed shear connectors (similar to the redistribution of load in plastic design). At ultimate loading, all shear connectors are equally loaded. Under certain loading conditions, however, a closer spacing is required. Equation 10-8 is a check for this. The q values given in Table 10-1 should not be confused with VQ/I shear connection values discussed in Chapter 4.

The allowable loads for the various types of connectors are as follows: for stud connectors, proportional to the square of the stud diameter and $(f_c')^{1/2}$; for channel connectors proportional to $(t_f + 0.5t_w)$, the bar diameter, d, and to $(f_c')^{1/4}$.

Allowable horizontal shear load values of mechanical connectors for other than those listed or for the case other than use with ASTM C33 or ASTM C330 produced by rotary kiln are required to be established by suitable tests.

The concrete cover, except for shear connectors installed in the ribs of formed steel decks (Section 10-12), must be a minimum of 1 in. lateral to the connector. Theoretically, there need not be any cover between the top of the connector and the top surface of the concrete slab.

Stud shear connectors connected by welds to a beam, at locations other than directly over the beam web, have a tendency to tear out from a thin flange prior to full achievement of q. To avoid this situation, the diameter of stud shear connectors $\leq 2.5t_f$, where t_f is the thickness of the flange to which the connector is welded for connectors other than those welded directly over the beam web.

The limiting center-to-center spacings of stud shear connectors on the beam shall be as indicated in Fig. 10-10. At times, a uniform connector spacing is not possible (e.g., a uniform spacing of connectors may

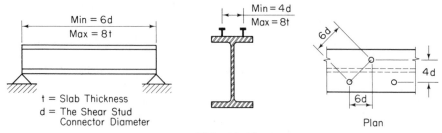

t = Slab Thickness
d = The Shear Stud
    Connector Diameter

Plan

FIG. 10-10

place them too close together; placement of connectors in metal deck rib flutes limits uniform placement of connectors). For these cases it is suggested that the remaining or additional studs should be placed symmetrically about the supports where the horizontal shear is of the greatest magnitude.

The plan view of the top flange of the beam in Fig. 10-10 shows the application of the minimum stud spacing rules. The maximum spacing rules apply in a similar manner.

AWS (American Welding Society) D1.1 (Structural Welding Code), Part F, discusses the welding of studs, manufacturing tolerances, mechanical requirements (ASTM A108, Grades 1010 through 1020), welder workmanship, quality control, inspection requirements, and so on. Of interest to the designer is the longitudinal and lateral spacing requirements between studs and between stud and edge of beam or beam flange. The major requirements are as follows:

(1)   Minimum spacing center to center, 2.5 in.
(2)   Minimum distance from edge of stud base to edge of flange $= d +$ $\frac{1}{8}$ in. $\geq 1\frac{1}{2}$ in.

The weld from stud shear connector to beam flange is of equivalent strength to the shank of the stud and is automatically performed with good assurance of quality. There is no need to design or specify its size.

10-9   SHORED AND NONSHORED CONSTRUCTION

Two methods of composite construction are used, shored and unshored construction. The forms for the floor slabs are usually hung from the steel beams and the floor beams may be shored from the floor below or may carry the dead load themselves. Which is used is a design decision based on economy and on the magnitude of dead-load deflection. Shoring may also be required to limit the steel stress to a value below initial yielding,

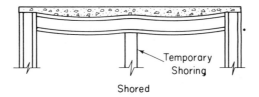

Shored

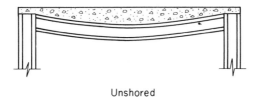

Unshored

FIG. 10-11

as will be seen. Both shored and unshored construction are shown in Fig. 10-11.

The AISC Specification permits both shored and unshored composite construction with mechanically anchored slab-to-beam design as well as encased-in-concrete-beam type of assumption. The method of construction, shored or unshored, does not influence the ultimate strength of the composite member. The method of construction does affect the steel stresses in the member.

For encased-in-concrete composite beams with no temporary shores,

$$f_b = \frac{M_{DL}}{S_s} + \frac{M_{LL}}{S_{tr}} \le F_b \qquad (10\text{-}9)$$

For encased-in-concrete composite beams with temporary shores,

$$f_b = \frac{M_{DL} + M_{LL}}{S_{tr}} \le F_b \qquad (10\text{-}10)$$

For mechanically connected composite beams with or without temporary shores,

$$f_b = \frac{M_{DL} + M_{LL}}{S_{tr}} \le F_b \qquad (10\text{-}11a)$$

(*Note:* For unshored construction, Eq. 10-12 must also be met.)

Equation 10-9 says that, for beams encased in concrete without temporary shoring, the steel beam must be proportioned to support all the dead load without assistance from the concrete. The superimposed live load can be assumed to be taken by the composite section (steel + concrete). For encased composite beams with temporary shores, the composite section may be relied upon to take both the dead and live load (Eq. 10-10).

For composite design utilizing mechanical shear connectors, the assumption is made that, for both shored and unshored construction, the composite section takes the dead and live loads. The liberalization (for both shored and unshored construction) is based on an ultimate-strength concept (although the proportioning of the members is based on the elastic section modulus of the transformed cross section, S_{tr}).

To assure that the maximum bending stress in the steel beam under actual (service) loading conditions is kept a substantial amount below the initial yielding level for all ratios of M_{LL}/M_{DL}, S_{tr} in the tension (bottom) flange of the steel beam, for unshored construction, should be limited to

$$S_{tr} = \left[1.35 + 0.35 \left(\frac{M_{LL}}{M_{DL}} \right) \right] S_s \qquad (10\text{-}12)$$

As an alternative (for the case of mechanically connected unshored composite sections), Eq. 10-11b may be used.

$$f_b = \frac{M_{DL}}{S_s} + \frac{M_{LL}}{S_{tr}} \leq 1.35 F_b \qquad (10\text{-}11b)$$

In Eqs. 10-9 through 10-12, M_{LL} is the moment produced by live load (loads applied after the concrete has reached 75% of its required strength), M_{DL} is the moment produced by dead load (loads applied before the concrete has reached 75% of its required strength), and, S_s is the section modulus of the steel beam used in the composite section referred to the flange, where the stress is being computed.

In Eq. 10-12, S_s should be computed for the steel section's tension flange at sections subject to positive moment and to both flanges (tension and compression) in negative-moment areas. By using the S_s value referred to the appropriate flange of the steel section, the corresponding stresses can be computed. These stresses in the steel beam alone should not exceed the appropriate allowables (Chapter 6). The usual $33\frac{1}{3}\%$ increase in allowable bending stress due to wind and/or seismic loads acting alone or in conjunction with gravity loading should not be applied in the negative-moment area. S_{tr} is the section modulus of the transformed composite cross section referred to the bottom flange based upon the effective width b. f_b is the computed bending stress in the steel section.

Equation 10-12 is arrived at as follows: setting the maximum computed bending stress in the composite section equal to $0.66F_y$,

$$f_b = \frac{(M_{DL} + M_{LL})}{S_{tr}} = 0.66F_y \qquad (A)$$

For unshored construction the maximum computed bending stress in the composite section after the concrete has set was assumed to be $0.89F_y$; therefore,

$$f_b = \frac{M_{DL}}{S_s} + \frac{M_{LL}}{S_{tr}} \leq 0.89F_y \qquad (B)$$

$0.66F_y(1.35) \approx 0.89F_y$, and combining Eqs. A and B,

$$\frac{M_{DL}}{S_s} + \frac{M_{LL}}{S_{tr}} \leq \frac{1.35(M_{DL} + M_{LL})}{S_{tr}} \qquad (C)$$

and

$$\frac{M_{DL}}{S_s} \leq \frac{(1.35M_{DL} + 1.35M_{LL} - M_{LL})}{S_{tr}}$$

$$\frac{M_{DL}}{S_s} = \frac{(1.35M_{DL} + 0.35M_{LL})}{S_{tr}}$$

Solving for S_{tr},

$$S_{tr} \leq \left[1.35 + 0.35\left(\frac{M_{LL}}{M_{DL}}\right) \right] S_s \qquad (10\text{-}12)$$

In computations for mechanically connected composite sections in positive-moment areas, the steel section does not have to adhere to the compact criteria, b, of Eqs. 6-7a, 6-7b, and 6-9a. $F_b = 0.66F_y$ regardless of the width-to-thickness criteria of unstiffened or stiffened compression elements or compression flange lateral support.

For mechanically connected composite sections, the flexural stress in the concrete slab is determined by $(M_{DL} + M_{LL})$ and M_{LL} for shored and unshored construction, respectively. If the actual S_{tr} exceeds the value computed from Eq. 10-12, shored construction must be used. The actual S_{tr} (not S_{tr} computed by Eq. 10-12) should be used to compute the actual concrete-slab flexural stresses. The stress in the concrete $\leq 0.45f_c'$.

For unshored construction, the concrete compressive stress will seldom control, and the bottom flange steel stress will usually control strength design. Adequate lateral support for the compression flange of the steel section will be provided by the hardened concrete slab. During construction (before hardening of the concrete slab), adequate lateral support must be provided or the working stresses reduced accordingly. Usually, steel deck with adequate connection to the steel beam compression flange or properly constructed concrete forms provide the necessary lateral support. For full encased-in-concrete type of composite construction, special attention should be made to provide proper lateral support during the construction stage.

For fully encased beams in composite construction, the AISC Specification suggests a limiting F_b of $0.66F_y$ because the encased beam is restrained from local and lateral buckling. In lieu of computing the bending stress on the basis of Eq. 10-9 or 10-10 and the composite section properties (for encased beams), an allowable stress $F_b = 0.76F_y$ for the steel beam section alone can be applied for the steel section resisting unassisted the positive moment produced by all dead and live loads. If the section is proportioned in this manner, no temporary shoring is required. If the beam checks on this account, it will check by Eqs. 10-9 and 10-10 also.

10-10 PARTIAL COMPOSITE ACTION

Sections 10-2 and 10-7 made mention of partial composite action relative to definition and horizontal shear, V_h', respectively. In cases where partial composite action is assumed, the effective section modulus is determined by

$$S_{eff} = S_s + \left(\frac{V_h'}{V_h}\right)^{1/2} (S_{tr} - S_s) \qquad (10\text{-}13)$$

where V_h and V_h' are horizontal shear as defined in Section 10-7. For V_h' in Eq. 10-13, a value of q times the number of connectors used between the point of maximum moment and the nearest point of zero moment should be used. S_s is the section modulus of the steel beam referred to the bottom flange. S_{tr} is the section modulus of the transformed composite section referred to the bottom flange of the steel beam.

Equation 10-13 provides for a parabolic variation from no composite action (the strength of the beam is that of the bare steel beam) to full composite action (the strength of the beam is that of the full transformed section). S_{eff} is less than the section modulus for fully effective composite action, S_{tr}, but is more than the section modulus of the steel beam alone, S_s. The V_h'/V_h expression in Eq. 10-13 is the degree of shear connection. The parabolic relationship is conservative.

For partial composite action (mechanically connected), Eq. 10-11a becomes

$$f_b = \frac{(M_{DL} + M_{LL})}{S_{eff}} \leq F_b \qquad (10\text{-}14a)$$

and Eq. 10-11b (alternative approach for unshored construction) becomes

$$f_b = \left(\frac{M_{DL}}{S_s}\right) + \left(\frac{M_{LL}}{S_{eff}}\right) \le 1.35 F_b \qquad (10\text{-}14b)$$

10-11 USE OF COVER PLATES

Designers are often asked whether cover plates should be used. The cover-plated beam will always result in a lighter section, but, as mentioned before (relative to plate girders), the lighter section is not necessarily the least expensive section. Fabrication costs have to be compared as well as material costs.

Where deflections are not critical, the use of bottom cover plates is quite effective in strengthening or reducing the depth of the composite beam. Attaching a thick plate to the bottom flange costs approximately the same as attaching a thin plate. Therefore, to facilitate and economize on fabrication, it would be preferable to use a narrow and thicker plate rather than a wider and thinner one. The narrower cover plate better lends itself to automatic welding procedures with minimum handling (Fig. 10-12). One further comment: when the cover-plated section is exposed to the elements, the narrower cover plate does not tend to trap water.

As a general guideline (and this may vary from one area of the country to another), when choosing between a cover-plated and a non-cover-plated section:

(1) If the cover-plated section saves less than 7 plf, do not use a cover plate.

(2) If the cover-plated section saves more than 12 plf, use a cover plate.

(3) Between a savings of 7 and 12 plf, minor savings may result either way.

The choice relates to cost savings of material versus increased cost of fabrication and is dependent on reference time, location of site and plant, and the size of the project. All these facts should be studied.

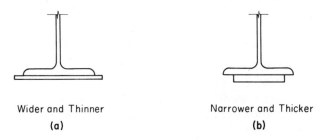

Wider and Thinner   Narrower and Thicker
(a)   (b)

FIG. 10-12

10-12   END REACTIONS

The web and end connections of the steel beam as per the AISC Specification should be designed to carry the total end reaction. If the end
reactions, R, are not shown on the design drawings, R can be computed
by Eq. 10-15, which is accurate for uniformly loaded simple beams that
are fully stressed. The equation is conservative for most other types of
loading.

$$R = \frac{0.33 S_{tr} F_b}{L} \qquad\qquad (10\text{-}15)$$

where L is the span (ft) and F_b is the allowable bending stress in the
absence of axial load (ksi).

10-13   DEFLECTIONS

Deflections of composite beams are generally from 50% to 65% of the
deflection of noncomposite beams. Composite beams are stiffer than noncomposite beams having equal loading, span length, and depth. In practice, shallower composite sections are used, and deflections of the steel
section alone under construction loads should be checked.

The AISC Commentary to the Specification suggests a depth-to-
span ratio for fully stressed beams to prevent deflection problems of
$F_y/800$ (when members of less depth are used, the allowable stress should
be decreased in the same ratio that the recommended depth is decreased).
This guide gives a depth-to-span ratio of $\frac{1}{22}$ and $\frac{1}{16}$ for $F_y = 36$ and 50 ksi,
respectively. The depth used in these ratios is from top of concrete slab
to bottom of steel beam.

To minimize transient vibrations due to pedestrian footfall for composite beams supporting large, open, partitionless areas (no source of
damping), the radio d/l should be limited to $\frac{1}{20}$, where d is the depth of
the bare steel beam. This depth would apply to noncomposite sections
as well.

Deflections for simple-span uniformly loaded composite sections
can be quickly computed:

$$\Delta = \frac{M L^2}{160 S y} \qquad\qquad (10\text{-}16a)$$

where Δ = deflection, in.
$\quad\ \ M$ = moment, kip-ft
$\qquad S$ = section modulus, in.^{3} ($S = I/y$ or $Sy = I$)
$\qquad y$ = distance from bottom of steel to neutral axis, in.
$\qquad L$ = span length, ft

For unshored construction (mechanically connected and encased composite),

$$\Delta_{DL} = \frac{M_{DL}L^2}{160I_s} \qquad (I_s = S_s y_{bs}) \qquad (10\text{-}16b)$$

Short-term,

$$\Delta_{LL} = \frac{M_{LL}L^2}{160I_{tr}} \qquad (I_{tr} = S_{tr} y_b]$$

$$\Delta_{\text{total}} = \Delta_{DL} + \Delta_{LL} \qquad (10\text{-}16c)$$

For shored construction (mechanically connected and encased composite),

$$\Delta_{DL} = \frac{M_{DL}L^2}{160I_{tr}} \qquad (10\text{-}16d)$$

Short-term,

$$\Delta_{LL} = \frac{M_{LL}L^2}{160I_{tr}}$$

$$\Delta_{\text{total}} = \Delta_{DL} + \Delta_{LL} \qquad (10\text{-}16e)$$

For the usual span of composite beam, Δ_{DL} is usually limited (rule of thumb) to $1\frac{1}{2}$ in. and Δ_{LL} is limited to $\frac{1}{360}$ (plastered ceilings).

For long-term creep (the "flow" of material over a period of time) deflection (when the sustained load is 50% of the total load), S_{tr} and y_b should be based upon a modular ratio, n, double the value in the Composite Design Properties Tables (Section 10-15). Using S_{tr} and y_b from the partial slab tables for beams with a full slab gives an estimate of this deflection. Full-width slab and partial slab, y_b and y_{bs}, are as defined in Fig. 10-13. This nomenclature is used in the Composite Design Properties Tables (AISC Manual, Seventh Edition).

As stated before, for stress computations the compression area of lightweight or normal-weight (stone) concrete can be converted to an equivalent area (transformed area) of steel by dividing this compression area by n, the modular ratio (e.g., bt/n). In this case n is the modular ratio for normal-weight concrete of the strength specified. However, for deflection computations, bt/n is based upon the actual modular ratio, n, for the strength and weight of concrete used.

The effective moment of inertia, I_{eff}, for deflection computations is determined by Eq. 10-17 for partial composite action (accounts for decrease in stiffness due to partial composite action):

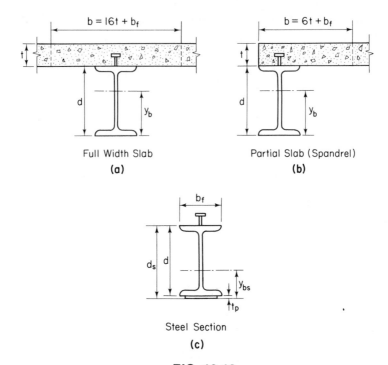

Full Width Slab
(a)

Partial Slab (Spandrel)
(b)

Steel Section
(c)

FIG. 10-13

$$I_{\text{eff}} = I_s + \left(\frac{V_h'}{V_h}\right)^{1/2} (I_{tr} - I_s) \qquad (10\text{-}17)$$

where I_s = moment of inertia of the bare steel beam, in.[4]
        I_{tr} = moment of inertia of the transformed compsoite section, in.[4]

Referring to Fig. 10-14, the transformed area properties for deflection are
simplified. Specifically, the computations for locating the neutral axis, y_b,
and for the transformed moment of inertia, I_{tr}, are

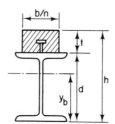

FIG. 10-14

ΣM about bottom of steel section:

$$\frac{bt}{n}\left(h - \frac{t}{2}\right) + A_s\left(\frac{d}{2}\right) = \left(\frac{bt}{n} + A_s\right)y_b$$

Assume that $\phi = bt/nA_s = A_c/nA_s$.

$$\phi A_s\left(h - \frac{t}{2}\right) + \frac{A_s d}{2} = (\phi A_s + A_s)y_b$$

$$A_s\left[\phi\left(h - \frac{t}{2}\right) + \frac{d}{2}\right] = A_s(\phi + 1)y_b$$

$$\frac{A_s[\phi(h - t/2) + d/2]}{A_s(\phi + 1)} = y_b$$

$$y_b = \left(\frac{\phi}{\phi + 1}\right)\left(h - \frac{t}{2}\right) + \left(\frac{d}{2}\right)\left(\frac{1}{\phi + 1}\right) \qquad (10\text{-}18)$$

Again, from Fig. 10-14, $I = \Sigma(I_0 + Ay^2)$,

$$I_{tr} = I_s + A_s\left[y_b - \frac{d}{2}\right]^2 + \frac{bt^3}{n(12)} + \frac{bt}{n}\left[d + \frac{t}{2} - y_b\right]^2 \qquad (10\text{-}19)$$

10-14   ECONOMY

In general, it is recommended for full or partial composite action that $6 \times 6 \times 10 \times 10$ welded wire mesh be used for transverse reinforcement. If a large slab force is anticipated owing to the use of high-strength steel beams, or full composite action, attention should be given to transverse reinforcement so that longitudinal shear in the slab is not critical. $6 \times 6 \times 10 \times 10$ wire mesh for these cases will still usually do the job. In general, composite design has reflected real economies in the typical office-type structure and has resulted in shallower beam depths.

For high-rise buildings with a tight construction schedule, shoring is not usually recommended. However, for the case of one- and two-story buildings or when metal deck is required to span long distances and loads are relatively light, shored construction should be carefully considered.

The designer knowing the economic and other factors concerned with the particular construction site should carefully consider the economics involved with the following questions: full composite, noncomposite, or partial composite? (Full-composite action overestimates the number of shear connectors required and is almost always uneconomical when compared to partial-composite action. This will be more obvious by examining the summary of results of the design examples of Section 10-16). Shored or unshored construction? Cover-plated or non-cover-plated? Mechanically shear-connected or encased composite? Many articles have been written about compositely designed structures that were built. These articles have reflected very significant economies, making it obvious that composite design can offer a savings.

10-15   DESIGN AIDS AND TABLES

There are numerous references on the subject of composite design, many of which present valuable design aids and tables. However, for the most part, discussions will be confined to those design aids and tables found in the AISC Manual.

It is worth mentioning that several of the steel-manufacturing companies, as well as the shear connector companies, produce many of these other valuable publications. Composite Steel-Concrete Construction, by Ivan M. Viest, Chairman of the Subcommittee on the State-of-the-Art Survey, of the Task Committee on Composite Construction, of the Committee on Metals, Structural Division, *Journal of the Structural Division, ASCE,* vol. 100, no. ST5, Proc. Paper 10561, May 1974, pp. 1085–1139, is an excellent report with supporting bibliography giving the current state of the art of composite construction.

Because composite construction usually involves relatively long beams spans and wide spacings of beams, the Specification provision for effective slab width which controls is the rule that the projection beyond the flange is $8t$ (Section 10-4). Additionally, slab thicknesses from 4 to $5\frac{1}{2}$ in. are most often used from the standpoint of both the required fireproofing and the wide spacing of the beams. For this reason, the AISC Manual (Seventh Edition), in its Properties of Composite Beams Tables, uses an effective slab width $b = 16t + b_f$ for full-width slabs and $b = 6t + b_f$ (t is the slab thickness and b_f is the flange width of the steel compression flange) for partial (spandrel beam) slabs for slab thicknesses of 4 to $5\frac{1}{2}$ in. (in $\frac{1}{2}$-in. increments) for sections without and with bottom cover plates. Figure 10-15* is a typical page from the AISC Manual tables for no cover plates; 91 different beam sections are listed. Similar tables are given for sections utilizing cover plates.

---

* Since these tables are from the AISC Manual, 7th ed., they include some of the older shapes that are no longer made. They do not include any of the new shapes currently available.

The following additional assumptions were made in preparing these tables: the concrete slab rests directly on the steel section, the concrete slab is connected to the steel beam by stud or channel shear connectors, $f_c' = 3.0$ ksi, $n = 9$, and steel beam depths are from 8 to 36 in. (sufficient for the usual 20- to 60-ft span for girders and longer spans for beams). For the usual building design, these assumptions will suffice. The afore-mentioned tabulated effective slab width should be checked with the other limits based on beam spacing and beam span. Should the effective slab width, b, be governed by these other limits, y_b, I_{tr}, S_{tr}, and S_{tr}/S_t values are affected. These values can be adjusted by referring to Eq. 10-18 for y_b, which is changed by a change in value of ϕ (which reflects any change in b) and Eq. 10-19 for I_{tr}, which is affected by a change in y_b and b. S_{tr} and S_t can be changed similarly. These adjustments should be made in the final design only. For determining a trial section, the tables may be utilized since slight changes in concrete area or b/n ratios are of little significance.

All data in the tables are applicable to any of the steel grades per-missible by the AISC Specification. However, the allowable web shear values, V, in the cover-plated beam tables apply only for A36 steel. The tables may be used for any value of n for stress computations because ultimate moment of the composite section is independent of n.

In these tables, values of b, the effective width of slab, are tabulated for each section separately for each slab thickness. Values for S_{tr}, S_t, S_s, I_{tr}, y_b, b, S_{tr}/S_t are listed (S_t is the section modulus for the transformed section referred to the top of concrete slab).

The S_{tr}/S_t tabulated values can be used in conjunction with footnote 2 to the table to determine if the concrete stress governs. Concrete stress usually controls for sections with heavy cover plates, partial width slabs used with high-strength beams. However, concrete stresses should be checked for all cases.

To prevent overstressing of the unshored steel beam, a proper value of S_{tr} should be used (a value equal to or less than Eq. 10-12). For all tabulated beams in the tables, $S_{tr} \leq 2.31 S_s$. Therefore (Eq. 10-12), when $M_{LL}/M_{DL} \geq 2.74$, Eq. 10-12 need not be checked (see footnote 3 to tables). Footnote 5 to the tables lists values of S_{tr}/S_t at balanced design where, for values equal to or less than these values, the allowable concrete design will not be exceeded for shored construction. These values are listed for F_y values of 36, 42, 45, 50, 55, and 60 ksi and for f_c' values of 3.0, 3.75, and 4.0 ksi. Values of I_{tr} and y_b given in the tables are used to check other tabular data or to calculate S_j, the transformed section modulus at the top of the steel beam.

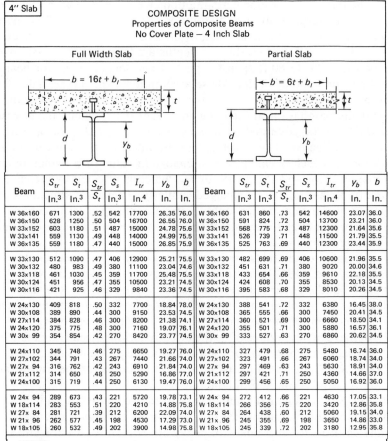

| 4" Slab | | COMPOSITE DESIGN
Properties of Composite Beams
No Cover Plate — 4 Inch Slab | | | | | | | | | | | | |

	Full Width Slab							Partial Slab						
Beam	S_{tr}	S_t	$\dfrac{S_{tr}}{S_t}$	S_s	I_{tr}	y_b	b	Beam	S_{tr}	S_t	$\dfrac{S_{tr}}{S_t}$	S_s	I_{tr}	y_b	b
	In.³	In.³		In.³	In.⁴	In.	In.		In.³	In.³		In.³	In.⁴	In.	In.
W 36×160	671	1300	.52	542	17700	26.35	76.0	W 36×160	631	860	.73	542	14600	23.07	36.0
W 36×150	628	1250	.50	504	16700	26.55	76.0	W 36×150	591	824	.72	504	13700	23.21	36.0
W 33×152	603	1180	.51	487	15000	24.78	75.6	W 33×152	568	775	.73	487	12300	21.64	35.6
W 33×141	559	1130	.49	448	14000	24.99	75.5	W 33×141	526	739	.71	448	11500	21.79	35.5
W 36×135	559	1180	.47	440	15000	26.85	75.9	W 36×135	525	763	.69	440	12300	23.44	35.9
W 33×130	512	1090	.47	406	12900	25.21	75.5	W 33×130	482	699	.69	406	10600	21.96	35.5
W 30×132	480	983	.49	380	11100	23.04	74.6	W 30×132	451	631	.71	380	9020	20.00	34.6
W 33×118	461	1030	.45	359	11700	25.48	75.5	W 33×118	433	654	.66	359	9610	22.18	35.5
W 30×124	451	956	.47	355	10500	23.21	74.5	W 30×124	424	608	.70	355	8530	20.13	34.5
W 30×116	421	925	.46	329	9840	23.36	74.5	W 30×116	395	583	.68	329	8010	20.26	34.5
W 24×130	409	818	.50	332	7700	18.84	78.0	W 24×130	388	541	.72	332	6380	16.45	38.0
W 30×108	389	890	.44	300	9150	23.53	74.5	W 30×108	365	555	.66	300	7450	20.41	34.5
W 27×114	384	828	.46	300	8200	21.38	74.1	W 27×114	360	521	.69	300	6660	18.50	34.1
W 24×120	375	775	.48	300	7160	19.07	76.1	W 24×120	355	501	.71	300	5880	16.57	36.1
W 30× 99	354	854	.42	270	8420	23.77	74.5	W 30× 99	333	527	.63	270	6860	20.62	34.5
W 24×110	345	748	.46	275	6650	19.27	76.0	W 24×110	327	479	.68	275	5480	16.74	36.0
W 27×102	344	791	.43	267	7440	21.66	74.0	W 27×102	323	491	.66	267	6060	18.74	34.0
W 27× 94	316	762	.42	243	6910	21.84	74.0	W 27× 94	297	469	.63	243	5630	18.91	34.0
W 21×112	314	650	.48	250	5290	16.86	77.0	W 21×112	297	421	.71	250	4360	14.66	37.0
W 24×100	315	719	.44	250	6130	19.47	76.0	W 24×100	299	456	.65	250	5050	16.92	36.0
W 24× 94	289	673	.43	221	5720	19.78	73.1	W 24× 94	272	412	.66	221	4630	17.05	33.1
W 18×114	283	553	.51	220	4210	14.88	75.8	W 18×114	266	356	.75	220	3420	12.86	35.8
W 27× 84	281	721	.39	212	6200	22.09	74.0	W 27× 84	264	438	.60	212	5060	19.15	34.0
W 21× 96	262	577	.45	198	4530	17.29	73.0	W 21× 96	245	355	.69	198	3650	14.86	33.0
W 18×105	260	532	.49	202	3900	14.98	75.8	W 18×105	245	339	.72	202	3180	12.95	35.8

NOTES:
1. Tables are based on specified concrete strength $f'c$ = 3.0 ksi and modular ratio n = 9.
2. Proper selection of S_{tr} value assures that stress in steel beam does not exceed allowable.
   Concrete stress due to bending does not exceed allowable under the following conditions:

F_y = 36 ksi	Condition	F_y = 50 ksi
S_{tr}/S_t	f'_c = 3.0 ski	S_{tr}/S_t
≤ .51	Shored construction	≤ .37
≤ 1.01	Unshored const. with M_L/M_D < 1.0	≤ .74
≤ .76	Unshored const. with M_L/M_D < 2.0	≤ .55
≤ .67	Unshored const. with M_L/M_D < 3.0	≤ .49

3. For unshored construction see discussion that follows (Sect. 10-15)

FIG. 10-15   *Courtesy of American Institute of Steel Construction*

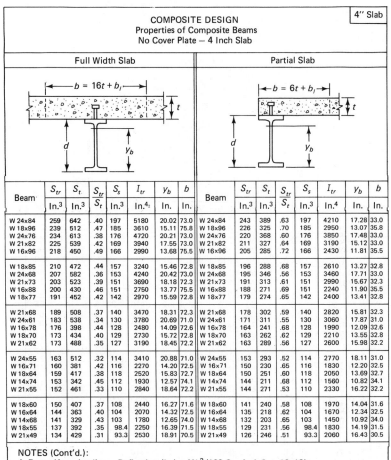

COMPOSITE DESIGN
Properties of Composite Beams
No Cover Plate — 4 Inch Slab

4″ Slab

Full Width Slab $b = 16t + b_f$

Partial Slab $b = 6t + b_f$

Beam	S_{tr} In.³	S_t In.³	$\dfrac{S_{tr}}{S_t}$	S_s In.³	I_{tr} In.⁴	y_b In.	b In.	Beam	S_{tr} In.³	S_t In.³	$\dfrac{S_{tr}}{S_t}$	S_s In.³	I_{tr} In.⁴	y_b In.	b In.
W 24x84	259	642	.40	197	5180	20.02	73.0	W 24x84	243	389	.63	197	4210	17.28	33.0
W 18x96	239	512	.47	185	3610	15.11	75.8	W 18x96	226	325	.70	185	2950	13.07	35.8
W 24x76	234	613	.38	176	4720	20.21	73.0	W 24x76	220	368	.60	176	3850	17.48	33.0
W 21x82	225	539	.42	169	3940	17.55	73.0	W 21x82	211	327	.64	169	3190	15.12	33.0
W 16x96	218	450	.49	166	2990	13.68	75.5	W 16x96	205	285	.72	166	2430	11.81	35.5
W 18x85	210	472	.44	157	3240	15.46	72.8	W 18x85	196	288	.68	157	2610	13.27	32.8
W 24x68	207	582	.36	153	4240	20.42	73.0	W 24x68	195	346	.56	153	3460	17.71	33.0
W 21x73	203	523	.39	151	3690	18.18	72.3	W 21x73	191	313	.61	151	2990	15.67	32.3
W 16x88	200	430	.46	151	2750	13.77	75.5	W 16x88	188	271	.69	151	2240	11.90	35.5
W 18x77	191	452	.42	142	2970	15.59	72.8	W 18x77	179	274	.65	142	2400	13.41	32.8
W 21x68	189	508	.37	140	3470	18.31	72.3	W 21x68	178	302	.59	140	2820	15.81	32.3
W 24x61	183	538	.34	130	3780	20.69	71.0	W 24x61	171	311	.55	130	3060	17.87	31.0
W 16x78	176	398	.44	128	2480	14.09	72.6	W 16x78	164	241	.68	128	1990	12.09	32.6
W 18x70	173	434	.40	129	2730	15.72	72.8	W 18x70	163	262	.62	129	2210	13.55	32.8
W 21x62	173	488	.35	127	3190	18.45	72.2	W 21x62	163	289	.56	127	2600	15.98	32.2
W 24x55	163	512	.32	114	3410	20.88	71.0	W 24x55	153	293	.52	114	2770	18.11	31.0
W 16x71	160	381	.42	116	2270	14.20	72.5	W 16x71	150	230	.65	116	1830	12.20	32.5
W 18x64	159	417	.38	118	2520	15.83	72.7	W 18x64	150	251	.60	118	2050	13.69	32.7
W 14x74	153	342	.45	112	1930	12.57	74.1	W 14x74	144	211	.68	112	1560	10.82	34.1
W 21x55	152	461	.33	110	2840	18.64	72.2	W 21x55	144	271	.53	110	2330	16.22	32.2
W 18x60	150	407	.37	108	2440	16.27	71.6	W 18x60	141	240	.58	108	1970	14.04	31.6
W 16x64	144	363	.40	104	2070	14.32	72.5	W 16x64	135	218	.62	104	1670	12.34	32.5
W 14x68	141	329	.43	103	1780	12.65	74.0	W 14x68	132	203	.65	103	1450	10.92	34.0
W 18x55	137	392	.35	98.4	2250	16.39	71.5	W 18x55	129	231	.56	98.4	1830	14.19	31.5
W 21x49	134	429	.31	93.3	2530	18.91	70.5	W 21x49	126	246	.51	93.3	2060	16.43	30.5

NOTES (Cont'd.):
4. For uniform loading: Deflection (in.) = $ML^2/160\,Sy_b$ (ref. Sec. 10–13)
      End reaction (kips) = $0.33S_{tr}F_b/L$ (ref. Sec. 10–12)
       where M = kip-ft., L = ft., S = in.³, y_b = in., S_{tr} = in.³, F_b = ksi
5. Ratio of S_{tr}/S_t at balanced design:

F_y (ski)	$f'c = 3.0$	3.75	4.0
36	.51	.59	.61
42	.44	.50	.52
45	.41	.47	.48
50	.37	.42	.44
55	.34	.38	.40
60	.31	.35	.36

Courtesy of American Institute of Steel Construction.

FIG. 10-15 *(Continued)*

4″ Slab

COMPOSITE
Properties of composite beams
With Cover Plate — 4 Inch Slab

$b = 16t + b_l$

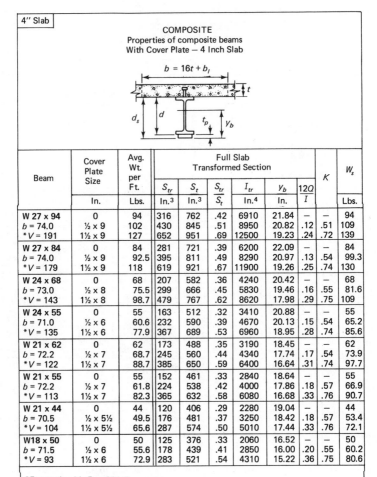

Beam	Cover Plate Size	Avg. Wt. per Ft.	Full Slab Transformed Section						K	W_s
			S_{tr}	S_t	$\dfrac{S_{tr}}{S_t}$	I_{tr}	y_b	$\dfrac{12Q}{I}$		
	In.	Lbs.	In.³	In.³		In.⁴	In.			Lbs.
W 27 x 94	0	94	316	762	.42	6910	21.84	—	—	94
b = 74.0	½ x 9	102	430	845	.51	8950	20.82	.12	.51	109
*V = 191	1½ x 9	127	652	951	.69	12500	19.23	.24	.72	139
W 27 x 84	0	84	281	721	.39	6200	22.09	—	—	84
b = 74.0	½ x 9	92.5	395	811	.49	8290	20.97	.13	.54	99.3
*V = 179	1½ x 9	118	619	921	.67	11900	19.26	.25	.74	130
W 24 x 68	0	68	207	582	.36	4240	20.42	—	—	68
b = 73.0	½ x 8	75.5	299	666	.45	5830	19.46	.16	.55	81.6
*V = 143	1½ x 8	98.7	479	767	.62	8620	17.98	.29	.75	109
W 24 x 55	0	55	163	512	.32	3410	20.88	—	—	55
b = 71.0	½ x 6	60.6	232	590	.39	4670	20.13	.15	.54	65.2
*V = 135	1½ x 6	77.9	367	689	.53	6960	18.95	.28	.74	85.6
W 21 x 62	0	62	173	488	.35	3190	18.45	—	—	62
b = 72.2	½ x 7	68.7	245	560	.44	4340	17.74	.17	.54	73.9
*V = 122	1½ x 7	88.7	385	650	.59	6400	16.64	.31	.74	97.7
W 21 x 55	0	55	152	461	.33	2840	18.64	—	—	55
b = 72.2	½ x 7	61.8	224	538	.42	4000	17.86	.18	.57	66.9
*V = 113	1½ x 7	82.3	365	632	.58	6080	16.68	.33	.76	90.7
W 21 x 44	0	44	120	406	.29	2280	19.04	—	—	44
b = 70.5	½ x 5½	49.5	176	481	.37	3250	18.42	.18	.57	53.4
*V = 104	1½ x 5½	65.6	287	574	.50	5010	17.44	.33	.76	72.1
W18 x 50	0	50	125	376	.33	2060	16.52	—	—	50
b = 71.5	½ x 6	55.6	178	439	.41	2850	16.00	.20	.55	60.2
*V = 93	1½ x 6	72.9	283	521	.54	4310	15.22	.36	.75	80.6

*For steels with F_y = 36 ksi.

NOTES:
1. Tables are based on specified concrete strength $f'c$ = 3.0 ksi and modular ratio n = 9.
2. Proper selection of S_{tr} value assures that stress in steel beam does
   not exceed allowable. Concrete stress due to bending does not
   exceed allowable under the following conditions:

F_y = 36 ksi	Condition	F_y = 50 ksi
S_{tr}/S_t	$f'c$ = 30 ksi	S_{tr}/S_t
⩽ .51	Shored construction	⩽ .37
⩽ 1.01	Unshored const. with M_L/M_D < 1.0	⩽ .74
⩽ .76	Unshored const. with M_L/M_D < 2.0	⩽ .55
⩽ .67	Unshored const. with M_L/M_D < 3.0	⩽ .49

3. For unshored construction see discussion that followed Fig. 10-15.

Courtesy of American Institute of Steel Construction.

FIG. 10-16

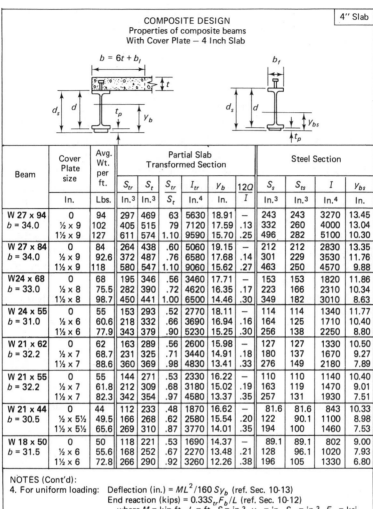

COMPOSITE DESIGN
Properties of composite beams
With Cover Plate — 4 Inch Slab

4″ Slab

$b = 6t + b_f$

Beam	Cover Plate size	Avg. Wt. per ft.	Partial Slab Transformed Section						Steel Section			
			S_{tr}	S_t	$\frac{S_{tr}}{S_t}$	I_{tr}	y_b	$\frac{12Q}{I}$	S_s	S_{ts}	I	y_{bs}
	In.	Lbs.	In.³	In.³		In.⁴	In.		In.³	In.³	In.⁴	In.
W 27 x 94	0	94	297	469	63	5630	18.91	—	243	243	3270	13.45
b = 34.0	½ x 9	102	405	515	79	7120	17.59	.13	332	260	4000	13.04
	1½ x 9	127	611	574	1.10	9590	15.70	.25	496	282	5100	10.30
W 27 x 84	0	84	264	438	.60	5060	19.15	—	212	212	2830	13.35
b = 34.0	½ x 9	92.6	372	487	.76	6580	17.68	.14	301	229	3530	11.76
	1½ x 9	118	580	547	1.10	9060	15.62	.27	463	250	4570	9.88
W24 x 68	0	68	195	346	.56	3460	17.71	—	153	153	1820	11.86
b = 33.0	½ x 8	75.5	282	390	.72	4620	16.35	.17	223	166	2310	10.34
	1½ x 8	98.7	450	441	1.00	6500	14.46	.30	349	182	3010	8.63
W 24 x 55	0	55	153	293	.52	2770	18.11	—	114	114	1340	11.77
b = 31.0	½ x 6	60.6	218	332	.66	3690	16.94	.16	164	125	1710	10.40
	1½ x 6	77.9	343	379	.90	5230	15.25	.30	256	138	2250	8.80
W 21 x 62	0	62	163	289	.56	2600	15.98	—	127	127	1330	10.50
b = 32.2	½ x 7	68.7	231	325	.71	3440	14.91	.18	180	137	1670	9.27
	1½ x 7	88.6	360	369	.98	4830	13.41	.33	276	149	2180	7.89
W 21 x 55	0	55	144	271	.53	2330	16.22	—	110	110	1140	10.40
b = 32.2	½ x 7	61.8	212	309	.68	3180	15.02	.19	163	119	1470	9.01
	1½ x 7	82.3	342	354	.97	4580	13.37	.35	257	131	1930	7.51
W 21 x 44	0	44	112	233	.48	1870	16.62	—	81.6	81.6	843	10.33
b = 30.5	½ x 5½	49.5	166	268	.62	2580	15.54	.20	122	90.1	1100	8.98
	1½ x 5½	65.6	269	310	.87	3770	14.01	.35	194	100	1460	7.53
W 18 x 50	0	50	118	221	.53	1690	14.37	—	89.1	89.1	802	9.00
b = 31.5	½ x 6	55.6	168	252	.67	2270	13.48	.21	128	96.1	1020	7.93
	1½ x 6	72.8	266	290	.92	3260	12.26	.38	196	105	1330	6.80

NOTES (Cont'd):
4. For uniform loading: Deflection (in.) = $ML^2/160\,Sy_b$ (ref. Sec. 10-13)
End reaction (kips) = $0.33S_{tr}F_b/L$ (ref. Sec. 10-12)
where M = kip-ft., L = ft., S = in.³, y_b = in., S_{tr} = in.³, F_b = ksi
5. Ratio of S_{tr}/S_t at balanced design:

F_y (ksi)	$f'c = 3.0$	3.75	4.0
36	.51	.59	.61
42	.44	.50	.52
45	.41	.47	.48
50	.37	.42	.44
55	.34	.38	.40
60	.31	.35	.36

FIG 10-16 *(Continued)*

The partial slab listings are useful for the case of spandrel beams where the slab is on one side only and also as an aid to the designer in interpolating for narrow flange (b < full slab) composite beams. Properties from these tables are also used to estimate long-term creep deflections for full concrete slab composite sections (see Section 10-13).

The same types of tables for beams with cover plates are shown in Fig. 10-16.* These tables also list the average weight per foot; the maximum weight in pounds per linear foot at the center of the beam, including the cover plate, W_s; K, the theoretical cover-plate length factor; a $12Q/I$ factor; and the properties of the bare steel section alone.

K is a factor used to determine the theoretical length of cover plate required. Its value is exact for uniformly loaded simple beams. The theoretical cutoff point (end of cover plate) for any type of loading occurs where $M = M_{max}(S_{tr})_{\text{non-cover-plated}}/(S_{tr})_{\text{cover-plated}}$. According to the AISC Specification, L_{cp} actual, the length of cover plate required, equals the theoretical length, L_{cp}, plus a length as described in Section 7-4. For a uniformly loaded simple span, the theoretical length of cover plate in feet, $L_{cp} = KL$, where L is the span in feet.

$12Q/I$ tabulated values are used to compute the total force (kips) to be developed by the cover-plate end welds:

$$F = \left(\frac{12Q}{I}\right)M \qquad\qquad (10\text{-}20)$$

where F = total force to be developed by the cover plate end welds, kips
M = moment at the theoretical cutoff point, kip-ft
Q = statical moment of the cover plate area about the neutral axis of the transformed section, in.^{3}

$(12 Q/I)(V/12)$ gives the horizontal shear (klf) of beam to be developed by intermediate welds. V is the vertical shear (kips) at the theoretical cutoff point.

Properties for the bare steel section are given to aid in computation of deflections and construction load stresses. They are also used to interpolate for properties of a trial section where b or f_c' are different from those assumed to establish the tabular values.

Properties of Composite Beams with Cover Plate Tables can be

---

* These tables are from the AISC Manual, 7th ed. They include some of the older shapes that are no longer made. They do not include any of the new shapes currently available.

used (the errors are negligible) to obtain interpolated values in $\frac{1}{8}$-in. increments between listed cover-plate thicknesses. In fact, interpolation between tabulated slab widths and thicknesses for all tables may be used with negligible error. The effect of a slight change in b/n (concrete areas) is not significant (a reduction in concrete area of 60% reduces the moment of inertia by only 10 to 15%). Therefore, the tables may be used to arrive at a trial section for other values of f'_c, t, and b.

As a review, the following computed stress equations may be used: at the bottom of steel for the transformed section,

$$f_b = \frac{12M}{S_{tr}} \qquad\qquad (10\text{-}21a)$$

At the top of steel for the transformed section,

$$f_b = \frac{12M}{S_j} \qquad\qquad (10\text{-}21b)$$

At the top of concrete slab,

$$f_b = \frac{12M}{nS_t} \qquad\qquad (10\text{-}21c)$$

The AISC Manual, Seventh Edition, also presents a series of Composite Beam Selection Tables* (Fig. 10-17) for 24 beam sections with and without cover plates for full-width slabs only. Separate tables are given for slab thicknesses of 4, $4\frac{1}{2}$, 5, and $5\frac{1}{2}$ in. S_{tr} is listed in descending values. These tables are used to select a trial section. After trial selection, the designer uses the Properties of Composite Beams Tables to obtain the necessary design parameters to complete the design computations. Actually, the Properties of Composite Beams Tables can also be used to select a trial section since S_{tr} is listed there (for non-cover-plated sections) in descending order. When selecting a section, the designer should scan the sections listed above the required S_{tr} for a possible lighter steel shape. The Properties of Composite Beams Tables (Figs. 10-15 and 10-16) for non-cover-plated and cover-plated sections and Composite Beam Selection Tables (Fig. 10-17) from the AISC Manual, Seventh Edition, have been retained because of their usefulness. The Eighth Edition of the AISC Manual does not include Properties of Composite Beams Tables and includes Composite Beam Selection Tables for non-

---

* Since these tables are from the AISC Manual, 7th ed., they include some of the older shapes that are no longer made. They do not include any of the new shapes currently available.

4" Slab

COMPOSITE DESIGN
Composite Beam Selection Table
4 Inch Slab (Full Width)

S_{tr}	Section Beam	Section Cover Plate	Avg. Wt. per Foot	S_{tr}	Section Beam	Section Cover Plate	Avg. Wt. per Foot
In.³		In. x In.	Lb.	In.³		In. x In.	Lb.
1180	W 36 x 170	1½ x 10	202	281	W 27 x 84	0	84
1100	W 36 x 150	1½ x 10	184	271	W 18 x 45	1½ x 6	68.3
1030	W 36 x 135	1½ x 10	170	245	W 21 x 62	½ x 7	68.7
957	W 33 x 130	1½ x 10	165	238	W 16 x 40	1½ x 6	64.0
908	W 33 x 118	1½ x 10	154	232	W 24 x 55	½ x 6	60.6
872	W 36 x 170	½ x 10	177	224	W 21 x 55	½ x 7	61.8
788	W 36 x 150	½ x 10	158	220	W 18 x 35	1½ x 5	54.9
755	W 30 x 108	1½ x 9	140	207	W 24 x 68	0	68
722	W 30 x 99	1½ x 9	132	204	W 16 x 36	1¼ x 6	55.7
720	W 36 x 135	½ x 10	143	178	W 18 x 50	½ x 6	55.6
713	W 36 x 170	0	170	176	W 21 x 44	½ x 5½	49.5
678	W 27 x 102	1½ x 9	134	173	W 21 x 62	0	62
663	W 33 x 130	1½ x 10	138	166	W 18 x 45	½ x 6	50.7
652	W 27 x 94	1½ x 9	127	163	W 24 x 55	0	55
628	W 36 x 150	0	150	152	W 21 x 55	0	55
619	W 27 x 84	1½ x 9	118	143	W 14 x 30	1 x 5½	44.0
612	W 33 x 118	½ x 10	127	142	W 16 x 40	½ x 6	46.1
559	W 36 x 135	0	135	133	W 16 x 26	1 x 4½	37.4
513	W 30 x 108	½ x 9	116	132	W 16 x 36	½ x 6	42.3
512	W 33 x 130	0	130	131	W 18 x 35	½ x 5	40.0
479	W 24 x 68	1½ x 8	98.7	125	W 18 x 50	0	50
479	W 30 x 99	½ x 9	107	120	W 21 x 44	0	44
461	W 33 x 118	0	118	112	W 18 x 45	0	45
457	W 27 x 102	½ x 9	110	105	W 14 x 22	1 x 4	32.2
430	W 27 x 94	½ x 9	102	104	W 14 x 30	½ x 5½	35.8
395	W 27 x 84	½ x 9	92.5	96.3	W 16 x 26	½ x 4½	30.8
389	W 30 x 108	0	108	92.8	W 16 x 40	0	40
385	W 21 x 62	1½ x 7	88.7	86.2	W 18 x 35	0	35
367	W 24 x 55	1½ x 6	77.9	82.8	W 16 x 36	0	36
365	W 21 x 55	1½ x 7	82.3	75.8	W 12 x 19	1 x 3	26.3
354	W 30 x 99	0	99	75.6	W 14 x 22	½ x 4	26.3
344	W 27 x 102	0	102	63.4	W 14 x 30	0	30
316	W 27 x 94	0	94	59.5	W 16 x 26	0	26
299	W 24 x 68	½ x 8	75.5	56.5	W 12 x 19	½ x 3	22.0
287	W 21 x 44	1½ x 5½	65.6	46.4	W 14 x 22	0	22
283	W 18 x 50	1½ x 6	72.9	36.8	W 12 x 19	0	19

Courtesy of American Institute of Steel Construction.

FIG. 10-17

cover-plated sections only. Should the reader decide to use the tables presented in this text, a preliminary check as to availability should be made (i.e., check AISC Manual, Eighth Edition, Part 1, Shape Dimensions and Properties Tables). If the shape is no longer made, a new shape may be chosen and its composite properties may be calculated or obtained from other sources.

For full composite action: for $V_h = A_s F_y / 2$ (Eq. 10-2), $V_h / q = N_s$ = number of mechanical connectors necessary for full composite action between the point of maximum moment and zero moment based on steel section.

$$N_s = \frac{W_s F_y (144)}{490(q)2} = W_s \left(\frac{0.147 F_y}{q} \right) = U_s W_s \qquad (10\text{-}22a)$$

where W_s is the maximum weight per foot of the steel section in pounds. For cover-plated sections this is the weight per foot (plf) at the center of the beam, including the cover plate (W_s is listed in the Properties of Composite Beams Tables). A_s in in.^{2} = $W_s(144)/490$ (490 pcf is the weight of steel). q is the allowable horizontal shear (kips) as determined by Table 10-1 and/or Table 10-2.

For $V_h = 0.85 f_c' A_c / 2$ (Eq. 10-13), $V_h / q = N_c$ = number of mechanical connectors necessary for full composite action between point of maximum moment and zero moment based on concrete section. A_c is the effective concrete flange area = bt (in.^{2}):

$$N_c = \frac{0.85 f_c' A_c}{2} = \frac{A_c (0.425 f_c')}{q} = U_c A_c \qquad (10\text{-}22b)$$

Using Eqs. 10-22a and b and the values of q obtained from Table 10-1 and/or Table 10-2, the coefficients U_s and U_c can be evaluated (Table 10-3). To determine the number of connectors to be used between the point of maximum moment and the point of zero moment for full composite action, the least value of $W_s U_s$ or $A_c U_c$ is used. U_s and U_c are coefficients used to obtain N_s and N_c, respectively.

10-16 DESIGN EXAMPLES

The design examples presented illustrated the previously discussed design procedures and the use of the various design aids and tables. Two types of composite construction, encased and mechanically connected, will be discussed along with full and partial composite action. A summary at the end of the section will make apparent the economies involved in the basic assumptions by listing the steel shape designation and the number of shear connectors utilized for each design assumption.

Table 10-3
MECHANICAL SHEAR CONNECTOR COEFFICIENTS
(FOR STONE CONCRETE)

Connector	U_s						U_c		
	F_y = 36 Ksi			F_y = 50 Ksi			All values of F_y		
	$f'c$ (Ksi)			$f'c$ (Ksi)			$f'c$ (Ksi)		
	3.0	3.5	⩾4.0	3.0	3.5	⩾4.0	3.0	3.5	⩾4.0
$\frac{1}{2}$ in. diam. x 2 in. hooked or headed stud	1.038	0.963	0.897	1.442	1.337	1.246	0.250	0.270	0.288
$\frac{5}{8}$ in. diam. x $2\frac{1}{2}$ in. hooked or headed stud	0.662	0.616	0.575	0.919	0.855	0.799	0.160	0.173	0.185
$\frac{3}{4}$ in. diam. x 3 in. hooked or headed stud	0.461	0.424	0.398	0.639	0.588	0.553	0.111	0.119	0.128
$\frac{7}{8}$ in. diam. x $3\frac{1}{2}$ in. hooked or headed stud	0.339	0.315	0.294	0.471	0.438	0.408	0.082	0.089	0.094
C3 x 4.1*	1.231	1.126	1.058	1.710	1.564	1.469	0.297	0.316	0.340
C4 x 5.4*	1.150	1.058	0.998	1.597	1.469	1.386	0.277	0.298	0.320
C5 x 6.7*	1.080	0.998	0.945	1.500	1.386	1.312	0.260	0.281	0.304

Notes: 1. For *lightweight concrete* divide any of the tabular values by the corresponding reduction coefficient obtained from **Table 10-2**.
2. *multiply tabular values by the length of channel shear connector in inches to obtain the corrected Mechanical Shear Connector Coefficient.
3. The tabular values can be used for studs longer than shown.
4. $N_s = U_s W_s$; $N_c = U_c A_c = U_c(\text{bt})$; and N_1, the number of shear connectors required between point of maximum moment and point of zero moment, is the lesser value of N_s and N_c.

EXAMPLE 10-1

For the typical bay warehouse framing plan shown in Fig. 10-18, design the interior beam, *AB,* for the following loading conditions: LL = 300 psf; partition = 0 psf (normally ≈20 psf); ceiling = 6 psf; and mechanical = 4 psf. Assume that w = 110 pcf, f'_c = 3.0 ksi lightweight concrete (rotary-kiln-produced aggregates conforming to ASTM C330), and a 4-in. slab, and design for each of the following cases for encased composite action: **(1)** shored, F_y = 36-ksi steel, **(2)** shored, F_y = 50-ksi steel, **(3)** unshored, F_y = 36-ksi steel, and **(4)** unshored, F_y = 50-ksi steel.

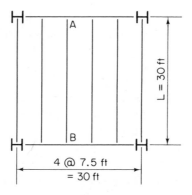

FIG. 10-18

solution:

See Eq. 10-1: $E_c = w^{1.5}(33)(f_c')^{1/2} = (110)^{1.5}(33)(3000)^{1/2} = 2{,}085{,}276$ psi $[(110)^{1.5} = (1.10 \times 10^2)^{1.5} = 1.10^{1.0} \times 1.10^{0.5} \times 10^3 = 1153.6896]$, and $n = E_s/E_c = 29{,}000{,}000/2{,}085{,}276 = 14$.

(1) Shored, $F_y = 36$-ksi steel:
$F_b = 0.76F_y$ (Section 10-9) and $S_s = M_{max}/F_b$.
Loads applied before concrete hardens:

4-in. lightweight concrete slab	
$= 0.110(7.5)0.33$	$= 0.272$
assume steel and concrete	
encasement	$= \underline{0.275}$
	0.527 kip/ft

$$M_{DL} = \frac{0.527(30)^2}{8} = 61.5 \text{ kip-ft}$$

Loads applied after concrete hardens:

$$LL = 0.300 \text{ kip/ft}^2$$
$$\text{partitions} = 0.0$$
$$\text{ceiling} = 0.006$$
$$\text{mechanical} = \underline{0.004}$$
$$0.310 \text{ kip/ft}^2$$

$$M_{LL} = \frac{0.310(7.5)(30)^2}{8} = 261.6 \text{ ft-kips}$$

$$M_{LL} + M_{DL} = 323 \text{ kip-ft}$$

See Eq. 10-10: $S_{s\,req} = 12(323)/0.76(36) = 142$ in.^{3}.
From the AISC Manual ASD Selection Tables, choose
W 24 × 68 ($S_x = 154$ in.^{3} > 142 in.^{3}). If the section is
okay for $0.76F_y$, no shoring is required.

(2) Shored, $F_y = 50$-ksi steel:
$S_{s\,req} = 142(36/50) = 102$ in.^{3}. Choose W 24 × 55 ($S_x
= 114$ in.^{3} > 102 in.^{3}). If the section is okay for $0.76F_y$,
no shoring is required.

(3) Unshored, $F_y = 36$-ksi steel:
To obtain a trial section: $S_{s\,req} = 142$ in.^{3} (design case
1). From the ASD Selection Table, choose a beam with
slightly less S_s value; *try* W 24 × 62 ($S_s = 131$ in.^{3}).
Effective width, b: $2(8t) + 12 = 64 + 12 = 76$ in.
(governs); $l/4 = 12(30)/4 = 90$ in.; $2(78)/2 + 12 = 90$
in. $A_s = 18.2$ in.^{2}; $b_f = 7.040$ in.; $d = 23.74$ in.; $t_f =
0.590$ in.; $I_s = 1550$ in.^{4}; $n = 14$.

Neglecting any area of concrete stem, 12a (see
Fig. 10-19), and finding y_b from the table below, where
y is the distance from the section's center-of-gravity
axis to the top of the concrete slab, $y_b = Ay/A =
286.7/39.91 = 7.18$ in.

Section	A	y	Ay
Beam	18.2	13.37	243.3
Slab	21.71	2.0	43.4
	39.91		286.7

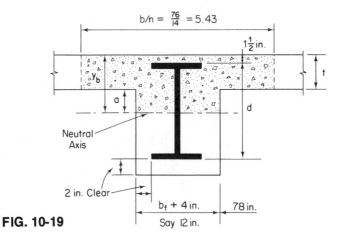

FIG. 10-19

Only the compression area of concrete (above the neutral axis) is considered in stress computations, although to compute the live-load deflection, the entire cross section may be used. $a = y_b - t = 7.18 - 4.0 = 3.18$ in. $A_{stem\,tr} = 12a/n = 2.7$ in.^{2}. Finding I_{tr} and S_{tr}:

Section	A	$\bar{y}$	$\bar{y}^2$	$A\bar{y}^2$	I_0
Beam	18.2	6.19	38.32	697.4	1550
Slab	21.71	5.18	26.83	582.5	29
Stem	2.73	1.59	2.53	6.90	32
				1286.8	1611
				$I_{tr} = 2888$ in.^{4}	

$\bar{y}$ is the distance from the center-of-gravity axis of the element (section) to the neutral axis of the transformed section $(y - y_b)$. I_0 is the moment of inertia of the element (section) about its own center-of-gravity axis. S_{tr} at top slab $= I_{tr}/y_b = 402.2$ in.^{3}. S_{tr} at bottom steel $= I_{tr}/(d + 1\frac{1}{2} - y_b) = 159.9$ in.^{3}.

From Eq. 10-9,

$$f_b = \frac{M_{DL}}{S_s} + \frac{M_{LL}}{S_{tr}} \leq 0.66F_y$$

$$= \frac{12(61.5)}{131} + \frac{12(261.6)}{159.9}$$

$$= 5.63 + 19.63 = 25.26 \text{ ksi} > 24 \text{ ksi}$$

$$f_b \text{ slab} = \frac{M_{LL}}{nS_{tr}} \text{ at top slab} = \frac{12(261.6)}{402.2(10)}$$

$$= 0.781 \text{ ksi} < 0.45f'_c = 1.35\text{ksi}$$

Use W 21 × 68 to avoid an overstress.

Case 3, an encased beam without shores with $F_y = 36$ ksi, was intentionally checked using Eq. 10-9, utilizing the actual geometric properties to compute I_{tr} (moment-of-inertia method). A more economical beam design and section selected with a lesser S_s value is apparent. If the encased section checks, using an allowable bending stress of $0.76F_y$ for the steel section alone $(M_{DL} + M_{LL}/S_s)$, it will check using the less conservative approach of Eq. 10-9. The same would hold true for the shored case except that the long method of computation would involve Eq. 10-10.

(4) Unshored, F_y = 50-ksi steel:
Same design as case 2 (short method of computation). *Use* W 24 × 55.
The following table summarizes Example 10-1:

Case	F_y	Steel Section	Savings
1	36 shored	W 24 × 68[a]	
2	50 shored	W 24 × 55	13 p/f
3	36 unshored	W 21 × 68	3 in. depth
4	50 unshored	W 24 × 55	13 p/f

[a] W 24 × 62 required (1% overstress) using Eq. 10-10.

EXAMPLE 10-2

Same data as given in Example 10-1. Design a flat soffit slab (no haunch as in section 1–1 of Fig. 10-1) utilizing mechanical shear connectors with unshored construction, if possible, for the following cases: **(1)** full composite, F_y = 36-ksi steel; **(2)** full composite, F_y = 50-ksi steel; **(3)** partial composite, F_y = 36-ksi steel; and **(4)** partial composite, F_y = 50-ksi steel.

solution:

Loads applied before concrete hardens:

$$4\text{-in. lightweight concrete slab} = 0.272$$
$$\underline{\text{Assume steel section} = 0.080}$$
$$0.352 \text{ kip/ft}$$

Loads applied after concrete hardens:

$$0.310 \text{ kip/ft}^2 \quad \text{(Example 10-1)}$$
$$M_{DL} = \frac{0.352(30)^2}{8} = 39.6 \text{ kip-ft}$$
$$M_{LL} = 261.6 \text{ kip-ft (Example 10-1)}$$
$$M_{DL} + M_{LL} = 301.2 \text{ kip-ft}$$
$$n = 14 \text{ (Example 10-1)}$$

(1) Full composite, F_y = 36 ksi:
  (a) Section choice: from Eq. 10-11b, for unshored construction, the required section moduli are

$$S_s = \frac{12M_{DL}}{0.66F_y} = \frac{12(39.6)}{24} = 19.8 \text{ in.}^3$$

$$S_{tr} = \frac{12(M_{DL} + M_{LL})}{0.66F_y} = \frac{12(301.2)}{24} = 150.6 \text{ in.}^3$$

For a 4-in. slab, from Fig. 10-15:
Select W 21 × 55. From AISC Manual, 8th ed., Dimensions and Properties Tables, this section is no longer rolled. Therefore, select W 24 × 55: S_{tr} = 164 in.^{3} (computed from Eqs. 10-18 and 10-19) > 150.6 in.^{3}; S_s = 114 in.^{3} > 19.8 in.^{3}.

(b) Check concrete stress: M_{LL}/M_{DL} = 261.6/39.6 = 6.61 ≮ 3.0. Therefore, the concrete stress-check tables at the bottom of Fig. 10-15 cannot be used. For a W 24 × 55, S_t = 504 in.^{3} (computed) and $(f_b)_{concrete}$ = 12(M_{DL} + M_{LL})/$S_t n$ = 12(301.2)/504(9) = 0.797 ksi < 0.45f_c' = 1.35 ksi (n = 9 for strength may be used).

(c) Check for shoring requirement (Fig. 10-15): M_{LL}/M_{DL} = 6.61 > 2.74 and Eq. 10-12 need not be checked (e.g., S_{tr} ≤ [1.35 + 0.35(6.61)]114 = 417.6 in.^{3} > 152 in.^{3}). Therefore, no temporary shoring is required.

(d) Deflection (Fig. 10-20): from Properties for Designing Tables, AISC Manual: For a W 24 × 55: A = 16.2 in.^{2} = A_s; d = 23.57 in.; b_f = 7.005 in.; t_f = 0.505 in.; I = 1350 in.^{4} = I_s.

For the effective width of the concrete slab, b: $L/4$ = 12(30)/4 = 90 in.; spacing = 12(7.5) = 90 in.; 2(8t) + b = 64 + 7.005 = 71.0 in. (governs).

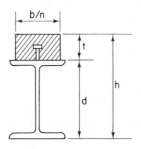

FIG. 10-20

From Section 10-13, Eqs. 10-18 and 10-19:

$$\phi = \frac{bt}{nA_s} = \frac{71.0(4)}{14(16.2)} = 1.25$$

$$y_b = \left(\frac{\phi}{\phi + 1}\right)\left(h - \frac{t}{2}\right) + \left(\frac{d}{2}\right)\left[\frac{1}{\phi + 1}\right]$$

$$= \left(\frac{1.25}{2.25}\right)(27.57 - 2.0) + \frac{11.79(1)}{2.25}$$

$$= 14.21 + 5.24 = 19.45 \text{ in.}$$

$$I_{tr} = I_s + A_s\left[y_b - \frac{d}{2}\right]^2 + \frac{bt^3}{12n}$$

$$+ \frac{bt}{n}\left[d + \frac{t}{2} - y_b\right]^2$$

$$= 1350 + 16.2(19.45 - 11.79)^2$$

$$+ \frac{71.0(4)^3}{12(14)} + \frac{71.0(4)}{14}.$$

$$[(23.57 + 2.0 - 19.45)]^2$$

$$= 1350 + 950 + 27 + 760 = 3087 \text{ in.}^4$$

The live-load deflection (Eq. 10-16c) is

$$\Delta_{LL} = \frac{M_{LL}L^2}{160 I_{tr}} = \frac{261.6(30)^2}{160(3087)}$$

$$= 0.477 \text{ in.} < \frac{L}{360} = 1.0 \text{ in.}$$

Since in a warehouse, the sustained load is likely to be larger than 50% of the total load, long-term creep deflection is significant. I_{tr} should be based on a value of $2n$ for this case. For $n = 28$, $\phi = 0.625$:

$$y_b = \left(\frac{0.625}{1.625}\right)(27.57 - 2.0) + 11.79\left(\frac{1}{1.625}\right)$$

$$= 9.83 + 7.26 = 17.09 \text{ in.}$$

$$I_{tr} = 1350 + 16.2(17.09 - 11.79)^2$$

$$+ 14 + \frac{71.0(4)}{28}.$$

$$[(23.57 + 2.0 - 17.09)]^2$$

$$= 1350 + 455 + 14 + 729 = 2534 \text{ in.}^4$$

$$\Delta_{LL} = 0.477 \left(\frac{3087}{2534} \right) = 0.581 \text{ in.}$$

$$d \not< \left(\frac{F_y}{800} \right) L = \left(\frac{36}{800} \right) 360$$

$$= 16.2 < 23.57 + 4$$

The dead-load deflection (Eq. 10-16b) is

$$\Delta_{DL} = \frac{M_{DL} L^2}{160 I_s} = \frac{39.6(30)^2}{160(1350)}$$
$$= 0.165 \text{ in.} < \text{say 1.5 in.}$$

(e) Shear connectors/connector load values: *select* for use $\frac{3}{4}$-in.-diameter × 3-in.-headed stud shear connectors and from Tables 10-1 and 10-2 for w = 110 pcf and $f_c' = 3.0$ ksi: $q = 11.5(0.83) = 9.55$ kips/stud.

The maximum diameter for studs located other than directly over the beam web (Section 10-8) is $2.5(0.505) = 1.263$ in. $> \frac{3}{4}$ in. diameter.

(f) Horizontal shear (full composite action): from Eqs. 10-2 and 10-3, respectively:

$$V_h = \frac{16.2(36)}{2} = 291.6 \text{ kips} \quad \text{(governs)}$$

$$V_h = \frac{0.85(3.0)(71.0)4}{2} = 362.1 \text{ kips}$$

(g) Number of studs from point of maximum moment (center line) and zero moment (supports): V_h/q = 291.6/9.55 = 30.5; *use* 31 studs each half of beam, a total of 62 studs (Fig. 10-21c).

Spacing (see Fig. 10-10, Section 10-8):

(i) One stud across, Fig. 10-21a: longitudinal = 12(30)/61 = 5.90 in. $< 8t > 6d$.

(ii) Two studs across, Fig. 10-21b (as an alternative): longitudinal = 12(30)/30 = 12.0 in. $< 8t > 6d$; transverse = 7.055 − 3.0 = 4.055 in. $< 8t > 4d$.

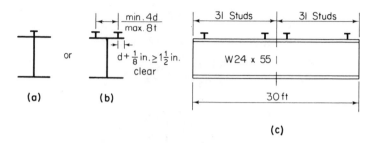

FIG. 10-21

Note: Computing (Eq. 10-19), $I_{tr} = 3368$ in.[4]
and from Eq. 10-19 (step d), $I_{tr} = 3087$ in.[4]. Therefore, lightweight concrete ($w = 110$ pcf, $f'_c = 3.0$
ksi) reduces the stiffness of the transformed section by 10%.

(2) Full composite, $F_y = 50$ ksi:

(a) Section choice: from Eq. 10-11b, for unshored
construction, the required section moduli are S_s
$= 14.4$ in.^{3}; $S_{tr} = 109.5$ in.^{3}.

Select, from Composite Beam Selection Ta-
bles, Fig. 10-17, W 21 × 44: $S_{tr} = 120$ in.$^3 > 109.5$
in.^{3}; $S_s = 81.6$ in.$^3 > 14.4$ in.^{3}.

(b) Check concrete stress: $M_{LL}/M_{DL} = 6.61 \nless 3.0$; for
a W 21 × 44, $S_t = 406$ in.^{3}, and f_b concrete $= 0.989$
ksi $< 0.45f'_c = 1.35$ ksi.

(c) Check for shoring requirement: $M_{LL}/M_{DL} = 6.61 >$
2.74 and Eq. 10-12 need not be checked. No temporary shoring is required.

(d) Deflection: for a W 21 × 44, $A = 13.0$ in.$^2 = A_s$; d
$= 20.66$ in.; $b_f = 6.500$ in.; $t_f = 0.450$ in.; $I = 843$
in.$^4 = I_s$, $b = 64 + 6.5 = 70.5$ in. $\phi = 1.55$; $y_b =$
$13.77 + 4.05 = 17.82$ in.

$$I_{tr} = 843 + 13.0(17.82 - 10.33)^2$$

$$+ \frac{70.5(4)^3}{12(14)} + \frac{70.5(4)}{14}$$

$$[(20.66 + 2.0 - 17.82)]^2$$

$$= 843 + 729 + 27 + 472 = 2071 \text{ in.}^4$$

$\Delta_{LL} = 0.710$ in. < 1.0 in.
$\Delta_{DL} = 0.264$ in. < 1.5 in.

For $n = 28$, $\phi = 0.775$:

$$y_b = \left(\frac{0.775}{1.775}\right)(24.66 - 2.0) + 10.33\left(\frac{1}{1.775}\right)$$

$$= 9.89 + 5.82 = 15.71 \text{ in.}$$

$$I_{tr} = 843 + 13.0(15.71 - 10.33)^2$$

$$+ 14 + \left[\frac{70.5(4)}{28}\right](20.66 + 2.0 - 15.71)^2$$

$$= 843 + 376 + 14 + 486 = 1719 \text{ in.}^4$$

$$\Delta_{LL} = 0.855 \text{ in.} < 1.0 \text{ in.}$$

$$\Delta_{DL} = 0.264 \text{ in.}$$

(e) Shear connectors/connector load values: $q = 9.55$ kips/stud; $2.5(0.450) = 1.13$ in. $> \frac{3}{4}$ in. diameter.

(f) Horizontal shear (full composite action):

$$V_h = \frac{13.0(50)}{2} = 325 \text{ kips} \quad \text{(governs)}$$

$$V_h = \frac{0.85(3.0)(70.5)4}{2} = 359.6 \text{ kips}$$

(g) Number of studs from point of maximum moment (center line) and zero moment (supports): $V_h/q = 325/9.55 = 34.03$ kips; use 35 studs each half of beam, a total of 70 studs. Spacing:

    (i) One stud across: longitudinal $= 12(30)/69 = 5.22$ in. $< 8t > 6d$.

    (ii) Two studs across: longitudinal $= 12(30)/34 = 10.58$ in. $< 8t > 6d$; transverse $= 6.50 - 3 = 3.50$ in. $< 8t > 4d$.

    Note: From the Composite Design Properties Tables, $I_{tr} = 2280$ in.^{4} and from Eq. 10-19 (step d), $I_{tr} = 2071$ in.^{4}.

    Lightweight concrete reduced the stiffness of the transformed section by $\approx 10\%$.

(3) Partial composite, $F_y = 36$ ksi: W 24 × 55 (see case 1):

(a) Horizontal shear: from Eq. 10-13: $S_{eff} = S_s + [V_h'/V_h]^{1/2}(S_{tr} - S_s)$ or $V_h' = V_h[(S_{eff} - S_s)/(S_{tr} - S_s)]^2$; $V_h = 291.6$ kips (step 1f); $S_{eff} = S_{trreq} = 150.6$

in.³ (step 1a); $S_{tr\ actual} = 164$ in.³ (step 1a); $S_{s\ actual}$ $= 114$ in.³ (step 1a).

$$V_h' = 291.6 \left[\frac{(150.6 - 114)}{(164 - 114)} \right]^2 = 156.2 \text{ kips}$$

$V_h' > 0.25 V_h = 72.9$ kips (Eqs. 10-6 and 10-7)

(b) Shear connectors: number of studs = 156.2/9.55 = 16.4. Use 17 studs each half of the beam, 34 total. When compared to case 1, there is a savings of 28 studs for the beam span, indicating the savings with partial composite action.

Spacing:

(i) One stud across: longitudinal = 12(30)/33 = 10.91 in. < 8t > 6d.

(ii) Two studs across: longitudinal = 12(30/16 = 22.5 in. < 8t > 6d; transverse: 4.055 in. < 8t > 4d.

(c) Deflection (steps 1d, f): from Eq. 10-17:

$$I_{eff} = I_s + \left(\frac{V_h'}{V_h} \right)^{1/2} (I_{tr} - I_s)$$

$$= 1350 + \left(\frac{156.2}{291.6} \right)^{1/2} (3087 - 1350)$$

$$= 1350 + 1271 = 2621 \text{ in.}^4$$

$$\Delta_{LL} = 0.477 \left(\frac{3087}{2621} \right) = 0.562 \text{ in.} < 1.0 \text{ in.}$$

I_{eff} reduced from I_{tr} (stiffness reduced) by about 15% or Δ_{LL} increased by about 15%.

(4A) Partial composite, $F_y = 50$ ksi: W 21 × 44 (see case 2)

(a) Horizontal shear: from Eq. 10-13: $V_h' = V_h[(S_{eff} - S_s)/(S_{tr} - S_s)]^2$; $V_h = 325$ kips (step 2f); $S_{eff} = S_{tr\ req} = 109.5$ in.³ (step 2a); $S_{tr\ actual} = 120$ in.³ (step 2a). $S_{s\ actual} = 81.6$ in.³ (step 2a).

$$V_h' = 325 \left[\frac{(109.5 - 81.6)}{(120 - 81.6)} \right]^2 = 171.6 \text{ kips}$$

$V_h' > 0.25 V_h = 81.3$ kips

(b) Shear connectors: number of studs = 171.6/9.55 = 18.0. *Use* 18 studs each half of the beam, for a total of 36. When compared to case 2, there is a savings of 34 studs for the beam span, indicating a very definite savings with partial composite action.

Spacing:

(i) One stud across: longitudinal = 12(30)/35 = 14.4 in. < $8t$ > $6d$.

(ii) Two studs across: longitudinal = 12(30)/17 = 21.1 in. < $8t$ > $6d$; transverse = 6.50 − 3.0 = 3.5 in. < $8t$ > $4d$.

(c) Deflection (steps 2d, f): from Eq. 10-17:

$$I_{eff} = 843 + \left(\frac{171.6}{325}\right)^{1/2}(2071 - 843)$$

$$= 843 + 892 = 1735 \text{ in.}^4$$

$$\Delta_{LL} = 0.710\left(\frac{2071}{1735}\right) = 0.847 \text{ in.} < 1.0 \text{ in.}$$

I_{eff} decreased by 16% from I_{tr} for full composite action and Δ_{LL} increased by about the same percentage.

(4B) Select a deeper shape: W 24 × 55: S_{tr} = 164 in.³; S_s = 114 in.³; I_{tr} = 3368 in.⁴; I_s = 1350 in.⁴; A_s = 16.2 in.²; b_f = 7.005 in.

(a) Horizontal shear: V_h = 16.2(50)/2 = 405 kips. b = 71.0 in.; V_h = 0.85(3.0)(71.0)4/2 = 362.1 kips (governs).

From Eq. 10-13 (step 4Aa): V_h' = 362.1 [(109.5 − 114)/(164 − 114)]² = negative value. *Use* 0.25 (362.1) = 90.5 kips = V_h' (Eq. 10-7).

(b) Number of connectors: 90.5/9.55 = 9.5

Use 10 headed ¾-in.-diameter stud connectors per half of the beam span, for a total of 20. By using the deeper steel shape (W 24 × 55), an additional 16 connectors are saved, reflecting additional economy over W 21 × 44.

The only other advantage over the shallower (W 21 × 44, F_y = 50 ksi) section would be a decrease in the deflection. The weight is increased along with an increase of 3 in. in depth. There seems to be some advantage in using this larger depth.

The following table summarizes Example 10-2:

Case	Beam	F_y	No. Studs	% Composite	Δ_{DL}	Δ_{LL}
1	W 24 × 55	36	62	100	0.165	0.477
2	W 21 × 44	50	70	100	0.264	0.710
3	W 24 × 55	36	34	54	0.165	0.562
4A	W 21 × 44	50	36	53	0.264	0.847
4B	W 24 × 55	50	20	25	—	—

The summary clearly shows the advantages and disadvantages (deflection mainly) of partial composite action and the advantages and disadvantages of 50-ksi versus 36-ksi material.

EXAMPLE 10-3

Design the beam *AB* in Example 10-1 using a mechanically connected cover-plated beam. Assume the same loading, $F_y = 36$ ksi, and full composite action.

solution:

(a) Required section moduli: from Example 10-2: $S_s = 19.8$ in.^{3}; $S_{tr} = 150.6$ in.^{3}; $S_{ts} = 19.8$ in.^{3}.

(b) Section selection: from Section 10-13: from the top of the concrete to the bottom of the steel, $d/L = \frac{1}{22}$; $d \geq 12(30)/22 = 16.4$ in. or ≥ 12.4 steel section.

From Fig. 10-17, select the lightest section having $S_{tr} \geq 150.6$ in.^{3} and $S_s \geq 19.8$ in.^{3}, which is W 21 × 44 with PL $\frac{1}{2}$ × $5\frac{1}{2}$ ($S_{tr} = 176$ in.^{3}; 49.5 plf).

From Fig. 10-16, by interpolation, using W 21 × 44 with PL $\frac{3}{8}$ × $5\frac{1}{2}$ (48.1 plf): $S_{tr} = 162$ in.^{3}; $S_{tr}/S_t = 0.35$; $y_b = 18.57$ in.; $K = 0.43$ (coefficient to determine length of cover plate); $W_s = 51.1$ plf (maximum weight per foot); $S_s = 111.9$ in.^{3}; $S_{ts} = 88.0$ in.^{3}; $12Q/I = 0.14$; $y_{bs} = 9.32$ in.; $b = 70.5$ in.; $S_t = 462$ in.^{3}.

The allowable web shear, from Fig. 10-16: $V = 104$ kips > 40.2 kips (maximum shear in span); $S_{tr} = 162$ in.$^3 > 150.6$ in.^{3}; $S_{ts} = 88.0$ in.$^3 > 19.8$ in.^{3}.

(c) Check concrete stress (Example 10-2): for a W 21 × 44 with PL $\frac{3}{8} \times 5\frac{1}{2}$, $S_t = 462$ in.^{3} and

$$(f_b)_{concrete} = \frac{12(M_{DL} + M_{LL})}{S_t n}$$

$$= \frac{12(301.2)}{462(9)} = 0.869 \text{ ksi} < 0.45 F_c' = 1.35 \text{ ksi}$$

(d) Check for shoring requirement (Example 10-2): $M_{LL}/M_{DL} = 6.61 > 2.74$. No temporary shoring is required.

(e) Deflection: $M_{DL} = 39.6$ kip-ft; $M_{LL} = 261.6$ kip-ft.

$$\Delta_{DL} = \frac{M_{DL}L^2}{160 I_s} = \frac{M_{DL}L^2}{160 S_s y_{bs}}$$

$$= \frac{(39.6)(30)^2}{160(111.9)9.32} = 0.214 \text{ in.} < 1.50 \text{ in.}$$

$$\Delta_{LL} = \frac{M_{LL}L^2}{160 I_{tr}} = \frac{M_{LL}L^2}{160 S_{tr} y_b}$$

$$= \frac{261.6(30)^2}{160(162)18.57} = 0.489 \text{ in.} < 1.0 \text{ in.}$$

(f) Shear connectors (full composite action):

Use $\frac{3}{4}$-in.-diameter × 3-in.-headed studs; $t_f = 0.450$ in.; $2.5 t_f = 1.13$ in. $> \frac{3}{4}$ in.

Using the Table 10-3 method, for $F_y = 36$ ksi, $f_c' = 3.0$ ksi; $U_s = 0.461$ (stone concrete).

From Table 10-2 (lightweight): $U_s = 0.461/0.83 = 0.555$.

From Eq. 10-22a, $N_s = U_s W_s = 0.555(51.1) = 28.4$.

From Table 10-3, $U_c = 0.111$ (stone concrete); from Table 10-2, for lightweight concrete, $U_c = 0.111/0.83 = 0.134$; From Eq. 10-22b, $N_c = 0.134(4)70.5 = 37.7$ (N_s governs).

Use 58$\frac{3}{4}$-in.-diameter × 3-in.-headed studs (29 each point of maximum moment).

Spacing:

(i) One stud across (Fig. 10-21a): longitudinal $= 12(30)/57 = 6.3$ in. $< 8t > 6d$.

(ii) Two studs across (Fig. 10-21b): longitudinal $= 12(30)/28 = 12.9$ in. $< 8t > 6d$; transverse $= b_f - 3.00 = 6.5 - 3.0 = 3.5$ in. $< 8t > 4d$.

(g) Cover plate: KL = theoretical length = 0.43(30) = 12.9 ft; x = distance from end support to theoretical cutoff point = (30 − 12.9)/2 = 8.55 ft.

From Example 10-2, $w_{DL} + w_{LL}$ = 0.352 + 0.310(7.5) = 2.677 klf; M at theoretical cutoff point = [(2.677)8.55/2] (30 − 8.55) = 244.6 kip-ft; $F = (12Q/I)M$(Eq. 10-20), where F is the total force to be developed by the cover-plate end welds (kips). M is the moment at the theoretical cutoff point of the cover plate (kip-ft); $12Q/I$ is as tabulated in Fig. 10-16. F = 244.6(0.14) = 34.2 kips.

The AISC Specification for a minimum-size fillet weld: thicknesses: cover plate = $\frac{3}{8}$ in.; flange = 0.450 in. = $\frac{7}{16}$ in.

Therefore, minimum size = $\frac{3}{16}$-in. fillet. Using E70 electrodes, a $\frac{3}{16}$-in. fillet has a capacity of 0.3(70)0.707(0.1875) = 2.78 kips/in.

The weld length required = F/capacity = 34.2/2.78 = 12.3 in. (6.5 in. each side) each end of cover plate.

The weld length, per Fig. 7-5, is $a' = 2b_p$ = 2(5.5) = 11.0 in. each side at each end of cover plate, where b_p = width of cover plate (in.).

Use 11.0 in. each side at each end of cover plate.

For the intermittent welds on the cover plate (AISC Specification): minimum length = 4 (weld size) ≥ $1\frac{1}{2}$ in. *Use* $1\frac{1}{2}$-in. or $\frac{3}{16}$-in. fillet × $1\frac{1}{2}$ in.

$$V_c = \text{vertical shear at cutoff} = (w_{DL} + w_{LL})\left[\frac{L}{2} - x\right]$$

$$= 2.677(15 - 8.55) = 17.27 \text{ kips}$$

For the horizontal shear (kips/in.), $(12Q/I)/(V_c/12)$ = 0.14(17.27/12) = 0.20 kip/in. Required spacing of intermittent cover-plate welds = 2 (capacity of weld size/horizontal shear) = 2(2.78)1.5/0.20 = 41.7 in.

The maximum spacing permitted (AISC Specification) = $24t_p$ ≤ 12 in. = 24(0.375) = 9 in. < 12 in. (governs), where t_p = thickness of cover plate (actually thinnest of plates joined).

Use 9-in. spacing each side of cover plate at each end of cover plate. (Refer to Fig. 10-22 for details of the bottom cover plate.)

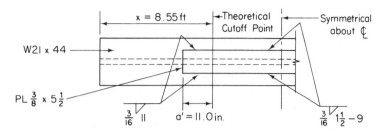

FIG. 10-22

Note: Step (g) of the design example deals with weld design. For a more detailed discussion of weld design, refer to Chapter 11.

The following table summarizes and compares full composite designs (Example 10-3):

Section	plf	Example	Case	F_y	No. Studs	Δ_{DL}	Δ_{LL}
W 24 × 55	55	10-2	1	36	62	0.165	0.477
W 21 × 44 plus PL $\frac{3}{8}$ × $5\frac{1}{2}$ × 14-9	51.5	10-3	—	36	58	0.214	0.489

Conclusion: The cover-plated beam reflects a 9% savings in weight. This savings in weight should be compared with the added costs of welding, fabrication, and handling to determine the true economies involved.

EXAMPLE 10-4

Assume the loading shown in Fig. 10-23 to be the total loading on an A36 fully composite steel beam. Assuming $f_c' = 3.0$-ksi stone concrete and a 4-in. concrete slab ($n = 9$), design the beam using mechanical shear connectors.

solution:

(a) Required section modulus: $M_{max} = 212.5$ kip-ft; $S_{tr} = 12M_{max}/0.66F_y = 12(212.5)/24 = 106.3$ in.^{3}. From Fig. 10-17, Composite Beam Selection Table, *select* W 21 × 44: $S_{tr} = 120$ in.$^3 > 106.3$ in.^{3}. Assume that S_s for the

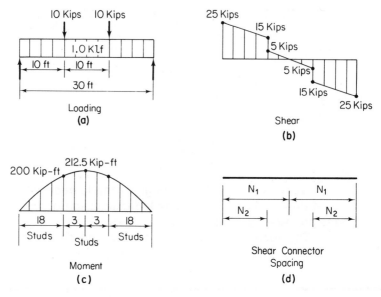

FIG. 10-23

section chosen is ample. Further assume that the concrete stress and deflections are within allowable limits and that no shoring is required. $S_s = 81.6$ in.³

(b) Shear connectors: assume controlling $b = 16t + b_f = 64 + 6.5 = 70.5$ in. From Table 10-1, for ¾-in.-diameter × 3-in.-headed shear connectors, $q = 11.5$ kips. The maximum diameter $= 2.5t_f = 2.5(0.450) = 1.13$ in. > ¾ in.

(c) Horizontal shear (full composite action); from Eqs. 10-2 and 10-3, $A_s = 13.0$ in.²

$$V_h = \frac{13.0(36)}{2} = 234 \text{ kips} \quad \text{(governs)}$$

$$V_h = \frac{0.85(3.0)\,(70.5)4}{2} = 359.6 \text{ kips}$$

(d) Number and spacing of connectors: $N_1 = V_h/q = 234/11.5 = 20.3$. *Use* 21. From Eq. 10-8, $N_2 = N_1 [(M\beta/M_{max}) - 1]/(\beta - 1)$; $M = 200$ kip-ft; $N_1 = 21$ − ¾-in.-diameter × 3-in. studs required between M_{max} and supports: $\beta = S_{tr}/S_s = 120/81.6 = 1.47$; $M_{max} = 212.5$ kip-ft. $N_2 = 21 [(200) (1.47/212.5) - 1]/(1.47 - 1.0) = 17.2$. *Use* 18.

The stud spacing must be adjusted to meet the N_2 requirement (number of studs between the concentrated load and end support). The balance, $N_1 - N_2$, is placed between the concentrated load and the point of maximum moment (the center line of the beam span). This spacing is shown in Fig. 10-23d.

Equation 10-8 checks whether N_1 (required to develop M_{max}) uniformly spaced connectors provide N_2 connectors between the concentrated load and the nearest point of zero moment. $M\beta/M_{max} = 1.384 \geq 1.0$ and $N_2 = 18 \leq N_1 = 21$. If $M\beta/M_{max} < 1.0$ or $N_2 \geq N_1$, Eq. 10-8 would not apply.

10-17 STEEL FORMED DECK: CRITERIA OF DESIGN/DESIGN EXAMPLES

Formed metal deck has been in common use as a component of floor systems for the past four decades. More recently, tests performed at Lehigh University have contributed to the development of metal deck in composite design. This composite system is composed of steel beams and concrete slabs with shear connectors welded through the deck to the beams. The corrugations of the steel deck can run parallel or perpendicular to the beam span. The latter case is the more common case in current design practice. The sketches in Fig. 10-24a and b show the system for ribs perpendicular and parallel to the span of the beam, respectively.

The general ground rules for the design criteria to follow have the following limitations:

(1) $h_r \leq 3.0$ in., where h_r is the rib height.

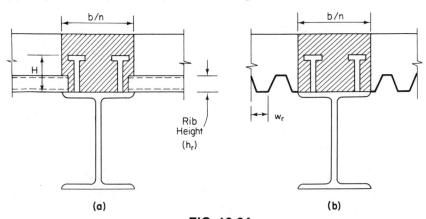

FIG. 10-24

(2) w_r minimum = 2.0 in., where w_r is the average rib width (for cal-
     culations ≤ width at top of deck).

(3) Stud shear connector diameter ≤ $\frac{3}{4}$ in. (welded directly through the
     metal deck or through prepunched holes in the metal deck).

(4) $H \geq h_r + 1\frac{1}{2}$ in.

(5) b, the effective width of concrete slab, should be computed by using
     the total thickness of concrete slab, t (including ribs).

(6) The slab thickness above the steel deck ≥ 2 in.

For case (a) of Fig. 10-24 (ribs normal to beam span):

(1) In Eq. 10-3, A_c, the area of concrete, should be computed by ne-
     glecting the area of concrete below the top of the steel deck. This
     area of concrete shall also be neglected when determining section
     properties. In other words, for the case where the ribs are normal
     to the beam span, only the concrete above the metal deck is con-
     sidered structurally effective.

(2) The allowable horizontal shear loads for stud shear connectors, as
     determined by Tables 10-1 and 10-2, shall be further reduced by
     multiplying the value obtained from Eq. 10-23a (value not to exceed
     1.0).

$$\left(\frac{0.85w_r}{h_r n^{1/2}}\right)\left[\frac{H}{h_r} - 1.0\right] \leq 1.0 \qquad (10\text{-}23a)$$

where H = height of the steel connector (in.) ($H \leq h_r + 3$ in. in
              Eq. 10-23 computations); actual H may > $h_r + 3$
      h_r = rib height, in.
      w_r = average width of rib, in.
      n = number of connectors per rib (for calculations ≤ 3);
              actual n may > 3

(3) The slab should be anchored to the steel beam to resist uplift by
     welded studs or welded studs and puddle welds at an interval not
     to exceed 16 in. Maximum stud spacing shall be 32 in. along the
     beam length.

For case (b) of Fig. 10-24 (ribs parallel to beam span):

(1) In Eq. 10-3, A_c, the area of concrete, should be computed by con-
     sidering all concrete below the top of the steel deck to be effective.
     The same area shall be used for determining section properties.

(2) It is preferred that the metal deck shall be split over the supporting
     steel beam to form a haunch (Fig. 10-24b). For nominal $h_r \geq 1\frac{1}{2}$ in.:

w_r of the haunch or rib over the beam ≥ 2 in. for the first stud in the tranverse row plus 4 stud diameters for each additional stud.

(3)   When $w_r/h_r > 1.5$, stud connector values as determined from Tables 10-1 and 10-2 may be used. When $w_r/h_r < 1.5$, the further reduction of stud connector values by multiplying by the value of Eq. 10-23b should be used.

$$\left(\frac{0.6w_r}{h_r}\right)\left[\frac{H}{h_r} - 1.0\right] \leq 1.0 \qquad\qquad (10\text{-}23b)$$

The use of metal deck in composite design with stone or lightweight concrete, with partial or full composite action, for shored or unshored construction, will reflect additional savings in the amount of stud shear connectors used. Its use usually does not significantly change strength or stiffness. This will be illustrated by the design example that follows.

Referring to Fig. 10-25, it is shown that for the case where the deck ribs are normal to the beam span (only the concrete above the deck is effective for composite action), Eqs. 10-18 and 10-19 adjust to become Eqs. 10-24 and 10-25, respectively.

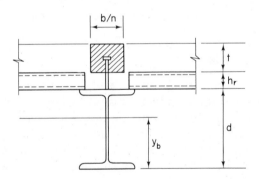

FIG. 10-25

ΣM about bottom of steel section:

$$\left(\frac{bt}{n}\right)\left(d + h_r + \frac{t}{2}\right) + A_s\left(\frac{d}{2}\right) = \left(\frac{bt}{n} + A_s\right)y_b$$

Assume that $\phi = bt/nA_s = A_c/nA_s$:

$$\phi A_s\left(d + h_r + \frac{t}{2}\right) + A_s\left(\frac{d}{2}\right) = (\phi A_s + A_s)y_b$$

$$A_s\left[\phi\left(d + h_r + \frac{t}{2}\right) + \frac{d}{2}\right] = A_s(\phi + 1)y_b$$

$$\phi\left(d + h_r + \frac{t}{2}\right) + \frac{d}{2} = (\phi + 1)y_b$$

$$y_b = \frac{(d/2)}{(\phi + 1)} + \left[\frac{\phi}{\phi + 1}\right]\left(d + h_r + \frac{t}{2}\right) \qquad (10\text{-}24)$$

Again, from Fig. 10-25,

$$I_{tr} = I_s + A_s\left[y_b - \left(\frac{d}{2}\right)\right]^2 + \frac{bt^3}{n(12)} + \frac{bt}{n}\left[d + h_r + \left(\frac{t}{2}\right) - y_b\right]^2 \qquad (10\text{-}25)$$

Property and selection tables for composite beams using steel formed deck are not given in the AISC Manual. Most manufacturers can provide tables of properties for beams utilizing metal deck. The metal deck may be composite or noncomposite with the slab. Both produce the same design composite beam properties. The section moduli for composite beams utilizing metal deck are about the same as those for a plain concrete slab of the total thickness. Thus, the plain slab selection tables, Fig. 10-17, can be used for preliminary estimates of S_{tr} when metal deck is used. Conversely, the various table available from manufacturers for composite design with metal deck can be used for plain slab composite design.

EXAMPLE 10-5

For the same data as given in Example 10-1, design a composite section utilizing headed stud-type mechanical shear connectors, unshored construction, and $1\frac{1}{2}$-in. metal deck spanning normal to the beam span for the following cases: (1) full composite, F_y = 36-ksi steel; (2) partial composite, F_y = 36-ksi steel. Neglect any slight change in dead load.

solution:

w = 100 pcf, f_c' = 3.0 ksi, n = 14, 4-in. slab. Assume that loading is the same as in Example 10-2: M_{DL} = 39.6 kip-ft, M_{LL} = 261.6 kip-ft.

(1)   Full Composite, F_y = 36-ksi steel:

    (a) Section choice [from Example 10-2, step (a)]:
           Select W 24 × 55; S_{tr} = 164 in.³; S_s = 114 in.³; b_f = 7.005 in.; t_f = 0.505 in.; I = 1350 in.⁴ = I_s; A = 16.2 in.² = A_s.

    (b) Strength properties (see Fig. 10-25): h_r = 1.5 in.; t = 2.5 in.; d = 23.57 in.; b = 71.0 in. [Example

10-2, step (d) and Section 10-17]; $\phi = bt/nA_s = 71.0(2.5)/9(16.2) = 1.22$ ($n = 9$ for all weights of concrete for strength calculations).
From Eq. 10-24,

$$y_b = \frac{(23.57/2)}{2.22} + \left(\frac{1.22}{2.22}\right) 26.32$$

$$= 5.31 + 14.46 = 19.77 \text{ in.}$$

From Eq. 10-25

$$I_{tr} = 1350 + 16.2(19.77 - 11.79)^2 + \frac{71.0(2.5)^3}{9(12)}$$

$$+ \left[\frac{71.0(2.5)}{9}\right] (23.57 + 1.5 + 1.25 - 19.77)^2$$

$$= 1350 + 1032 + 10 + 846 = 3228 \text{ in.}^4$$

At the bottom of the steel section:

$$S_{tr} = \frac{I_{tr}}{y_b} = \frac{3228}{19.77} = 163.3 \text{ in.}^3$$

At the top of the concrete slab:

$$S_t = \frac{I_{tr}}{(d - y_b + h_r + t)} = \frac{3228}{7.80} = 413.8 \text{ in.}^3$$

(c) Flexural stresses: (i) Concrete (unshored): S_{tr}/S_t = 164/413.8 = 0.396; M_{LL}/M_{DL} = 6.61 ≮ 3.0; therefore, the concrete check tables at the bottom of Fig. 10-15 cannot be used.
    $(f_b)_{concrete}$ = $12(M_{DL} + M_{LL}/S_t n)$ = 12(331.2)/ 413.8(9) = 1.07 ksi < 0.45f_c' = 1.35 ksi. (ii) Steel: S_{tr} = 163.3 in.^{3} > 150.6 in.^{3} [Example 10-2, step (1a)]; S_s = 114 in.^{3} > 19.8 in.^{3} [Example 10-2, step (1a)].
(d) Deflection: using $n = 14$, $\phi = 0.784$: from Eq. 10-24:

$$y_b = \frac{11.785}{1.784} + \left(\frac{0.784}{1.784}\right) (23.57 + 1.5 + 1.25)$$

$$= 6.61 + 11.57 = 18.18$$

From Eq. 10-25

$$I_{tr} = 1350 + 16.2(18.18 - 11.785)^2 + \frac{71.0(2.5)^3}{14(12)}$$

$$+ \left[\frac{71.0(2.5)}{14}\right] (23.57 + 1.5 + 1.25 - 18.18)^2$$

$$= 1350 + 663 + 7 + 366 = 2386 \text{ in.}^4$$

From Eq. 10-16c,

$$\Delta_{LL} = \frac{M_{LL}L^2}{160 I_{tr}} = \frac{261.6(30)^2}{160(2386)} = 0.617 \text{ in.} < \frac{L}{360} = 1.0 \text{ in.}$$

Neglect long-term creep in this problem. $\Delta_{DL} = 0.165$ in. $<$ say 1.5 in. [Example 10-2, step (1d)].

(e) Shear connectors/connector load values: from Eq. 10-23a (reduction factor to be applied to shear connector values obtained from Tables 10-1 and 10-2): $h_r = 1.5$ in.; assume that $w_r = 2.0$ (2.0 in. is minimum, Section 10-17); $H \leq h_r + 3.0$ in. $= 4.5$; $H \geq h_r + 1\frac{1}{2}$ in. $= 3.0$ in.; $d \leq \frac{3}{4}$ in.

Use $\frac{3}{4}$-in.-diameter $\times$ 3-in.-headed-stud shear connectors. Assume 2 connectors per rib.

$$\frac{0.85 w_r}{h_r n^{1/2}}\left(\frac{H}{h_r} - 1.0\right) \leq 1.0$$

$$\frac{0.85(2.0)}{1.5(2)^{1/2}}\left(\frac{3.0}{1.5} - 1.0\right) = 0.80 \leq 1.0$$

For $\frac{3}{4}$-in.-diameter $\times$ 3-in.-headed studs, $f'_c = 3.0$ ksi, $w = 110$ pcf from Tables 10-1 and 10-2: $q = 11.5(0.83) = 9.55$ kips/stud. The reduced q value $= 9.55(0.8) = 7.64$ kips/stud.

The maximum diameter for studs located other than directly over the beam web (Section 10-8): $2.5(0.505) = 1.26$ in. $> \frac{3}{4}$ in. diameter.

(f) Horizontal shear (full composite action): from Eqs. 10-2 and 10-3, respectively:

$$V_h = \frac{16.2(36)}{2} = 291.6 \text{ kips}$$

$$V_h = \frac{0.85(3.0)(71.0)(2.5)}{2} = 226.3 \text{ kips} \quad \text{(governs)}$$

(g) Number of studs from point of maximum moment (center line) and zero moments (supports); V_h/q = 226.3/7.64 = 29.6.

Use 30 studs per half of the beam span.

(2) Partial composite, F_y = 36-ksi steel: W 24 × 55

(a) Number of studs: from Eq. 10-13,

$$V_h' = V_h \left[\frac{(S_{eff} - S_s)}{(S_{tr} - S_s)} \right]^2$$

$$= 226.3 \left[\frac{(150.6 - 114)}{(163.3 - 114)} \right]^2 = 124.7 \text{ kips} > 0.25 V_h$$

q = 7.64 kips/stud; 124.7/7.64 = 16.3.

Use 17 studs each half of the beam span (a savings of 26 studs/beam).

(b) Deflection: from Eq. 10-17.

$$I_{eff} = 1350 + \left(\frac{124.7}{226.3} \right)^{1/2} (2386 - 1350)$$

$$= 1350 + 769 = 2119 \text{ in.}^4$$

$$\Delta_{LL} = \left(\frac{2386}{2119} \right) (0.617) = 0.695 \text{ in.} < 1.0 \text{ in.}$$

The following table summarizes Example 10-5:

Case	Beam	No. Studs	% Composite	Δ_{LL}
1	W 24 × 55	60 (62)	100 (100)	0.617 (0.477)
2	W 24 × 55	34 (34)	55 (54)	0.695 (0.562)

Note: Numbers in parentheses indicate comparable results for flat soffit slabs. (F_y = 36 ksi; Example 10-2). Utilizing steel-formed deck does not significantly change strength or stiffness.

PROBLEMS

(10-1) Design an encased composite floor beam (interior) for a 4-in. slab (non-coverplated, unshored construction) to meet the following criteria: live load = 150 psf; partition plus ceiling load = 25 psf; beam spacing = 10 ft; simple beam span = 32 ft. Use A36 steel and 3000-psi stone concrete.

(10-2) Determine the dead load and live load deflections for the beam chosen in Problem 10-1.

(10-3) Design the composite beam in Problem 10-1 using a mechanically connected non-coverplated beam. Assume full composite action and use stud-type shear connectors. Design the connectors also.

(10-4) Modify your solution to Problem 10-3 by designing for partial composite action, as permitted by the AISC Specification.

(10-5) (a) Design the interior beam using mechanically connected non-coverplated beam for the warehouse floor shown. Assuming full composite action, indicate the type, number, and spacing of shear connectors required. Assume beams simply supported and no temporary shoring. Live load is 200 psf; slab is 4 in. thick. Use A36 steel and 3000-psi stone concrete.
(b) Design the beam assuming no composite action, but full lateral support.

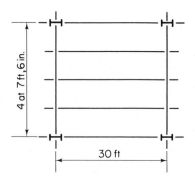

(10-6) Design a flat soffit slab using mechanical stud shear connectors in unshored construction to satisfy the following loading conditions. Assume full composite action, 4-in. slab, 3000-psi stone concrete, and A36 steel. The beam is a typical interior beam. Assume an effective slab width based on the slab thickness ($b = 16t + b_f$).

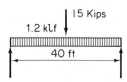

(10-7) Redesign the shear connectors in Problem 10-6 to provide "partial composite action" as permitted by the AISC specification.

(10-8) Using the same data as given in Problem 10-1, design a composite section using stud shear connectors in conjunction with a 2-in. metal deck spanning normal to the beam span. Assume unshored construction, full composite action, and A36 steel.

(10-9) (a) Is a W 21 × 57 supporting a 4-in. slab (3000-psi stone concrete) satisfactory if $M_D = 1100$ kip-in. and, $M_L = 2500$ kip-in.
Assume full composite action. The beams are 30 ft long and spaced at 8-ft intervals. Beams are simply supported. (A36 steel.)

(b) Determine the type and number of shear connectors needed to assure full composite action.

(c) What W shape would be required if there were no composite action? Assume full lateral support. (A36 steel.)

(10-10) Design a mechanically connected cover-plated beam (unshored) for the following loads and conditions: full composite action, A36 steel, and 3000-psi stone concrete. There is a 4-in. concrete slab. Beams are spaced at 7.5 ft and span is 28 ft. The loading is such that the dead load moment is 1800 kip-in. and the live load moment is 4800 kip-in.

11

connections

11-1   INTRODUCTION

We have already discussed the material and its use as structural elements (tension members, beams, columns, plate girders, etc.). The last structural element, the connection, is as important, if not more so, than the actual supporting structural members. Poorly or wrongly designed connections can lead to a failure and can affect the stresses of the main structural members. They affect the overall behavior of the structure. They are the last link to structural adequacy.

Too often, structural engineers have little to do with the actual design of connections. They usually specify the type of connection to be used, as well as the type of fastener to be employed. For economy, for simply supported members, beam reactions should be shown on the contract drawings. If reactions are not shown, the connections are designed for one-half the total uniform load capacity of the beam span.

The effect of any concentrated load must be taken into account when designing connections. Further, all beam reactions must be shown on the contract drawings for compositely designed and continuous beam framing.

The AISC Specification states that all plans and drawings must indicate the type of construction and type of connection assumed. Further supplementary data, such as assumed loads, shears, moments, axial forces, and other forces, should be shown so that members and their connections can be designed. This will enable one to prepare the shop drawings.

The shop drawings provide complete information necessary to fabricate the structural elements of the structure. They provide member and connection location, type, and size. These drawings clearly distinguish between shop-installed connectors and field-installed connectors. Shop drawings and design drawings were discussed in Section 1-12.

All connections carrying stress, with the exception of lacing, sag bars, and girts, shall be designed to support a minimum load of 6 kips. Connections at the ends of tension or compression members in trusses should develop the force due to the design, which should not be less than half the effective strength of the member.

Part 4 of the AISC Manual devotes itself to the subject of connections, and it will be referred to throughout this chapter.

11-2 ENGINEER–STEEL DETAILER– FABRICATOR RELATIONSHIP

The more cooperative with each other and the more considerate of the other, the more smoothly the design–detailing–fabricating operations will run and the more readily and economically the structure will be built.

The engineer should provide ample design information for the steel detailer so that he may prepare proper shop drawings. In turn, the detailer must convey ample information to the fabricator to facilitate the fabrication of the steel members and connections. In addition, the detailer and the fabricator should be granted freedom of choice of connection or connector to yield maximum economy.

11-3 SIMPLE (GRAVITY)

There are basically three types of construction or connections associated with various design criteria, as discussed in the AISC Specification. The first type will be discussed in two parts, the first part in this section and the second part in the next section. The first type is termed simple framing or type 2. *Simple framing* is unrestrained at the beam ends and in simple framing (gravity) this means that when a beam is subject to vertical load-ing (gravity) the ends of the beam are free to rotate under this gravity load. Its ends are connected for shear only (no moment capacity of the connection), and connection flexibility is a must. The column to which the beam is connected is not required to be designed for moment. The beam-to-column end connection must be capable of accomodating simple beam rotation, ϕ, without failure of the connection. Some examples of simple beam-to-column connections are shown in Fig. 11-1. The connec-tors are not shown in the connection because these may be bolted, riveted (although riveting is rarely used today), welded, or a combination of both welding and bolting. The examples shown are all assumed to be flexible connections. The beam-end rotation is achieved by the deformation of the angles. In bolted connections, additional flexibility is gained by elon-gation of the bolts after initial prestress is overcome.

In the seated connections, the top angle contributes lateral stability only and is sized accordingly, so as not to contribute to rigidity or stiff-ness of the connection and minimize rotational capabilities also. The angle used should be the thinnest possible.

For the case of riveted or bolted angle-frame connections (Fig. 11-1a), g, the gage on the outstanding angle legs should be as large as is practicable. For welded angle-frame connections, using longer outstand-ing angle legs accomplishes the same effect. For seated connections the same principles are applied.

In summary, for maximum flexibility, use minimum angle thickness, maximum g (for rivets and bolts), or maximum length of outstanding angle legs (for welded construction).

For end-plate shear connections, values of $g = 3\frac{1}{2}$ in. and $5\frac{1}{2}$ in. with a relatively thin plate (Fig. 11-1d) can provide ample rotational ca-

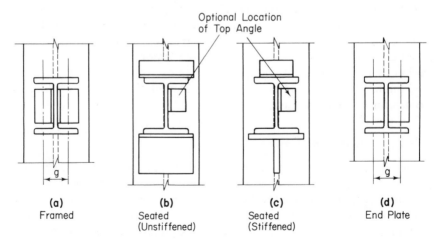

FIG. 11-1

pacity (comparable to that in Fig. 11-1a). The edge distance (from center line of fastener to edge of plate) should be more than $1\frac{1}{4}$ in., because if this is not the case, prying action on the bolt should be investigated (see Section 11-16). Prying action is an action that occurs in bolted connections when loaded in tension, causing an additional tensile force in the bolt, owing to the tendency of the connection material to pull away (pry) from the column flange.

11-4   SIMPLE (WIND)

The AISC Specification permits the designer to use simple connections (type 2) when wind load is present in addition to gravity loads. The wind moment is distributed to selected bents or frames where moment connections are added to resist the bending due to the wind. These connections are so designed that, when subject to the larger moments induced by gravity loads, they are relieved by the connection material deforming (yielding) without overstressing the welds (when welds are used). Figure 11-2 shows typical type 2 (simple) connections for gravity loads with moment connections for wind.

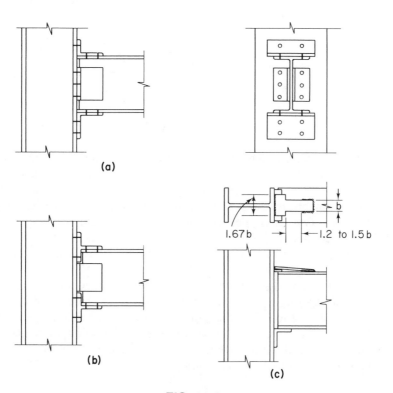

FIG. 11-2

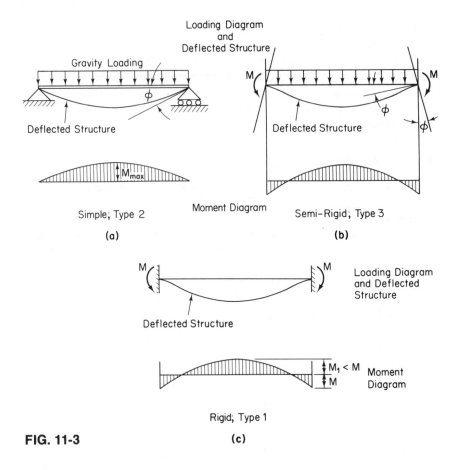

FIG. 11-3

All these connections could also be used as semirigid (type 3), but the prying action for the tees (Fig. 11-2b) may have to be looked into. The connection using the tees could also be used as rigid (type 1), but the tees are unable to develop the full moment capacity of the beam. Therefore, this type of connection used as type 1 would not be economical. The tee connection serves well as a "soft" wind moment connection when used in conjunction with type 2 simple (flexible) connections for shear.

Figure 11-2c employs a type of "soft" wind connection that obtains its rotational capacity by leaving a portion of the top moment plate free of weld (usually 1.2 to 1.5b in length). This weldless portion allows free deformation of the plate when subject to the applied wind moments.

Figure 11-3a illustrates free rotation at the ends of type 2 simple framing when subject to gravity loading.

11-5   SEMIRIGID

Somewhere between type 2 (simple) and type 1 (rigid) framing is the type 3 category of framing (semirigid). Its behavior is somewhere between free rotation and zero rotation and is illustrated in Fig. 11-3b. Type 3 framing assumes that the connections possess some known and dependable moment capacity. The AISC Specification will permit this type of framing when evidence is available that the connection used is capable of furnishing a predictable fraction of full end restraint as a minimum. The design of this connection shall be based upon this minimum. The behavior of these connections is involved and can only be determined by suitable testing.

The design of these connections is accomplished by the simplified assumption that the web angles take the shear and that the top and bottom angles, plates, or tees take the moment. The behavior of type 3 is similar to the behavior of type 2 for gravity loads with wind moment connections. Typical type 3 connections are similar to those shown in Fig. 11-2 for type 2 with wind moment connections.

Figure 11-2b and c would be used for larger moments or to increase rigidity. The angles shown in Fig. 11-2a would be replaced by tees bolted to the column flange or by flexible plates shop-welded to the column flange in this case. If tees are used, it may be necessary to look into the prying-action phenomenon.

11-6   RIGID

Rigid framing (type 1) assumes that the beam-to-column connections are rigid enough to hold unchanged the original angle made by the intersecting members. Type 1 framing is the most predictable in terms of performance. When properly designed, the joints will behave very closely to the theoretical behavior. Type 1 may be all bolted, riveted, or welded, or they may be a combination of bolts and welds. Figure 11-4 shows several types of rigid connections. The text to follow discusses the advantages and disadvantages of each type.

One of the simplest and probably the most economical of type 1 connections is shown in Fig. 11-4a. A single plate is shown, which is shop-welded to the column flange and field-bolted to the beam web. This plate takes the shear of the beam and also serves to hold the beam in place prior to welding. The bolts used are of the friction type (this will be defined and discussed in Sections 11-10 and 11), as required by the AISC Specification (bolts used in combination with welds may be considered to share the stress if they are designed under the friction-type assumption). Top and bottom beam flanges are beveled and field-welded

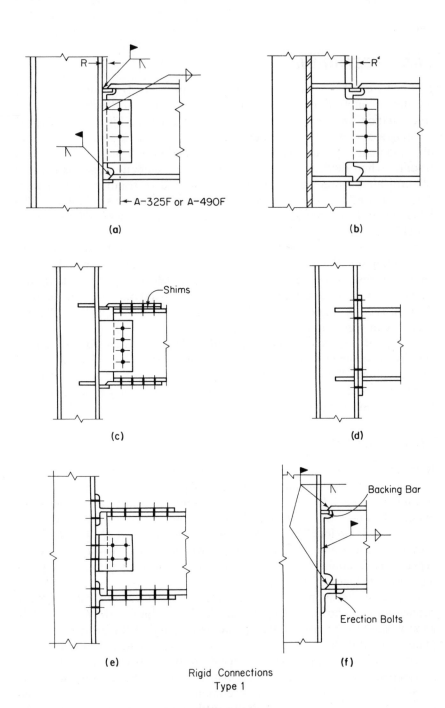

(a)

A-325F or A-490F

(b)

Shims

(c)

(d)

(e)

Backing Bar

Erection Bolts

(f)

Rigid Connections
Type 1

FIG. 11-4

372

($\curlywedge$ symbol). This type of connection necessitates a close tolerance of beam length to assure proper fit. This problem can be lessened by opening the weld root opening, R, and using more weld material in the field. If R were lessened, the beam would have to be ordered long to cover the case of beam-length underrun (within tolerances). When beams are ordered long and there is not an overrun, the fabricator must cut some beams in the shop, thereby increasing costs.

Figure 11-4b shows a variation of this type of connection that is usually employed when the beam is connected to the column web. For this case, as before, the web plate takes the shear and also serves to hold the beam in position prior to welding. In addition, the bolts employ the friction-type assumption. Horizontal plates are shop-welded between column flanges. These plates extend beyond the column flanges.

Figure 11-4c and d shows additional variations of type 1 connections usually employed where field-welding is not desired. The three plate connections shown in Fig. 11-4c use shims below the top flange plate. The shims take care of under- and overrun in the beam depth, as well as the tendency of the top flange to cock. It is a popular connection that requires less fabrication. Figure 11-4d employs an end plate and has also been used. In the latter case, tolerances relative to beam length are more critical. Once again, beams are required to be ordered long to take care of potential underrun. This requires cutting in the shop for the cases of no underrun prior to installation of the end plate. Both details have performed well. Column web stiffeners may or may not be required, and design investigation to determine this should be pursued. Finally, prying action of the bolts for the case employing an end plate should be checked.

The split-tee case shown in Fig. 11-4e is a valid type 1 connection, but it uses too much connection material. It is also often unable to develop the full beam-moment capacity, making it an uneconomical choice. Figure 11-4f is a common detail for all welded construction.

11-7   BEAM-LINE CONCEPT

Relative flexibility of the various categories of joints or connections can be compared by use of the *beam-line concept*. This concept utilizes the moment-rotation characteristics of a given connection superimposed upon a plot of the moment-rotation characteristics of the end of a particular beam under load. The moment rotation or $M-\phi$ characteristics of a particular connection are quite involved and are impossible to represent analytically. They must be determined in the laboratory by testing.

A typical $M-\phi$ curve of a certain connection is shown in Fig. 11-5a. The steeper the slope of the connection curve, the less flexible the connection. It is the slope of the curve that indicates the degree of flex-

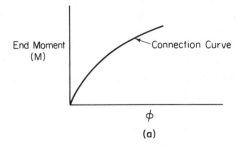

(a)

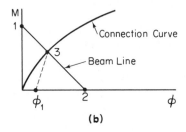

(b)

FIG. 11-5

ibility or rigidity of the connection. When the M–ϕ curve is determined in the laboratory, the beam line for a particular beam under load can then be superimposed as in Fig. 11-5b. The beam line is represented by straight line 1–2. Point 1, plotted on the ordinate axis, is the fixed-end moment of the beam with full rigidity ($\phi = 0$), while point 2, plotted on the abscissa axis, is the free-ended rotation of this beam with no end restraints (hinged ends, $M = 0$). The M–ϕ curve is found by testing, while the beam line is determined by analyses. Point 3, the intersection of the two, is descriptive of the actual connection behavior for a particular beam under load.

   If the beam represented in Fig. 11-5b were unloaded, the connection behavior would be represented by the dashed line 3–ϕ_1, where 3–ϕ_1 would be parallel to the initial slope of the connection curve. ϕ_1 represents the permanent (plastic) deformation of the connections. For all loadings to follow, the connection would behave elastically between ϕ_1 and point 3.

   For a fixed-ended uniformly loaded beam of span length L, point 1 of Figure 11-5b would represent the beam's fixed-end moment, $wL^2/12$, corresponding to a value of $\phi = 0$. Point 2 would represent the rotation of the beam at zero moment equal to $2f_bL/3Ed$.

   A direct comparison of the behavior of connections can be made

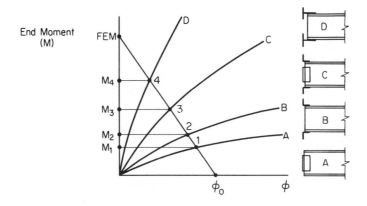

FIG. 11-6

by superimposing the moment-rotation curve of each of the connections under comparison on one plot. This plot is shown in Fig. 11-6 for the various type 2 and type 3 connections shown in the figure. Curve A is determined by test for a double-angle web connection. It is a type 2 simple connection designed for shear only. The intersection with the beam line, point 1, indicates the amount of moment, M_1, that the connection will absorb. This moment, M_1, is relatively small, and for design purposes (type 2, simple), it is assumed to be zero. The actual magnitude of M_1 depends on many variables, such as the column flange section size, the connection angle thickness, F_y of the angles, the beam size and its properties, and the bolt gage for bolted connections. Exerting proper control over these variables keeps M_1 small.

Curve B is the $M-\phi$ curve for a type 2 seat angle with top-angle clip connection. It is seen to be slightly higher than curve A, which indicates a slightly higher degree of rigidity (less flexibility) of joint. The end-plate type 2 shear connection shown in Fig. 11-1d would also fall somewhere slightly above curve A.

Curve C can represent type 2 with wind-moment connections or a type 3 semirigid connection. Curve C has a substantial amount of rigidity and absorbs a sizable moment, M_3. Design procedures are simplified by the assumption that the beam web angles take shear, while the top and bottom beam flange angles take the moment.

Although type 2 with wind-moment connections has a $M-\phi$ curve similar to type 3, the actual performance requirements differ, making both connections different. In type 3, moment and shear are applied simultaneously and are inseparable. The moment resistance part of the connection limits the connection's moment capacity and thereby defines the rigidity of the connection. The moment and shear for type 2 with wind-moment connections are separate and may or may not act simultaneously. In this connection, the moment connection (beam flange angles) does not limit the moment but rather is designed for the wind moment. This wind moment is independent of the gravity-beam moment. Further, the wind-moment connection must act as a truly flexible wind connection in the presence of gravity load only. The connection must not resist the rotation required to make this a true type 2 connection for gravity loading in the absence of wind.

Connection D, as shown in Fig. 11-6, is a type 3 semirigid connection. If beam web angles had been added, the connection could function as either type 3 or type 2 with wind-moment connections.

After selecting the type of construction (type 1, type 2 simple gravity, type 2 simple wind, or type 3), the designer must select an appropriate connector (rivet, weld, high-strength bolt, common bolt, etc). The variables that should be considered before making this decision were discussed in Section 2-4.

In the three sections that follow, the most commonly used structural mechanical fasteners will be discussed: rivets, common bolts, and high-strength bolts. Section 11-11 will discuss the highlights of high-strength-bolt specification. Section 11-12 will discuss welding.

11-8   RIVETS

Rivets used to be the most popular mechanical fastener, both in the field and in the shop. However, since the development of the high-strength and common bolt, rivets are no longer used in the field and are rarely used in the shop. Some fabricating shops still prefer riveting and, as mentioned in an earlier chapter, the option should be the fabricator's.

The AISC Specification requires that rivets conform to the provisions of ASTM A502, Grade 1, 2, or 3 (Steel Structural Rivets). Rivets are manufactured with cylindrical bodies and a head at one end, as shown in Fig. 11-7a. They are available from $\frac{1}{2}$ to $1\frac{1}{2}$ in. diameters.

When making a connection, the rivet is usually heated, placed in a prepunched hole, and then driven by means of a power hammer, which is equipped with a die to form a second head. The first head is held under pressure. The rivet after being driven is shown in Fig. 11-7b. Upon cooling, the rivet shrinks longitudinally, producing a clamping force on the connected material. Grade 1 is more widely used than the higher-strength grade 2 or 3 rivets. Grade 3 has enhanced corrosion-resistant characteristics. The more common head is the button head. Other heads are the high button (acorn), cone, pan, and countersunk.

The replacement of rivets by bolts has resulted from the noneconomy involved with large and skilled rivet crews, time-consuming place-

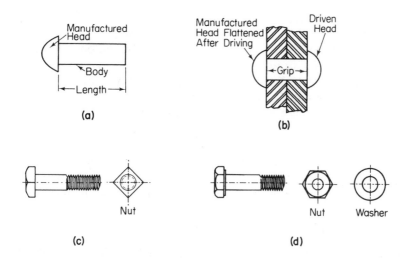

FIG. 11-7

ment and inspection, installation noise, and the potential fire hazard involved.

Allowable tension F_t, and shear stresses, F_v, on rivets (as well as bolts and threaded parts) are shown in Table 11-2. These allowable stresses are in ksi of nominal area of rivet prior to driving (or on the unthreaded nominal body area of bolts and threaded parts, unless otherwise noted in the table).

11-9   COMMON BOLTS

Common bolts are also referred to as A307, machine, unfinished, or rough bolts. These bolts should meet ASTM A307 (Specification for Low-Carbon Steel Externally and Internally Threaded Standard Fasteners). This type of bolt is significantly cheaper than high-strength bolts. They generally have square heads and nuts with no marking on the head surface (hexagon-shaped heads and nuts are less common). The body of the bolt is cylindrical and partially threaded, as shown in Fig. 11-7c. Although ASTM A307 does not specify a minimum yield-stress value, typically a well-defined yield stress is exhibited. This value is approximately $F_y = 55$ ksi (minimum typical tensile strength is 60 ksi).

Where permissible, common bolts should be used because they are more economical and easier to install. There is a definite tendency to specify the more expensive high-strength bolt for connections even when common bolts would do the job. Common bolts should be used in secondary or minimum connections where strength is not of prime consid-

eration; where cyclic, vibratory, or stress reversal loading is not anticipated; and where slip can be tolerated (i.e., bracing and secondary floor
beams, roof purlins, girts, etc.). The nuts on common bolts have a tendency to back off with time, for the aforementioned conditions.

F_t and F_v values for common bolts will be found in Table 11-2. F_t
and F_v values are applied to the nominal area of the bolts based on their
nominal (major) diameter. The bolts are available in $\frac{1}{4}$ to 4 in. diameters
inclusive.

Threads are unified coarse thread series (UNC Series), class 2A
(see ANSI B1.1, Unified Screw Threads). Common bolts are easily tightened by using spud wrenches. The tension induced by turning is usually
low, and it is usually considered that no clamping force is developed.

11-10   HIGH-STRENGTH BOLTS

Currently, the most often used mechanical fastener, by far, is the high-
strength bolt (HSB). The AISC Manual uses the notation shown in Table
11-1 for the three ways in which either of the two types of high-strength
bolts (ASTM A325 and A490) may be used. Each of the bolts designated
in Table 11-1 may be used with differing allowable stresses, as shown in
Table 11-2. The –N represents bolts assuming a bearing connection
where the bolt threads are included in the shear plane. The –X represents
bolts assuming a bearing connection where the bolt threads are excluded
from the shear plane. The shear plane is considered to be the plane
formed by the contact of two elements of the connection.

Table 11-1

Bolts	Application
A325-F, A490-F	Friction-type connection
A325-N, A490-N	Bearing-type connection with threads in the shear plane
A325-X, A490-X	Bearing-type connection with threads excluded from the shear plane

Bearing connections are merely connections in which it is assumed
that slippage occurs, after which the bolts are in bearing, whereas in a
friction connection this slippage is assumed not to occur. The word
"assumed," which was used twice, is important to understand. Although
the allowable tensile stress, F_t, in either the A325 or A490 bolt does not
vary when the bolt threads are in or out of the shear plane, the allowable
shear stress, F_v, does. The allowable shear stress is a maximum for
bearing-type connections when the threads are excluded from the shear

plane. Therefore, the A325-X and A490-X use is the most efficient and economical when this assumption is possible. Bearing-type connections should be used in all cases where slippage can be tolerated. Further, bearing-type connections cannot be used in combination with welds (if bolts are used in combination with welds, they cannot be considered to share the stress with the welds). Shear connections subjected to severe stress fluctuation, stress reversal where slippage is not desired, should be friction-type connections. Bolts using the friction-type connection assumption can be considered to share the stress with the welds. When making welded alterations to structures, any existing rivets or high-strength bolts can be assumed to carry existing dead-load stress where the new welding should be designed to carry any additional stress. Rivets and high-strength bolts in friction-type connections may be considered as sharing the dead- and live-load stress. The AISC Specification states where rivets, high-strength bolts, or welds shall be used in field connections.

A325 or A490 bolts are high-strength bolts using heavy hexagonal nuts and having heavy hexagonal heads. They have shorter thread lengths than other bolts. The A325 bolts come in three types. Type 1 is produced from a medium-carbon steel (available sizes are from $\frac{1}{2}$ to $1\frac{1}{2}$ in. in diameter). Type 2 is produced from low-carbon martensite steel and is limited to $\frac{1}{2}$ to 1 in. diameter in size. This type should not be hot-dipped galvanized. Type 3 is produced from steels with self-weathering characteristics comparable to ASTM A588 and A242 steels (available sizes are from $\frac{1}{2}$ to $1\frac{1}{2}$ in. in diameter). The A490 bolt is the stronger of the two and is produced from alloy steel.

Identifying marks on the heads of each of the three types of A325 bolts distinguish them. Type 1, at the option of the manufacturer, are identified by the mark "A325" and the manufacturer's symbol or as an option, and in addition, by three radial lines 120° apart. Type 2 is identified by three radial lines 60° apart. Type 3 is identified by the mark "A325" underlined and, at the option of the manufacturer, any other additional marks to identify the bolt as a self-weathering type. A490 bolts are marked by "A490" and the manufacturer's symbol.

The heavy hexagonal nuts for A325 bolts are similarly marked for identification on at least one face. These marks are the manufacturer's symbol and the number "2" or "2H," by three equally spaced circumferential lines or by the mark "D" or "DH." The nuts for A325 type 3 bolts are marked on one face by three circumferential marks and the number "3" in addition to any other marks desired by the manufacturer. A490 nuts are marked by "2H" and the manufacturer's mark or by "DH." Washers for A325 type 3 bolts bear the mark "3" near the outer edge of one face and any other marks desired by the manufacturer.

Table 11-2
ALLOWABLE WORKING STRESS FOR FASTENERS (ksi)[a]

Fastener [b]	F_t [c]	F_v [c]
A502, grade 1, hot-driven rivets	23.0 [d]	17.5 [e,f]
A502, Grades 2 and 3, hot-driven rivets	29.0 [d]	22.0 [e,f]
A307 bolts [g]	20.0 [d]	10.0 [e,f]
A449 bolts and threaded parts of steel [g,h,i]	$0.33F_u$ [d,j]	$0.17F_u$ [e,j,k]
A325F (standard holes)	44.0 [l]	17.5 [m]
A325F (oversized and short slotted holes)	44.0 [l]	15.0 [m]
A325F (long slotted holes)	44.0 [l]	12.5 [m]
A325N (standard and slotted holes)	44.0 [l]	21.0 [f]
A325X (standard and slotted holes)	44.0 [l]	30.0 [f]
A490F (standard holes)	54.0 [l]	22.0 [m]
A490F (oversized and short slotted holes)	54.0 [l]	19.0 [m]
A490F (long slotted holes)	54.0 [l]	16.0 [m]
A490N (standard and slotted holes)	54.0 [l]	28.0 [f]
A490X (standard and slotted holes)	54.0 [l]	40.0 [f]

[a] Any increase in allowable stresses due to wind or seismic stresses does not apply to fasteners subject to fatigue conditions.

[b] For fasteners in oversized and slotted holes, see Sec. 11-11.1.

[c] F_t and F_v are applied to the nominal area of the bolts and threaded parts based on their nominal (major) diameter (for rivets, apply to area of rivets prior to driving).

[d] Static load only.

[e] Bearing-type connection.

[f] In bearing-type connections used to splice tension members whose length between extreme fasteners measured parallel to the line of axial force > 50 in., use 0.80 F_v and 0.80 F_t.

[g] Threads permitted in shear plane.

[h] In the case of upset rods, the area of the body times $0.60F_y$ < F_t on the threaded portion.

[i] A449 bolts may be used for anchor bolts or threaded rods or for tension and bearing-type high-strength structural joints requiring diameters > $1\frac{1}{2}$ in., where high-strength material is desired. An A325 heavy hex nut and a hardened washer under the bolt head must be used for the tension and bearing-type shear high-strength structural joint application when the tightening re-

Table 11-2 (Continued)

quirements $> 0.5F_u$. Further, the same thread type and series should be used. Last, the quality assurance requirements are not as sophisticated nor as stringent as those called for in the production of A325 bolts. A449 bolts are quenched and tempered and have a strength grade about equal to A325 (proof load, tensile strength, reduction of area and elongation). There is no A449 equivalent to A325 type 3 bolts.

^{j} F_u is the lowest specified minimum tensile strength of the fastener (ksi).

^{k} $0.22F_u$ when threads are excluded from the shear plane.

^{l} See Table 11-5 for restrictions when bolts are subjected to fatigue loading in tension.

^{m} When specified by the designer, F_v for friction-type shear connections may have the applicable values in Table 11-3 for surface conditions other than for clean mill scale.

The high-strength bolt (HSB) is shown in Fig. 11-7d. All HSBs are heat-treated by quenching and tempering. HSBs are required to be tightened to a tension equal to or greater than 70% of the specified minimum tensile strength. This tension, induced by nut rotation, produces a high clamping force, which allows the contact surfaces to carry loads solely by friction. With this high preload, when the bolt is subject to additional external tension load, there will be little or no increase in internal bolt tension and the bolt can develop the full tensile strength. Tightening is accomplished by long-handled torque wrenches or powered impact wrenches. A nut will never back off on a properly installed high-strength bolt. The tightening procedures will be discussed in Section 11-11.2.

All dimensions of HSBs must conform to the American National Standard for Square and Hex Bolts and Screws (ANSI B18.2.1) and the heavy hex-nut dimension must conform to ANSI B18.2.2. Threads must be Unified Coarse Thread Series as specified by ANSI B1.1 (American National Standard for Unified Screw Threads) and must have class 2A tolerances for bolts and class 2B tolerances for nuts. Dimensions of the washers should conform to those of the Specification for Structural Joints Using ASTM A325 or A490 Bolts issued by the Research Council on Riveted and Bolted Structural Joints of the Engineering Foundation. Unless otherwise specified, washers are circular.

The allowable bearing stress, F_p, in kips per square inch, on the projected area of bolts and rivets in shear connections is expressed by Eq. 11-1b. The minimum distance, L, in inches, measured in the line of the force from the centerline of a standard hole to an edge of a connected part shall be expressed by Eq. 11-1a or by the applicable table of minimum edge distances in the AISC Specification. The corresponding minimum distance, L, in inches, for oversized or slotted holes shall be that required for standard holes plus the applicable increment as per the AISC Specification.

$$L = \frac{2P}{F_u t} \qquad (11\text{-}1a)$$

where P is the force transmitted by one fastener to the critical connected part in kips, t is the thickness of the critical connected part in inches, and F_u is the lowest specified minimum tensile strength of the connected part in kips per square inch.

$$F_p = 1.5F_u \qquad (11\text{-}1b)$$

The minimum distance center to center of all types of fastener holes is expressed by the AISC Specification (Section 1.16.4.2).

Table 11-3
F_v (ksi) BASED ON SURFACE CONDITIONS FOR FRICTION-TYPE SHEAR CONNECTION WITH STANDARD HOLES[a]

Class[b]	Surface Condition	A325F	A490F
A	Clean mill scale	17.5	22.0
B	Blast-cleaned carbon and low-alloy steel	27.5	34.5
C	Blast-cleaned quenched and tempered steel	19.0	23.5
D	Hot-dip galvanized and roughened[c]	21.5	27.0
E	Blast-cleaned, organic-zinc-rich paint	21.0	26.0
F	Blast-cleaned, inorganic-zinc-rich paint	29.5	37.0
G	Blast-cleaned, metallized with zinc	29.5	37.0
H	Blast-cleaned, metallized with aluminum	30.0	37.5
I	Vinyl wash	16.5	20.5

[a] For oversized and slotted holes, see Table 11-4.

[b] Reference is made to the specification for Structural Joints Using ASTM A325 or A490 Bolts for the appropriate type of coating for contact surfaces.

[c] If loads causing actual stresses greater than 50% of the allowable stresses tabulated are sustained over a long period of time (e.g., gravity), slip into bearing may occur. If this slip is detrimental, these increased working stresses are not recommended.

Values of this Table are applicable only when ≤ the lowest appropriate allowable working stress for bearing-type connections, taking into account the position of the threads relative to shear planes and footnote (f) of Table 11-2.

For end connections bolted to the web of a beam and designed for beam shear reaction (no fastener eccentricity accounted for), the center of nearest standard hole to end of beam web distance shall be equal to or larger than

$$\frac{2P_R}{F_u t} \qquad\qquad (11\text{-}1c)$$

where P_R is the beam reaction divided by the number of bolts and F_u and t are as previously defined. The requirements for Eq. 11-1c may be waived when the bearing stress induced by the fastener $\leq 0.90\ F_u$.

11-11   A325/A490 BOLT SPECIFICATIONS

Two basic specifications cover all high-strength bolts. ASTM A325 and ASTM A490 (American Society for Testing and Materials) cover the chemical and mechanical requirements for HSBs intended for use in structural joints. In addition to the chemical and mechanical requirements, the types of each bolt are described along with their size availability; the type of suitable nut and washers; the manufacturing process of the steel materials; the dimensions (bolt body, nuts, thread type, washers, etc.); the methods of testing; the quality-assurance requirements; the required identification (marking) of the bolt, nut, and washers; inspection and rejection procedures; and so on. Some of these requirements have already been discussed.

The second basic specification is entitled Structural Joints Using ASTM A325 or A490 Bolts, which is approved by the Research Council on Riveted and Bolted Structural Joints (RCRBSJ) of the Engineering Foundation and endorsed by both the AISC and the Industrial Fasteners Institute (IFI). This specification covers the design and assembly of structural joints using the aforenamed HSB. The AISC Specification dealing with HSB conforms to the RCRBSJ Specification. The RCRBSJ Specification includes a Commentary that is a valuable guide to the actual application of the Specification proper.

The Specification covers the specification requirements for bolts, nuts, and washers; the dimensions of the bolts, nuts, and washers; bolted parts (hole sizes; Section 11-11.1); permissible joint surface coatings (paint is permitted without consideration as to type in bearing joints and certain contact surface coatings are permitted with friction joints); design stresses (Tables 11-2 through 11-5) for applied tension, shear, and bearing; acceptable installation procedures (Section 11-11.2), which includes the minimum tension corresponding to the size and grade of fastener; and the required inspection procedures (Section 11-11.3).

Table 11-4
ADJUSTMENT COEFFICIENTS[a] C_1 FOR OVERSIZED OR SLOTTED HOLES

Description of Hole in Bolted Parts (excluding washers)	C_1[b]
Standard holes (nominally $\frac{1}{16}$ in. larger than bolt)	1.00
Oversized and short-slotted holes (Section 11-11.1)	0.85
Long-slotted holes (Section 11-11.1)	0.70

[a] Same for A325 and A490 bolts and any method of installation recognized in the specification for Structural Joints Using ASTM A325 or A490 Bolts.

[b] F_v value for friction-type shear connections for oversized and slotted holes is obtained by multiplying the value obtained from Table 11-3 by the appropriate C_1 value. Use of C_1 approximately gives the values in Table 2a of the specification for structural Joints Using ASTM A325 or A490 Bolts.

Table 11-5
RESTRICTIONS WHEN A325 AND A490 BOLTS ARE SUBJECT TO FATIGUE[a] LOADING IN TENSION

Use F_t from Table 11-2 for applied plus prying loads for:
1. Connections subject to $\leq$ 20,000 cycles of combined direct external tension and prying forces.
2. Connections subject to > 20,000 cycles but $\leq$ 500,000 cycles of direct tension where the prying load on the bolts $\leq$ 10% of the externally applied load.
3. Connections subject > 500,000 cycles of direct tension where the prying load on the bolts $\leq$ 5% of the externally applied load.

When the prying load > the values in (2) and (3):
1. F_t = 0.60 (Table 11-2 F_t value) for applied load only for connections subject to > 20,000 cycles but $\leq$ 500,000 cycles of direct tension.
2. F_t = 0.50 (Table 11-2 F_t value) for applied load only for connections subject to > 500,000 cycles of direct tension.

[a] Fatigue is defined as the damage that may result in fracture after a sufficient number of fluctuations of stress. Few connections in ordinary buildings need be designed for fatigue, since most load changes occur a small amount of times with minor fluctuations in stress. Occurrence of full-design wind or earthquake loads is too infrequent to warrant consideration in fatigue design. However, crane runways and supporting structures for machinery and equipment may be subject to fatigue loadings.

A loading cycle of 20,000 is equivalent to two applications every day for 25 years. A loading cycle of 500,000 is equivalent to fifty applications every day for 25 years.

Connections subject to fatigue loading should be designed in accordance with Appendix B to the AISC Specification, which will not be discussed.

Range of shear stress need not be considered on properly tightened A325 and A490 bolts, but the maximum stress $\leq$ the values given in Table 11-2. Where stress reversal is encountered, the use of A307 and A449 bolts is not recommended.

11-11.1 Oversized and Slotted Holes

All standard holes for HSB should be $\frac{1}{16}$ in. greater than the nominal bolt diameter, d. The holes can be punched (provided the thickness of material $\le d + \frac{1}{8}$ in.), subpunched and reamed or drilled (if the thickness of material $> d + \frac{1}{8}$ in.). The fasteners in such holes shall be proportioned to meet the allowable stresses given in Tables 11-2, 11-3, and 11-5 as applicable. With the approval of the designer and when shown on the design drawings, oversized, short-slotted, and long-slotted holes may be used (not permitted for riveted connections). These fasteners shall be proportioned to meet the allowable stresses in Table 11-2 through 11-5, as applicable.

(1) *Oversized holes* in connections may have nominal diameters equal to or less than $d + \frac{3}{16}$ in. for $d \le \frac{7}{8}$-in. diameter; $d + \frac{1}{4}$ in. for $d = $ 1-in. diameter; $d + \frac{5}{16}$ in. for $d \ge 1\frac{1}{8}$-in. diameter. Oversized holes can be used in any or all plies of friction-type connections only. Hardened washers shall be used over exposed oversized holes in an outer ply.

(2) *Short-slotted holes* are nominally $\frac{1}{16}$ in. wider than d, having a maximum length determined by the oversized diameter (determined in step 1) + $\frac{1}{16}$ in. They may be used in any or all plies of both friction- and bearing-type connections. The slots may be used in friction-type connections regardless of direction of loading. In bearing-type connections, the slots should be normal to the direction of loading. Hardened washers must be used over exposed short-slotted holes.

(3) *Long-slotted holes* are also nominally $\frac{1}{16}$ in. wider than d, having a length larger than in step 2, $= 2\frac{1}{2} d$. These slots may be used in friction-type connections regardless of direction of loading. In bearing-type connections, the slots should be normal to the direction of loading. A plate washer or continuous bar (of $\ge \frac{5}{16}$ in. thickness for HSB) having standard holes completely covering the slot should be used (of structural material grade, which need not be hardened) where long-slotted holes are used on an outer ply (exposed). If hardened washers are required (Section 11-11.2), they must be placed over the outer surface of the plate washer or continuous bar. These slots may be used in only one of the connected parts of either friction- or bearing-type connections at an individual contact surface.

Table 11-6 summarizes and lists the maximum hole size in inches for oversized and slotted holes.

Table 11-6
OVERSIZE AND SLOTTED HOLES

Nominal Bolt Size, Inches	Maximum Hole Size (Nominal), Inches		
	Oversize Holes	Short Slotted Holes	Long Slotted Holes
$\frac{5}{8}$	$\frac{13}{16}$	$\frac{11}{16} \times \frac{7}{8}$	$\frac{11}{16} \times 1\frac{9}{16}$
$\frac{3}{4}$	$\frac{15}{16}$	$\frac{13}{16} \times 1$	$\frac{13}{16} \times 1\frac{7}{8}$
$\frac{7}{8}$	$1\frac{1}{16}$	$\frac{15}{16} \times 1\frac{1}{8}$	$\frac{15}{16} \times 2\frac{3}{16}$
1	$1\frac{1}{4}$	$1\frac{1}{16} \times 1\frac{5}{16}$	$1\frac{1}{16} \times 2\frac{1}{2}$
$1\frac{1}{8}$	$1\frac{7}{16}$	$1\frac{3}{16} \times 1\frac{1}{2}$	$1\frac{3}{16} \times 2\frac{13}{16}$
$1\frac{1}{4}$	$1\frac{9}{16}$	$1\frac{5}{16} \times 1\frac{5}{8}$	$1\frac{5}{16} \times 3\frac{1}{8}$
$1\frac{3}{8}$	$1\frac{11}{16}$	$1\frac{7}{16} \times 1\frac{3}{4}$	$1\frac{7}{16} \times 3\frac{7}{16}$
$1\frac{1}{2}$	$1\frac{13}{16}$	$1\frac{9}{16} \times 1\frac{7}{8}$	$1\frac{9}{16} \times 3\frac{3}{4}$

Courtesy American Institute of Steel Construction.

11-11.2 Installation Procedures

Both ASTM A325 or A490 HSB (bearing or friction-type) may be installed by any of three methods of installation. Tightening may be done by turning either the bolt head or the bolt nut and preventing the unturned element from turning. All fasteners in a connection should be tightened to the minimum tension called for in the RCRBSJ Specification.

The three acceptable methods of installation are:

(1)   Turn-of-nut tightening.
(2)   Calibrated wrench tightening.
(3)   Tightening by use of a direct tension indicator.

The requirements (or nonrequirements) of washers are given in Table 11-7.

Turn-of-Nut Tightening   Of the three methods, the *turn-of-nut method* is the "preferred" installation procedure, although the RCRBSJ, without preference, endorses all three methods. The turn-of-nut tightening method is a strain-control procedure, as contrasted with a torque-control procedure (torque wrench or calibrated wrench). The effectiveness of the method depends on the uniformity of the starting point from which rotations of the turned element (usually the nut) are measured. This starting point is called the *snug* position, as is indicated by the start of impact (slipping) of the impact wrench, or the full effort of a person using a common spud wrench. After the snug condition, further prescribed ro-

Table 11-7
HARDENED WASHER REQUIREMENTS[a]

Bolt	Method of Installation	Washer Requirements[b]
A325	Turn-of-nut tightening	None
	Calibrated wrench tightening	Washer under turned element
A490	Turn-of-nut tightening	Washer under turned element
	Calibrated wrench tightening and direct tension indicator	or two washers (one at each end of the bolt) if the material has an $F_y < 40$ ksi

[a]Beveled washers must be used where required when the slope of the outer face of the bolted parts > 1:20 with respect to the plane normal to the bolt axis.

[b]Except as required by Section 11-11.1.

tation results in bolt elongation or deformation, which produces the required clamping force. For uniformity purposes the same amount of rotation from snug position is recommended for both A325 and A490 HSBs. Tolerances on nut rotation are also stipulated in the specification. Bolts installed in accordance with the recommended rotation (± tolerance) will sustain additional direct tension loads without any reduction in ultimate strength. Because of this reserve strength, if the fastener does not fail during the installation process, it will not fail at any time thereafter provided that the additional loads have been designed for.

The procedure is to bring enough bolts into the snug position to make the connection parts have good contact. Then additional bolts are placed in any holes that remain, and these bolts are brought to the snug condition. Then all bolts are further tightened by the amount of rotation from the snug tight position as prescribed by the RCRBSJ specification. The tightening procedure should systematically progress from the most rigid part of the connection to its free edges. To retain a uniform level of reserve deformation for all bolts, the nut rotation is proportioned in the specification to the bolt length (up to 12 diameters).

Calibrated Wrench Tightening When using *calibrated wrenches* to tighten a connection, a minimum of three bolts of each diameter being used must be tightened at least once each working day in a calibrating device, which allows the tension in the bolt to be read. The wrench is adjusted to stall at a bolt tension that is at least 5% greater than the RCRBSJ tension shown in tabular form in the Specification (varies with diameter of bolt and with type of bolt, A325 or A490). This installation method is one of torque control. The tightening procedure is further checked by verifying during actual installation that the turned element rotation from snug position is not greater than the RCRBSJ prescribed

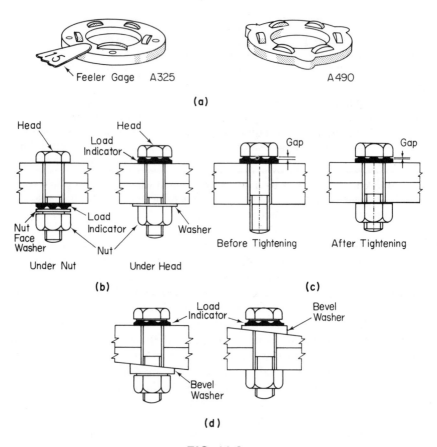

FIG. 11-8

amount. The identical recommended sequence of tightening designated for the turn-of-nut method is prescribed for this method of installation. In addition, it is recommended that the wrench be returned to touch up previously tightened bolts to the preset tension, which may have been loosened by the subsequent tightening of other bolts. This retightening is not considered to be reuse of the bolt. (A325 bolts may be reused, but A490 and galvanized A325 bolts may not be reused. In all cases, approval for reuse should be obtained from the engineer.)

Tightening by Use of a Direct Tension Indicator   An example of a *direct tension indicator* is the circular load indicator washer, which is manufactured to fit both A325 and A490 bolts. The washer is a hardened washer with embossed protrusions on one face as shown in Fig. 11-8a. The washer is installed under the unturned element such that the protrusions bear against the underside of the bolt head or bearing against a special nut-faced washer plate placed between the underside of the nut and the indicator washer (Fig. 11-8b). As the bolt is tightened, the gap is reduced as a result of the flattening of the protrusions caused by the induced clamping force. At a specified average remaining gap, which is measured by a feeler gage (by eye as experience is gained), the proper minimum bolt tension is achieved. Figure 11-8c shows a typical assembly in place before and after tightening. Gaps remaining after tightening are usually 0.015 in. for most A325 and A490 installations. The exceptions are for both A325 and A490 galvanized HSB (gap = 0.010 in.) and for the case of placement of the indicator under the nut for both A325 and A490 HSB (with nut-face washer plates, gap = 0.010 in.). Installation is achieved by ordinary tightening wrenches (calibration of a tightening device is unnecessary) and by the aforementioned recommended tightening procedures.

The load indicator devices for A325 bolts are used with types 1 and 2 and are usually maintained from $\frac{1}{2}$-in. to $1\frac{1}{4}$-in. diameters. The A490 indicator is made in diameters varying from $\frac{3}{4}$ to $1\frac{1}{4}$ in. The load capacity of each indicator matches each grade and diameter of bolt (minus zero and plus 15%). This is accomplished through certified quality control during the manufacturing process. Figure 11-8d shows how the load washer is used with tapered flanges.

The gap is not related to the bolt length. When using the indicator, an extra allowance of $\frac{1}{8}$ in. for bolt length must be made for the circular load indicator washer.

11-11.3 Inspection

The inspection of bolts installed by the turn-of-nut or calibrated wrench method is usually made at the time of installation to ensure that the proper procedure is used. When additional inspection procedures are needed, a suggested method is through the use of an inspecting wrench (a torque or power wrench), which can be adjusted in the same manner as described for calibrated wrench tightening. Three bolts of the same grade and size of representative length as used on the job are then placed in the calibration device, which is capable of indicating bolt tension. The surface under the turned element should be similar to that used in the actual structure (same grade of steel under turned element or use of a washer under the turned element, if this is the case in the field on the real structure).

When the inspecting wrench is a torque wrench, each of the test bolts is tightened initially in the calibration device (by any means) to 15% of the required bolt tension and then to the minimum specified tension. Tightening beyond the initial condition should not produce a nut rotation greater than that permitted. The inspection wrench is then calibrated by applying it to the tightened bolt, and the torque necessary to turn the nut or head 5 degrees (1 in. in a 12-in. radius) in the tightening direction is determined. The average torque so determined on the three test bolts is used as a reading to inspect the actual bolts on the structure in the manner described in the next paragraph. When the inspecting wrench is a power wrench, it should be adjusted in a manner such that each of the three test bolts will achieve a tightening tension from a minimum of 5% to a maximum of 10% greater than the specified minimum tension. This setting must not produce more than the specified nut rotation beyond the snug condition. This tested setting should be used as the setting for inspection of the actual bolts on the structure.

When either a precalibrated (as described) torque or calibrated inspection wrench is used for inspection, it shall be applied to 10% of the actual bolts on the structure (but not less than two bolts) randomly selected in each condition. If no nut or bolt head is turned by applying the preset setting for inspection, the connection is acceptable. If any nut or bolt head is turned by applying this preset inspection setting, this preset inspection setting should be applied to all bolts in the connection, and all the bolts that turned must be tightened and reinspected. As an alternative, all bolts in the connection may be retightened and submitted for the specified inspection.

With the direct tension indicator, inspection at the time of installation or after the time of installation is simpler. The inspector need only observe the installed bolts and satisfy himself that the indicators have closed to at least the average required gap. This can be checked through the use of a feeler gage (Fig. 11-8a). When the gage does not enter the gap (a gap being evident), the installation is considered a valid and proper one.

Visual inspection of the sides of bolt heads and nuts that were tightened with an impact wrench for peening indicates that the wrench was applied to the bolt. In bearing-type connections, no further inspection is really needed, since bolts in bearing are not dependent on high tension. The peening is an indication that the bolt has been tightened sufficiently to prevent nut loosening.

In a connection where washers were not used, torque readings will vary considerably and be relatively high. A torque multiplier device may then be necessary in this case. Further, when a washer has not been used under the turned element, heavy galling may occur, such that the required rotation or minimum tension was not attained. Yet a resistance to turning may occur, which can register a high torque reading on the inspection torque wrench. Because of this, some fabricators and erectors always use a hardened washer under the turned element, regardless of the tightening method used.

To equate torque values to the minimum bolt tensions specified for various bolt sizes is erroneous. All torque and impact wrenches used for inspection as well as installation should be precalibrated in a test device or cell capable of measuring the actual tension produced by the wrench on a representative bolt sample.

11-11.4 Economy of Material

The choice of which of the high-strength bolts to use is an economic decision. Cost comparisons between A325 and A490 connections depend upon more than the differential cost between the two bolt types (an approximate 50% material cost differential and a 30% installation cost differential). The use of A490 bolts would require fewer bolts than the use of A325 bolts for the same connection (therefore, there are fewer holes to be made in the connection material). However, the A490 bolts may require the use of washers while the A325 bolts may not require washers. The requirement of washers increases installation as well as material costs.

Further, since the A490 bolted connection requires fewer bolts, it requires less connection material. The type (grade) of connected material reflects the number of washers required and in this manner reflects upon final connection costs. The method of bolt installation dictates the use or nonuse of washers in addition. Under certain cases, different sizes of bolts will be required for each of the bolt grades (A325 and A490), and this has a bearing on cost.

A complete analysis of each connection, which would involve all material and labor, should be made to reach a decision as to which type of bolt to use, which size of bolt to use, what bolt action to assume (friction type or bearing type), and which method of installation to use.

Shop costs involved in hole preparation, field costs in physically filling holes with bolts, and the savings of time involved with fewer bolts in a connection could influence total completion time required for a structure. Lessening the construction time can have a tremendous impact on dollars (e.g., early occupancy or use of structure). A490 HSB, regardless of connection material, can conceivably reflect a significant savings in the connection. However, there are exceptions and for this reason a complete analysis is in order. A very excellent reference on this subject is an article by Frederick E. Graves, Cost Studies of High-Strength Bolted Connections, published in the October 1965 issue (vol. 2, no. 4) of the *AISC Engineering Journal,* pages 109–114. Graves presents numerous design examples to illustrate the analyses.

11-12   WELDING

Welding of steel is the process of joining or combining steel pieces by means of melting a similar metal into the joint, which joins the steel pieces. It has been said that, given a good welder and proper welding procedures and equipment, any two metals can be welded.

Welding can permit freedom and economy of design, a savings in fabrication costs, and aesthetic advantages. The weld itself is stronger than the parent (base) material, and the yield point of the weld metal is a multiple of the yield point of the base material. Two of the main reasons for this are that the core wire used in an electrode, for instance, is made of premium steel manufactured to much closer specifications than parent metal, and that during the welding process complete shielding of the molten weld metal, together with the use of various beneficial agents in the electrode coating, produces a superior steel (more uniform crystalline structure as well as higher physical properties).

Rather than treat the entire subject of welding, a concentrated discussion of electrodes, weld types, and weld strengths will be presented. For more details, the references listed in Section 11-20 should be consulted.

The AISC Specification parallels in its recommendations the American Welding Society (AWS) Specifications for symbols, welding electrodes for manual shielded metal-arc welding, submerged-arc welding, gas metal-arc welding, and flux cored-arc welding. Further, the AISC Specification recommends permissible stresses, and effective areas of weld metals for the various types of welds are defined. For Welder, Tacker and Welding Operator Qualifications the AWS Specifications should be referred to, as well as for the Qualification of Weld and Joint Details, required electrodes, and matching weld metal. All the welding processes (submerged-arc, gas metal-arc, flux cored-arc, electroslag, and electrogas) are found in an appropriate portion of the AWS D1.1 Specification. The proper electrodes or flux to be used are found there, together with associated design, workmanship, and so on, to ensure sound welds.

11-12.1 Electrodes

Optimally, the chemical and mechanical properties of the deposited weld metals should be as close as possible to those of the base metal. This necessitates a variety of electrodes to meet all the variations existing in a job and the variations involved with material types.

ASTM, in cooperation with AWS, has established a numbering system that classifies electrodes. The system takes the form of the prefix letter E followed by four or five digits. E stands for electrode. The first two (or three) digits designate the minimum tensile strength in kips per square inch. The third (or fourth) digit indicates the welding position (1 = all positions: flat, horizontal, vertical, and overhead; 2 = flat and horizontal position only; 3 = flat position only). The fourth (or fifth) digit designates the current supply and welding technique variables.

The first step in the selection of an electrode is to decide upon the minimum strength requirement. The next steps are the current supply and application and position of welding, respectively.

The AWS Specification states that E70XX electrodes should be used for all welds for A36 steel and high-strength steel up to A572 Grade 55; for A572 Grades 60 and 65, E80XX should be used; and for A514, E100XX (over $2\frac{1}{2}$ in. thick) and E110XX (for $2\frac{1}{2}$ in. and under thickness). These required electrodes are based upon the shielded metal-arc welding process (electrodes for the other welding processes are also given in the AWS Specification). Where different yield points of parent metal are encountered in any one connection, the electrode should be determined by the lower of the two yield points. These electrodes should be used for the making of groove welds (Section 11-12.2). Lower-strength weld metal may be used for other weld types.

The AISC and AWS specifications in tabular form, suggest the required weld strength level (in terms of the strength of the "matching" or base metal) for the various types of welds (complete-penetration groove welds, partial-penetration groove welds, fillet welds, and plug or slot welds) with various allowable stresses (i.e., in terms of base-metal allowable stress or a multiple of the weld-metal nominal tensile stress) of weld metal. Weld metal that is one strength level more than the base metal is permitted. Another table within the AWS Specification specifies the minimum preheat and interpass temperatures (°F) required for combinations of thickness of part joined, welding process, and the base metal type of material. Preheating and the proper interpass temperature reduce shrinkage stresses in the weld and adjacent base metal. Providing a slow rate of cooling prevents excessive hardening and lower ductility in both weld and adjacent base metal, and keeps welds from cracking.

11-12.2 Weld Types

Figure 11-9 shows the standard symbols for welds and weld designation. The single-pass (usually a $\frac{5}{16}$-in. fillet leg size can be achieved with a single pass) *fillet weld* is the most commonly used type of weld. Less labor and material per square inch of stress-carrying area are required. The size of a fillet weld is expressed as the length of the shorter of its two legs, as shown in Fig. 11-10a (leg size = a; effective throat = b). For efficiency and economy, equal leg fillet welds are usually used. The maximum size of fillet weld for design purposes along the edge of material $< \frac{1}{4}$ in. thick is the thickness of material, and for material $\geq \frac{1}{4}$ in. thick the size of fillet

WELDED JOINTS
Standard Symbols

Basic Weld Symbols									
Back or Backing	Fillet	Plug or Slot	Groove or Butt						
			Square	V	Bevel	U	J	Flare V	Flare Bevel
⌒	◺	▭	‖	⋁	⋁	⋃	⊬	⋁	⎰⎱

Supplementary Weld Symbols				
	Weld all Around	Field Weld	Contour	
			Flush	Convex
			For other basic and supplementary weld symbols, see AWS A2.4	
	○	▶	—	⌒

Standard Location of Elements of a Welding Symbol

Finish symbol

Contour symbol

Root opening, depth of filling for plug and slot welds

Size in inches

Reference line

Specification, process or other reference

Tail (may be omitted when reference is not used)

Basic weld symbol or detail reference

F
A
R
S
T
(Both sides)
(Arrow side)
(Other side)
L @ P
A
B

Groove angle or included angle of countersink for plug welds

Length of weld in inches

Pitch (c. to c. spacing) of welds in inches

Weld-all-around symbol

Field weld symbol

Arrow connects reference line to arrow side of joint. Use break as at A or B to signify that arrow is pointing to the grooved member in bevel or J-grooved joints.

Note:

Size, weld symbol, length of weld and spacing must read in that order from left to right along the reference line. Neither orientation of reference line nor location of the arrow alter this rule.

The perpendicular leg of ◺.⋁.⊬.⎰⎱ weld symbols must be at left.

Arrow and Other Side welds are of the same size unless otherwise shown.

Symbols apply between abrupt changes in direction of welding unless governed by the "all around" symbol or otherwise dimensioned.

These symbols do not explicitly provide for the case that frequently occurs in structural work, where duplicate material (such as stiffeners) occurs on the far side of a web or gusset plate. The fabricating industry has adopted this convention; that when the billing of the detail material discloses the identity of far side with near side, the welding shown for the near side shall be duplicated on the far side.

Courtesy of American Institute of Steel Construction.

FIG. 11-9

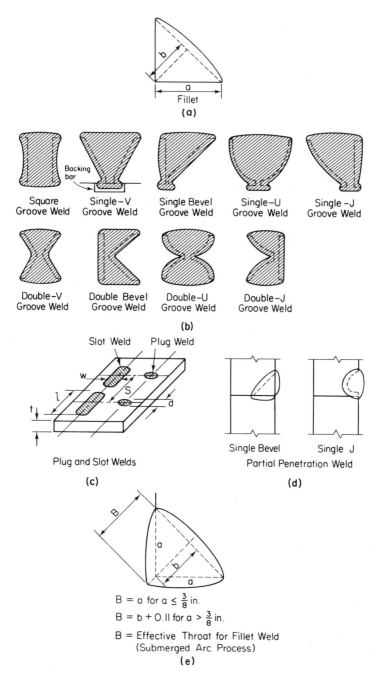

Fillet
(a)

Square Groove Weld   Single-V Groove Weld   Single Bevel Groove Weld   Single-U Groove Weld   Single-J Groove Weld

Backing bar

Double-V Groove Weld   Double Bevel Groove Weld   Double-U Groove Weld   Double-J Groove Weld
(b)

Slot Weld   Plug Weld

Plug and Slot Welds
(c)

Single Bevel   Single J
Partial Penetration Weld
(d)

$B = a$ for $a \leq \frac{3}{8}$ in.

$B = b + 0.11$ for $a > \frac{3}{8}$ in.

B = Effective Throat for Fillet Weld
(Submerged Arc Process)
(e)

FIG. 11-10

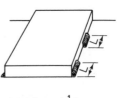

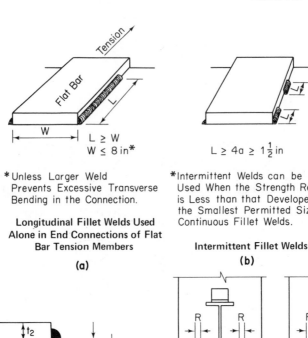

*Unless Larger Weld
Prevents Excessive Transverse
Bending in the Connection.

**Longitudinal Fillet Welds Used
Alone in End Connections of Flat
Bar Tension Members**

(a)

*Intermittent Welds can be
Used When the Strength Required
is Less than that Developed by
the Smallest Permitted Size of
Continuous Fillet Welds.

Intermittent Fillet Welds

(b)

$W \geq 5t_1 \geq 1$ in.

Lap Joints

(c)

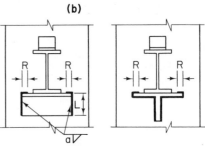

*R = Weld Return $\geq 2a$

* Applies to Welds Connecting
Brackets, Beam Seats, etc.

Side and Top Fillet Welds

(d)

FIG. 11-11

is $\frac{1}{16}$ in. less than the thickness of material (unless the weld is built out to full throat thickness). The minimum effective length $\geq 4a$; otherwise, $a \leq \frac{1}{4}$ (effective length) should be considered. Additional requirements for fillet welds are given in Fig. 11-11.

The minimum size of fillet weld in a connection joined by only fillet welds must be as in Table 11-8. a is determined by the thicker of the two parts joined and does not have to exceed the thickness of the thinner part jointed unless a larger size is required by stress computations. Table 11-8 is based upon the required minimum fillet weld size, a, providing sufficient heat input into the plate to give a desired slow rate of cooling.

Table 11-8
MINIMUM WELD SIZES

Thickness of Thicker Plate Joined (in.)	a_{min} for Fillet Welds; $b_{eff\,min}$ for Partial-Penetration Groove Welds
To $\frac{1}{4}$	$\frac{1}{8}$
$> \frac{1}{4}$ to $\frac{1}{2}$	$\frac{3}{16}$
$> \frac{1}{2}$ to $\frac{3}{4}$	$\frac{1}{4}$
$> \frac{3}{4}$ (to $1\frac{1}{2}$*)	$\frac{5}{16}$
$> 1\frac{1}{2}$ to $2\frac{1}{4}$*	$\frac{3}{8}$
$> 2\frac{1}{4}$ to 6*	$\frac{1}{2}$
> 6*	$\frac{5}{8}$

* For partial penetration welds *only.*

Thick plates offer greater restraint and produce a faster cooling rate for the welds. On multipass welds, the plate may have to be preheated to provide sufficient heat input. Table 11-8 also shows the minimum size of partial penetration groove welds.

One of the simplest and most efficient type of welds is the *butt weld,* usually used to join the edge of a plate to the edge of a plate. This type of weld usually requires some edge preparation of the plates, such as the beveling of the plate to form a groove. The weld metal is then placed in this groove and fused to the base metal. The groove may be a single or double bevel, or a single or double V, J, or U, depending on the base-metal thickness, position of welding, and accessibility for welding. Backing bars or strips are required for groove welds made from one side of the joint in order to attain the required amount of full penetration. In lieu of a backing bar, the root pass may be back-gouged or back-chipped with a torch and rewelded from the backside of the joint. *Groove welds* may be made as *full-* or *partial-penetration welds.* The single-V or double-V groove is the most commonly used. Butt welds with two square abutting ends may also be used. Figure 11-10b shows many of the standard groove welds and depicts the edge preparation of the two abutting plates. Partial-penetration welds are similar to those shown as full-penetration welds except that usually the joints are welded from one side. Additionally, Fig. 11-10d shows other partial-penetration welds. For a more detailed and more varied pictorial representation of full- and partial-penetration welds, reference is made to Welded Joints, Part 4, of the AISC Manual. This part of the Manual shows acceptable permissible welds without any further welding qualification under the AWS provisions. Acceptable joint forms, details, welding processes, and procedures are shown.

Plug or *slot welds* can be used to transmit shear and prevent buckling in a lap joint or to join elements of built-up members. They are used in circular holes or slotted holes with circular ends. Weld metal is deposited in the holes. Plug and slot welds are not usually used for strength connections but rather to stitch components together that are too wide to be joined efficiently by other types of welds. They are frequently used to fasten plates such as floor plates to avoid overhead (antigravity) welding. Plug and slot welds are shown in Fig. 11-10c.

The diameter of the holes for plug welds $d \geq t + \frac{5}{16}$ in. rounded to the next greater odd $\frac{1}{16}$ in., and in addition $d \geq 2\frac{1}{4}(t_w)$, where t_w is the thickness of the weld metal. S, the center-to-center spacing of plug welds $\geq 4d$. The thickness of plug or slot welds, t_w, must be equal to t for $t \leq \frac{5}{8}$ in. and $\geq t/2$, but $\geq \frac{5}{8}$ in. for $t > \frac{5}{8}$ in. (Fig. 11-10c). The length of a slot, l, of a slot weld $\leq 10 t_w$. The width of the slot $w \geq t + \frac{5}{16}$ in. rounded to the next greater odd $\frac{1}{16}$ in., and in addition $w \leq 2\frac{1}{4}(t_w)$. The slot ends should be semicircular or should have the corners rounded to a radius, $r \geq t$ (Fig. 11-10c).

The minimum center-to-center spacing of slot welds normal to their length should equal $4w$. The minimum center-to-center spacing, S, in the longitudinal direction on any line for slot welds should equal $2l$ (Fig. 11-10c).

When large or closely spaced plug or slot welds are used, there can be a large amount of shrinkage. For this case, fillet welds in the holes or slots may be used to transmit shear in lap joints or to prevent buckling or separation of lapped joints. They may also be used for joining elements of built-up members. Fillet welds in slots or holes are not considered to be plug or slot welds.

The AISC Specification gives additional rules pertaining to welded construction and should be referred to.

11-12.3 Weld Strength

All welds are proportioned to meet the allowable stresses given in Table 11-11. These allowable stresses are specified and vary with the type of weld (complete- or full-penetration groove welds, partial-penetration groove welds, fillet welds, or plug or slot welds), type (tension, compression, or shear), and direction of stress (normal to effective area, parallel to the weld axis, etc.). The allowable stresses are applied to the "effective area" of the type of weld listed. The effective areas of weld metal are as listed in Table 11-9. Effective throat thicknesses of the weld types are listed in Table 11-10. Table 11-11 lists the allowable weld stresses and Table 11-12 lists the "matching" weld metal (matched to the grade of base or parent metal).

Table 11-9
EFFECTIVE AREAS OF WELD METALS

Type of Weld	Effective Area	l_{eff}[a]
Groove	$l_{eff}b$[b]	Width of part joined
Fillet	$l_{eff}b$	Overall length + returns
Plug or slot	A[c]	
Fillets in holes or slots	$l_{eff}b$[d]	Length of center line of weld through the center of the plane through the throat

[a] Effective length of weld.

[b] b = effective throat thickness (Table 11-10).

[c] A = effective shearing cross-sectional area of the hole or slot in the plane of the faying surface.

[d] In case of overlapping fillet welds, the effective area ≤ the area of the hole or slot in the plane of the faying surface.

Table 11-10
EFFECTIVE THROAT THICKNESS OF WELDS

Type of Weld	Effective Throat Thickness
Fillet	Shortest distance from root to face of the diagrammatic weld (Fig. 11-10a) (0.707a for equal-leg fillet)
Fillet (submerged arc process)	$B = a$ for $a \leq \frac{3}{8}$ in.; $B = b + 0.11$ in. for $a > \frac{3}{8}$ in. (Fig. 11-10e)
Complete (full) penetration groove	Thickness of thinner element joined
Partial-penetration groove[a] (for all welding positions with shielded metal arc or submerged arc welding process)	Depth of groove (included angle ≥ 60°)[b]
	Depth of groove: $\frac{1}{8}$ in. (for included angle ≥ 45° and < 60°)[b]

[a] See AWS D1.1 and AISC for additional rules.

[b] Not less than $b_{eff min}$ shown in Table 11-8.

11-13 DESIGN AIDS

Part 4 of the AISC Manual covers the subject of connections. Many useful design aids are presented which, when used, shorten the design time significantly.

Table 11-11
ALLOWABLE WELD STRESSES

Type of Weld and Stress[1]	Allowable Stress	Required Weld Strength Level[2,3]
Complete Penetration Groove Welds		
Tension normal to the effective area	Same as base metal	"Matching" weld metal must be used; see Table 11-12.
Compression normal to the effective area	Same as base metal	Weld metal with a strength level equal to or less* than "matching" weld metal may be used.
Tension or compression parallel to axis of the weld[5]	Same as base metal	
Shear on the effective area	0.30 x nominal tensile strength of weld metal[4] (ksi), except shear stress on base metal shall not exceed 0.40 x yield stress of base metal	*One classification (10 ksi) for compression normal to effective area.
Partial Penetration Groove Welds		
Compression normal to effective area	Same as base metal	Weld metal with a strength level equal to or less than "matching" weld metal may be used.
Tension or compression parallel to axis of the weld[5]	Same as base metal[6]	
Shear parallel to axis of weld	0.30 x nominal tensile strength of weld metal[4] (ksi), except shear stress on base metal shall not exceed 0.40 x yield stress of base metal	
Tension normal to effective area	0.30 x nominal tensile strength of weld metal[4] (ksi), except shear stress on base metal shall not exceed 0.60 x yield stress of base metal	
Fillet Welds		
Shear on effective area	0.30 x nominal tensile strength of weld metal[4] (ksi), except shear stress on base metal shall not exceed 0.40 x yield stress of base metal	Weld metal with a strength level equal to or less than "matching" metal may be used.
Tension or compression parallel to axis of weld[5]	Same as base metal	
Plug and Slot Welds		
Shear parallel to faying surfaces (on effective area)	0.30 x nominal tensile strength of weld metal[4] (ksi), except shear stress on base metal shall not exceed 0.40 x yield stress of base metal	Weld metal with a strength level equal to or less than "matching" weld metal may be used.

[1] For definition of effective area see Table 11-9.

[2] For "matching" weld metal, see Table 11-12.

[3] Weld metal one strength level stronger than "matching" weld metal will be permitted.

[4] First two (or three) digits of electrode designation.

[5] Fillet welds and partial penetration groove welds joining the component elements of built-up members, such as flange-to-web connections, may be designed without regard to the tensile or compressive stress in these elements parallel to the axis of the welds.

[6] For joints designed to bear. For joints not designed to bear, 0.50 nominal tensile strength of weld metal (ksi), except stress on base metal ≯ 0.60 yield stress of base metal.

Courtesy of American Institute of Steel Construction.

Table 11-12
"MATCHING" WELD METALS

Base (Parent) Metal[3]	Welding Process[1,2]			
	Shielded Metal-Arc	Submerged-Arc	Gas Metal-Arc	Flux Cored-Arc
ASTM A36, A53 Gr. B, A500, A501, A529, and A570 Gr. D and E	AWS A5.1 or A5.5, E60XX or E70XX[3]	AWS A5.17 or A5.23 F6X or F7X-EXXX	AWS A5.18 E70S-X or E70U-1	AWS A5.20 E60T-X or E70T-X (except EXXT-2 and EXX-3)
ASTM A242, A441, A572 Grades 42 thru 55 and A588[4]	AWS A5.1 or A5.5, E70XX[5]	AWS A5.17 or A5.23 F7X-EXXX	AWS A5.18 E70S-X or E70U-1	AWS A5.20 E70T-X (except E70T-2 and E70T-3)
ASTM A572 Grade 60 & 65	AWS A5.5 E80XX[5]	AWS #A5.23, Grade F8X–EXXX	Grade E80S	Grade E80T
ASTM A514 over $2\frac{1}{2}$ in. thick	AWS A5.5 E100XX[5]	AWS A5.23 Grade F10X-EXXX	Grade E100S	Grade E100T
ASTM A514 $2\frac{1}{2}$ in. thick and under	AWS A5.5 E110XX[5]	AWS A5.23 Grade F11X-EXXX	Grade E110S	Grade E110T

Use of the same type filler metal having next higher mechanical properties is permitted.

[1] When welds are to be stress relieved the deposit weld metal shall not exceed 0.05 percent vanadium.

[2] See AWS for electroslag and electrogas weld metal requirements.

[3] On joints involving base metals of different yield strengths, filler metals applicable to the lower yield strength may be used.

[4] For architectural exposed bare unpainted applications requiring weld metal with atmospheric corrosion resistance and coloring characteristics similar to that of the base metal see AWS. The steel manufacturer's recommendation shall be followed.

[5] Low hydrogen classifications.

Courtesy of American Institute of Steel Construction.

11-13.1 Allowable Loads on Fasteners

Allowable tension, shear, and bearing values are listed for the various fasteners (threaded parts, rivets, common bolts, and high-strength bolts). Fasteners up to $1\frac{1}{2}$ in. in diameter are covered by these tables.

The Allowable Load in Shear Table (I-D) provides allowable shear values (kips) for bolts in standard, oversized, short-slotted, long-slotted, and long- or short-slotted holes normal to the load direction (as required in bearing-type connections). Values for rivets and threaded parts in standard holes are also given. Single and double shear allowables are given for both friction- and bearing-type connections. For bearing-type connections, values are given for threads included in and excluded from the shear plane.

The Allowable Load in Bearing Tables (I-E) provide allowable loads for fasteners in both friction- and bearing-type connections ($\frac{3}{4}$, $\frac{7}{8}$, and 1 in. diameters) for standard holes in material of various thicknesses. Values are given for the usual 3-in. center-to-center spacing, the minimum spacing to obtain full bearing (Eq. 11-1b, $F_p = 1.5F_u$), the preferred spacing ($3d$), the absolute minimum spacing ($2\frac{2}{3}d$), as well as for the preferred (1.50 in.) and absolute (1.25 in.) minimum end distances for different F_u values of the base material of the connected part.

The aforementioned values based upon various fastener spacing are arrived at by using the minimum distance between centers of standard holes (AISC Section 1.16.4.2), which is expressed as $(2P/F_ut) + (d/2)$, where P is the force transmitted by one fastener to the critical connected part (kips), F_u is the specified minimum tensile strength of the critical connected part (ksi), t is the thickness of the critical connected part (inches), and d is the diameter of the fastener (inches). Substitution of the appropriate minimum spacing and solving for P yields the value in the table. In a similar manner, using Eq. 11-1a (the minimum distance from the center of a standard hole to the edge of the connected part), these table values are calculated.

In addition, an Allowable Load in Bearing Table for Bolts and Rivets (I-F) is given (for one fastener, 1 in.-thick material) for various specified minimum tensile strengths of web material (F_u) for different values of edge distances l_v and l_h. l_v, the distance from center of hole to free edge of the connected part in the direction of the force, is expressed by Eq. 11-1a, and l_h, the distance from the center of hole to the end of the beam web, is expressed by Eq. 11.1c. Both equations are solved for P or P_R to obtain the tabular values.

Finally, a table giving Coefficients for Web Tear-out (Block Shear) for Bolts and Rivets in Bearing (I-G) (based on standard holes and a fastener spacing of 3 in.) is given. These coefficients, C_1 and C_2, when added and multiplied by F_ut, yield R_{BS}, the resistance to block shear in kips [i.e., $R_{BS} = (C_1 + C_2)F_ut$]. Coefficient values are tabulated for various combinations of l_v and l_h values (for C_1) and for combinations of a number of fasteners in a line and bolt diameters (for C_2). A complete explanation of the table is found in the notes at the bottom of the table.

Block shear (tearing failure) was discussed in Section 4-3 and will be discussed in more detail in this section. R_{BS} is actually the allowable reaction (maximum) to avoid this type of failure. For conditions that differ from those tabulated, the general equation shown at the bottom of the table may be used.

The AISC specification requires that the following conditions be checked to determine the allowable capacity of a framed beam connection: bolt shear, bolt bearing on the connecting material, beam web block shear, shear on the net area of the connection angles or plate, and local bending stresses.

Recent laboratory tests have indicated that at high-strength bolted beam end connections, where the top flange is coped and the web subject to high bearing stresses, failure may occur by shear along a plane through the vertical line of fasteners in the beam web or by a combination of this shear with tension along a plane perpendicular to the plane through the fasteners. F_v, the allowable shear stress, equals $0.30F_u$ and is applied on the area effective in resisting this tearing failure. The effective area is the minimum net failure surface bound by the bolt holes and the plane normal to the bolt holes. As an alternate, the tension and shear areas can be treated separately, and the allowable connection capacity (resistance to tearing failure) expressed as $0.30A_vF_u + 0.50A_tF_u$, where A_v and A_t are the net shear and net tension areas. A_v is the net area along a plane through the vertical line of web fasteners, and A_t is the net area along a plane normal to the vertical line of fasteners (through a fastener row). Similar connections where this type of failure is possible would be thin bolted gusset plates in double shear, for example. These situations should be investigated in a similar manner.

11-13.2 Framed Beam Connections (Bolted)

Allowable total loads in kips are given in the Framed Connections Tables for 2 to 10 rows of bolt fasteners (diameters vary: $\frac{3}{4}$, $\frac{7}{8}$, and 1 in.). Three tables are presented: Tables II-A, II-B, and II-C. Table II-A covers A307 bolts in standard or slotted holes, A325 and A490 bolts in friction-type connections (standard holes and class A, clean mill scale surface condition) and bolts (A325-N, A490-N, A325-X, A490-X) in bearing-type connections with standard or slotted holes. Table II-B covers bolts (A325-F, A490-F) in friction-type connections with long-slotted, oversized, and short-slotted holes with class A, clean mill scale surface condition. Table II-C gives the allowable shear in A36 connection angles and is used for checking the shear capacity on the net section of these angles.

Tabulated loads for Tables II-A and II-B are based on double shear of the bolts unless otherwise noted by footnote. Connection angle thicknesses in these tables were established to conform with an end distance of $1\frac{1}{4}$ in. for material with $F_u = 58$ ksi. The footnote mentioned indicates when certain tabulated values are governed by the bearing capacity of the angles or the net shear on the angles, and refers one to Table II-C.

When the beam flange is coped and standard holes at 3-in. spacing are used, the allowable load is found in Table II-A or II-C (as applicable) and, in addition, block shear must be checked by using the Coefficients for Web Tear-out (Block Shear) Table. When the beam flange is not coped and standard holes at 3-in. spacing are used, the allowable load is obtained from Table II-A or II-C, if the footnotes so indicate.

Table II-B gives the allowable load for bolts in friction-type connections with oversized or slotted holes and class A, clean mill scale surface condition. Again, if the beam is coped, block shear must be checked utilizing the Coefficients for Web Tear-out (Block Shear) Table (making appropriate adjustments to account for the length of slot as it affects l_v or l_h).

For bearing-type connections in slotted holes, the long dimension of the slot is required to be normal to the direction of the load. For noncoped beams with slotted holes in the web, allowable shear values are obtained from Table II-A and allowable bolt bearing values can be obtained from Table I-F by entering the table with the actual l_h reduced by the increment C_2 (AISC Specification Table 1.16.5.4). For coped beams, block shear must also be checked by using Table I-G by reducing l_h by the increment C_2. Table II-C is used to check the net shear capacity of the connection angles.

Table II-B is used for friction-type connections using slotted or oversized holes in the connection angles. When slotted holes are used in the beam web (long dimension of slotted hole normal to the direction of the load), bolt bearing values and block shear are obtained as discussed in the previous paragraph.

Table II-C values are based on an allowable shear value, $F_v = 0.3F_u$ (for A36 material) on the net section (taken vertically through the bolt holes) of the two angles for standard holes.

The length of the connection angles is dependent on the beam T-dimension for uncoped beams (the T-dimension is the depth of web clear of the flange to web fillet). For stability reasons during the erection process, the minimum length of connection angle should be at least $T/2$. End distances on the angles in these tables are $1\frac{1}{4}$ in. for standard holes. For oversized and slotted holes, the edge or end distance must be increased over that required for standard holes by the applicable increment, C_2 (AISC Specification, Table 1.16.5.4). Although the vertical spacing of fasteners is assumed to be 3 in. in these tables, the spacing may be varied in accordance with specifications. Supporting member gages should be chosen to meet the requirements of the specification provisions for minimum spacing, minimum edge distance, and maximum edge distance (not to exceed $12t$ or 6 in., where t is the thickness of the part). Clearances required for assembling the connection are necessary (refer to AISC Manual, Part 4). Framing angles with holes staggered in alternate angle legs are permitted and, when used, provide the necessary clearance with smaller gages on the legs of connection angles.

11-13.3 Framed Beam Connections (Welded)

The Framed Beam Connections Tables are so arranged as to permit the substitution of welds for bolts in the connections shown in the Tables, which fall within the weld capacities. In one case (Fig. 11-12, diagram top left) "Weld A" replaces the fasteners in the beam web angle legs. In another case (Fig. 11-12, diagram top right) "Weld B" replaces the fasteners in the outstanding angle legs.

To match the usual gages (see the last column of the AISC Manual, Seventh Edition, W Shapes Dimension for Detailing Table), the angle widths are generally 4 × 3½ in. with the 4-in. leg the outstanding leg (i.e., 3½-in. leg connects to the beam web). For the case with Weld A, the 3½-in. leg may be reduced to 3 in. For the case with Weld B, the outstanding leg may be reduced from 4 in. to 3 in. for values of angle lengths, L, equal to 5½ in. to 1 ft 5½ in. When the 3-in. legs are used, the tabulated capacities of Welds A and B are conservative. In bearing-type connections for the outstanding legs for bolts, the bearing capacity of the supporting member should be checked.

Angle thickness is equal to the larger weld size plus $\frac{1}{16}$ in. or the angle thickness used in the Frame Beam Connections, Bolted Tables. These tables are for the usual framed connection, which is welded with E70 electrodes in combination with the bolts used in bolted framed beam connections. The minimum web thickness for Welds A are tabulated for A36 and F_y = 50-ksi steel. The purpose of the minimum-web-thickness restriction is to prevent overstress of the beam web at the location of maximum weld stress. The minimum web thickness is calculated by equating the expression for allowable shear stress in the fillet welds to the allowable shear stress of the beam web and solving for the minimum web thickness.

The allowable capacity for Welds A in the table uses an instantaneous center solution (similar to that developed for Eccentric Loads on Weld Groups Tables, Fig. 11-20). This solution is discussed in Section 11-13.10. The capacity for Welds B is calculated utilizing traditional vector analysis techniques.

Although Welds A and B can be combined from Fig. 11-12, a more economical all-welded connection that provides greater flexibility in angle-length selection and connection capacity can be obtained through the use of the all-welded Framed Beam Connections shown in Fig. 11-13.

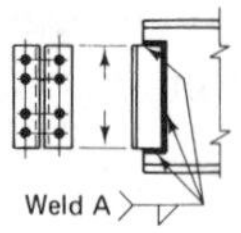

FRAMED BEAM CONNECTIONS

Welded—E70XX Electrodes

Allowable Loads in Kips

 Weld A

 Weld B

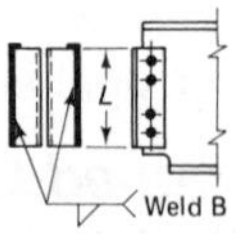

Weld A		Weld B		Angle Length L In.	[a] Minimum Web Thickness for Welds A		Maximum Number of Fasteners in One Vertical Row
Capacity, Kips	[b] Size, In.	[c] Capacity, Kips	Size In.		F_y = 36 ksi F_v = 14.5 ksi	F_y = 50 ksi F_v = 20 ksi	
275	5/16	296	3/8	29½	0.64	0.46	
220	1/4	247	5/16	29½	0.51	0.37	10
165	3/16	197	1/4	29½	0.38	0.28	
253	5/16	261	3/8	26½	0.64	0.46	
202	1/4	217	5/16	26½	0.51	0.37	9
152	3/16	173	1/4	26½	0.38	0.28	
230	5/16	223	3/8	23½	0.64	0.46	
184	1/4	186	5/16	23½	0.51	0.37	8
138	3/16	149	1/4	23½	0.38	0.28	
205	5/16	187	3/8	20½	0.64	0.46	
164	1/4	156	5/16	20½	0.51	0.37	7
123	3/16	125	1/4	20½	0.38	0.28	
183	5/16	152	3/8	17½	0.64	0.46	
146	1/4	126	5/16	17½	0.51	0.37	6
110	3/16	101	1/4	17½	0.38	0.28	
156	5/16	115	3/8	14½	0.64	0.46	
125	1/4	95.7	5/16	14½	0.51	0.37	5
93.5	3/16	76.6	1/4	14½	0.38	0.28	
128	5/16	80.1	3/8	11½	0.64	0.46	
102	1/4	66.9	5/16	11½	0.51	0.37	4
76.6	3/16	53.4	1/4	11½	0.38	0.28	
99.0	5/16	48.2	3/8	8½	0.64	0.46	
79.2	1/4	40.3	5/16	8½	0.51	0.37	3
59.4	3/16	32.2	1/4	8½	0.38	0.28	
61.8	5/16	21.9	3/8	5½	0.64	0.46	
49.5	1/4	18.3	5/16	5½	0.51	0.37	2
37.1	3/16	14.6	1/4	5½	0.38	0.28	

[a] When the beam web thickness is less than the minimum, multiply the connection capacity furnished by Weld A by the ratio of the actual web thickness to the tabulated minimum thickness. Thus, if 5/16" Weld A, with a connection capacity of 156 kips and a 14½" long angle, is considered for a beam of web thickness of 0.375" with F_y = 36 ksi, the connection capacity must be multiplied by 0.375/0.64, giving 91.4 kips.

[b] Should the thickness of material to which connection angles are welded exceed the limits set by AISC Specification (Table 11-8) for weld sizes specified, increase the weld size as required, but not to exceed the angle thickness.

[c] When welds are used on outstanding legs, connection capacity may be limited by the shear capacity of the supporting member.

Note1: Connection Angle: Two L 4 X 3½ X thickness X L; F_y = 36 ksi. See Sect. 11-13.3 for limiting values of thickness and optional width of legs.

Note 2: Capacities shown in this table apply only when the material welded is F_y = 36 ksi or F_y = 50 ksi steel.

Courtesy of American Institute of Steel Construction.

FIG. 11-12

11-13.4 Seated Connections (Bolted)

Seated unstiffened shear connections as shown in Fig. 11-1b should only be used when the beam is laterally supported by a top angle placed in either of the two locations shown. The AISC Manual has tables showing the allowable loads in kips of the outstanding legs of angles for various beam web thicknesses (F_y = 36 and 50 ksi) for outstanding angle legs of 4 in. (F_y = 36 ksi) of various thicknesses and of 6- or 8-in. lengths. The allowable loads are based upon a $\frac{3}{4}$-in. setback of beam from column face to provide for possible mill underrun in the beam length (nominal beam setback is $\frac{1}{2}$ in.).

A307 bolts can be used unless otherwise specified in the AISC Specification (see Field Connections). The allowable loads in Table A of the Seated Beam Connections in the AISC Manual are based on A36 steel for both beam and seat angle. These values are conservative when used with beams of greater strength. Table B of the Seated Beam Connections in the AISC Manual should be used when $F_y \geq$ 50 ksi for the beam. For beams of greater strength, the allowable loads in Table B will be conservative.

11-13.5 Seated Connections (Welded)

Seated unstiffened shear connections as shown in Fig. 11-1b can also be welded. Once again, the beam must be laterally supported by the top angle located in either of the two positions shown. Figure 11-14 shows allowable loads for all welded connections similar to those shown in the AISC Manual for bolted connections. These allowable loads are based upon the use of E70 electrodes. The loads for other electrodes may be found by multiplying the E70 tabulated loads by the ratio of the electrode strengths (i.e., for E80 electrodes, multiply the tabular values by $\frac{80}{70}$). Of course, the welds and base metal must meet the provisions of Tables 11-11 and 11-12.

The welds shown, which attach the beam to seat angle or beam to top angle, can be replaced by A307 bolts where permitted by the AISC Specification. The allowable loads in Tables A and B are based on $\frac{3}{4}$-in. beam setback from the column face, which provides for the possibility of mill underrun in the beam span (nominal beam setback is $\frac{1}{2}$ in.). The remarks in Section 11-13.4 relative to the allowable loads and Tables A and B are valid for these tables also.

The minimum size of welds and the maximum effective size of fillet welds as shown in Table 11-8 and discussed in Section 11-12.2, respectively, should be adhered to and, if necessary, the weld size or material thickness should be increased as required. No reduction of the tabulated weld capacities is required when unstiffened angle seats line up on opposite sides of the supporting web.

FRAMED BEAM CONNECTIONS

Welded—E70XX Electrodes

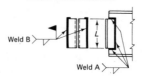

Weld A		Weld B		Angle Length L In.	Angle Size (F_y = 36 ksi)	[a]Minimum Web Thickness for Weld A	
Capacity Kips	[b]Size In.	[c]Capacity Kips	[b]Size In.			F_y = 36 ksi F_v = 14.5 ksi	F_y = 50 ksi F_v = 20 ksi
277	$\frac{5}{16}$	326	$\frac{3}{8}$	32	$4 \times 3 \times \frac{1}{2}$	0.64	0.46
221	$\frac{1}{4}$	271	$\frac{5}{16}$	32	$4 \times 3 \times \frac{3}{8}$	0.51	0.37
166	$\frac{3}{16}$	217	$\frac{1}{4}$	32	$4 \times 3 \times \frac{5}{16}$	0.38	0.28
269	$\frac{5}{16}$	302	$\frac{3}{8}$	30	$4 \times 3 \times \frac{1}{2}$	0.64	0.46
215	$\frac{1}{4}$	251	$\frac{5}{16}$	30	$4 \times 3 \times \frac{3}{8}$	0.51	0.37
161	$\frac{3}{16}$	201	$\frac{1}{4}$	30	$4 \times 3 \times \frac{5}{16}$	0.38	0.28
255	$\frac{5}{16}$	278	$\frac{3}{8}$	28	$4 \times 3 \times \frac{1}{2}$	0.64	0.46
204	$\frac{1}{4}$	231	$\frac{5}{16}$	28	$4 \times 3 \times \frac{3}{8}$	0.51	0.37
153	$\frac{3}{16}$	185	$\frac{1}{4}$	28	$4 \times 3 \times \frac{5}{16}$	0.38	0.28
240	$\frac{5}{16}$	254	$\frac{3}{8}$	26	$4 \times 3 \times \frac{1}{2}$	0.64	0.46
192	$\frac{1}{4}$	211	$\frac{5}{16}$	26	$4 \times 3 \times \frac{3}{8}$	0.51	0.37
144	$\frac{3}{16}$	169	$\frac{1}{4}$	26	$4 \times 3 \times \frac{5}{16}$	0.38	0.28
226	$\frac{5}{16}$	230	$\frac{3}{8}$	24	$4 \times 3 \times \frac{1}{2}$	0.64	0.46
181	$\frac{1}{4}$	191	$\frac{5}{16}$	24	$4 \times 3 \times \frac{3}{8}$	0.51	0.37
136	$\frac{3}{16}$	153	$\frac{1}{4}$	24	$4 \times 3 \times \frac{5}{16}$	0.38	0.28
211	$\frac{5}{16}$	206	$\frac{3}{8}$	22	$4 \times 3 \times \frac{1}{2}$	0.64	0.46
169	$\frac{1}{4}$	171	$\frac{5}{16}$	22	$4 \times 3 \times \frac{3}{8}$	0.51	0.37
127	$\frac{3}{16}$	137	$\frac{1}{4}$	22	$4 \times 3 \times \frac{5}{16}$	0.38	0.28
196	$\frac{5}{16}$	181	$\frac{3}{8}$	20	$4 \times 3 \times \frac{1}{2}$	0.64	0.46
157	$\frac{1}{4}$	152	$\frac{5}{16}$	20	$4 \times 3 \times \frac{3}{8}$	0.51	0.37
118	$\frac{3}{16}$	121	$\frac{1}{4}$	20	$4 \times 3 \times \frac{5}{16}$	0.38	0.28
181	$\frac{5}{16}$	157	$\frac{3}{8}$	18	$4 \times 3 \times \frac{1}{2}$	0.64	0.46
144	$\frac{1}{4}$	131	$\frac{5}{16}$	18	$4 \times 3 \times \frac{3}{8}$	0.51	0.37
108	$\frac{3}{16}$	105	$\frac{1}{4}$	18	$4 \times 3 \times \frac{5}{16}$	0.38	0.28
186	$\frac{5}{16}$	148	$\frac{3}{8}$	16	$3 \times 3 \times \frac{1}{2}$	0.64	0.46
149	$\frac{1}{4}$	123	$\frac{5}{16}$	16	$3 \times 3 \times \frac{3}{8}$	0.51	0.37
111	$\frac{3}{16}$	98.8	$\frac{1}{4}$	16	$3 \times 3 \times \frac{5}{16}$	0.38	0.28
147	$\frac{5}{16}$	124	$\frac{3}{8}$	14	$3 \times 3 \times \frac{1}{2}$	0.64	0.46
118	$\frac{1}{4}$	103	$\frac{5}{16}$	14	$3 \times 3 \times \frac{3}{8}$	0.51	0.37
88.3	$\frac{3}{16}$	82.5	$\frac{1}{4}$	14	$3 \times 3 \times \frac{5}{16}$	0.38	0.28
131	$\frac{5}{16}$	99.6	$\frac{3}{8}$	12	$3 \times 3 \times \frac{1}{2}$	0.64	0.46
105	$\frac{1}{4}$	83.1	$\frac{5}{16}$	12	$3 \times 3 \times \frac{3}{8}$	0.51	0.37
78.7	$\frac{3}{16}$	66.5	$\frac{1}{4}$	12	$3 \times 3 \times \frac{5}{16}$	0.38	0.28

FIG. 11-13 *Courtesy of American Institute of Steel Construction*

FRAMED BEAM CONNECTIONS

Welded—E70XX Electrodes

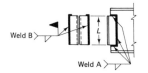

Weld A		Weld B		Angle Length L In.	Angle Size (F_y = 36 ksi)	[a] Minimum Web Thickness for Weld A	
Capacity Kips	[b] Size In.	[c] Capacity Kips	[b] Size In.			F_y = 36 ksi F_v = 14.5 ksi	F_y = 50 ksi F_v = 20 ksi
111	$\frac{5}{16}$	75.9	$\frac{3}{8}$	10	3 X 3 X $\frac{1}{2}$	0.64	0.46
89.1	$\frac{1}{4}$	63.3	$\frac{5}{16}$	10	3 X 3 X $\frac{3}{8}$	0.51	0.37
66.8	$\frac{3}{16}$	50.5	$\frac{1}{4}$	10	3 X 3 X $\frac{5}{16}$	0.38	0.28
101	$\frac{5}{16}$	64.3	$\frac{3}{8}$	9	3 X 3 X $\frac{1}{2}$	0.64	0.46
80.9	$\frac{1}{4}$	53.7	$\frac{5}{16}$	9	3 X 3 X $\frac{3}{8}$	0.51	0.37
60.7	$\frac{3}{16}$	42.9	$\frac{1}{4}$	9	3 X 3 X $\frac{5}{16}$	0.38	0.28
90.7	$\frac{5}{16}$	53.2	$\frac{3}{8}$	8	3 X 3 X $\frac{1}{2}$	0.64	0.46
72.6	$\frac{1}{4}$	44.4	$\frac{5}{16}$	8	3 X 3 X $\frac{3}{8}$	0.51	0.37
54.4	$\frac{3}{16}$	35.5	$\frac{1}{4}$	8	3 X 3 X $\frac{5}{16}$	0.38	0.28
80.1	$\frac{5}{16}$	42.5	$\frac{3}{8}$	7	3 X 3 X $\frac{1}{2}$	0.64	0.46
64.1	$\frac{1}{4}$	35.5	$\frac{5}{16}$	7	3 X 3 X $\frac{3}{8}$	0.51	0.37
48.1	$\frac{3}{16}$	28.3	$\frac{1}{4}$	7	3 X 3 X $\frac{5}{16}$	0.38	0.28
68.7	$\frac{5}{16}$	32.6	$\frac{3}{8}$	6	3 X 3 X $\frac{1}{2}$	0.64	0.46
54.9	$\frac{1}{4}$	27.1	$\frac{5}{16}$	6	3 X 3 X $\frac{3}{8}$	0.51	0.37
41.2	$\frac{3}{16}$	21.7	$\frac{1}{4}$	6	3 X 3 X $\frac{5}{16}$	0.38	0.28
57.2	$\frac{5}{16}$	23.4	$\frac{3}{8}$	5	3 X 3 X $\frac{1}{2}$	0.64	0.46
45.8	$\frac{1}{4}$	19.5	$\frac{5}{16}$	5	3 X 3 X $\frac{3}{8}$	0.51	0.37
34.3	$\frac{3}{16}$	15.7	$\frac{1}{4}$	5	3 X 3 X $\frac{5}{16}$	0.38	0.28
45.8	$\frac{5}{16}$	15.5	$\frac{3}{8}$	4	3 X 3 X $\frac{1}{2}$	0.64	0.46
36.6	$\frac{1}{4}$	12.9	$\frac{5}{16}$	4	3 X 3 X $\frac{3}{8}$	0.51	0.37
27.5	$\frac{3}{16}$	10.4	$\frac{1}{4}$	4	3 X 3 X $\frac{5}{16}$	0.38	0.28

[a] When the beam web thickness is less than the minimum, multiply the connection capacity furnished by Welds A by the ratio of the actual thickness to the tabulated minimum thickness. Thus, if $\frac{5}{16}$ " Weld A, with a connection capacity of 90.7 kips and an 8" long angle, is considered for a beam of web thickness 0.305" and F_y = 36 ksi, the connection capacity must be multiplied by 0.305/0.64, giving 43.2 kips.

[b] Should the thickness of material to which connection angles are welded exceed the limits set by AISC Specification (Table 11-8) for weld sizes specified, increase the weld size as required, but not to exceed the angle thickness.

[c] For welds on outstanding legs, connection capacity may be limited by the shear capacity of the supporting members.

Note 1:   Capacities shown in this table apply only when connection angles are F_y = 36 ksi steel and the material to which they are welded is either F_y = 36 ksi or F_y = 50 ksi steel.

Courtesy of American Institute of Steel Construction.

FIG. 11-13 (*Continued*)

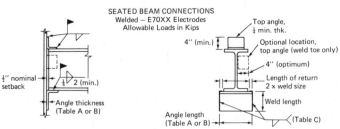

SEATED BEAM CONNECTIONS
Welded – E70XX Electrodes
Allowable Loads in Kips

TABLE A		Outstanding Leg Capacity, kips (based on OSL = $3\frac{1}{2}$ or 4 inches)										
	Angle Length		6 inches					8 inches				
	Angle Thickness, In.		$\frac{3}{8}$	$\frac{1}{2}$	$\frac{5}{8}$	$\frac{3}{4}$	1	$\frac{3}{8}$	$\frac{1}{2}$	$\frac{5}{8}$	$\frac{3}{4}$	1
F_y (ksi) 36	Beam Web Thickness In.	$\frac{3}{16}$	7.50	10.3	13.1	15.9	18.8	8.44	11.5	14.6	17.7	18.8
		$\frac{1}{4}$	9.57	13.0	16.3	19.7	26.2	10.6	14.4	18.1	21.8	26.2
		$\frac{5}{16}$	11.3	16.3	20.3	24.2	32.0	13.1	17.9	22.2	26.5	34.7
		$\frac{3}{8}$	12.4	19.3	24.3	28.7	37.6	14.3	21.5	26.3	31.2	40.9
		$\frac{7}{16}$	13.4	21.1	28.8	33.7	43.6	15.5	23.8	30.9	36.3	47.1
		$\frac{1}{2}$	14.3	22.8	31.6	39.2	50.0	16.5	25.7	35.1	41.8	53.6
		$\frac{9}{16}$	15.2	24.4	34.0	43.8	56.9	17.5	27.5	37.8	47.8	60.5

Note: Values above heavy lines apply only for 4-inch outstanding legs.

TABLE B		Outstanding Leg Capacity, kips (based on OSL = $3\frac{1}{2}$ or 4 inches)										
	Angle Length		6 inches					8 inches				
	Angle Thickness, in.		$\frac{3}{8}$	$\frac{1}{2}$	$\frac{5}{8}$	$\frac{3}{4}$	1	$\frac{3}{8}$	$\frac{1}{2}$	$\frac{5}{8}$	$\frac{3}{4}$	1
F_y (ksi) 50	Beam Web Thickness, in.	$\frac{3}{16}$	9.14	12.6	16.1	19.5	26.2	10.2	14.1	17.9	21.7	26.2
		$\frac{1}{4}$	11.9	16.1	20.3	24.5	32.9	13.1	17.7	22.4	27.0	36.2
		$\frac{5}{16}$	13.3	20.7	25.6	30.5	40.4	15.4	22.4	27.8	33.2	43.9
		$\frac{3}{8}$	14.6	23.3	31.1	36.7	47.9	16.9	26.3	33.4	39.5	51.7
		$\frac{7}{16}$	15.8	25.5	35.8	43.5	56.1	18.2	28.8	39.6	46.4	60.0
		$\frac{1}{2}$	16.9	27.7	39.0	50.6	64.8	19.5	31.1	43.2	53.9	68.8
		$\frac{9}{16}$	17.9	29.7	42.2	55.0	74.3	20.7	33.3	46.6	60.1	78.4

Note: Values above heavy lines apply only for 4-inch outstanding legs.

TABLE C	Weld Capacity, kips				
Weld Size In.	E70XX Electrodes				
	Seat Angle Size (long leg vertical)				
	$4 \times 3\frac{1}{2}$	$5 \times 3\frac{1}{2}$	6×4	7×4	8×4
$\frac{1}{4}$	11.5	17.2	21.8	28.5	35.6
$\frac{5}{16}$	14.3	21.5	27.3	35.6	44.5
$\frac{3}{8}$	17.2	25.8	32.7	42.7	53.4
$\frac{7}{16}$	20.1	30.1	38.2	49.8	62.3
$\frac{1}{2}$	–	34.4	43.6	56.9	71.2
$\frac{5}{8}$	–	43.0	54.5	71.2	89.0
$\frac{11}{16}$	–	47.3	60.0	78.3	–
$\frac{3}{4}$	–	–	–	–	–
Range of Available Seat Angle Thicknesses					
Minimum	$\frac{3}{8}$	$\frac{3}{8}$	$\frac{3}{8}$	$\frac{3}{8}$	$\frac{1}{2}$
Maximum	$\frac{1}{2}$	$\frac{3}{4}$	$\frac{3}{4}$	$\frac{3}{4}$	1

Courtesy of American Institute of Steel Construction.

FIG. 11-14

For the most economical seated connections, welded, bolted, or riveted, beam reactions should be shown on the contract drawings. If reactions are not shown, the connections should be required to be designed to support 50% of the total uniform load capacity (see AISC Manual, ASD Allowable Loads on Beams Tables, for laterally supported uniformly loaded shapes simply supported, and Section 6-14.1).

In computing the values tabulated in Fig. 11-14, the bending stress in the seat angle, as well as the crippling stress in the web of the supporting beam, was investigated. In addition, the distance from the outer face of the beam flange to the toe of the web/flange fillet, k, which is associated with the end reaction of a beam resting on a seat angle, was taken into account (Section 6-15).

11-13.6 Stiffened Seated Connections (Bolted)

When larger seats with larger capacities than those found in the Bolted Seated Beam Connections Tables (AISC Manual) and the Welded Seat Beam Connections Tables (Fig. 11-14) are necessary, stiffened seated connections should be used. Once again, as for unstiffened seated connections, stiffened seated connections should only be used when the beam is laterally supported by a top angle placed in either of the positions shown in Fig. 11-1c.

The allowable stiffener angle capacities shown in Table A of the Stiffened Seated Beam Connections Table (bolted) found in the AISC Manual are based upon allowable bearing stress on contact area (F_p) utilizing steel with $F_y = 36$ or 50 ksi. Fastener capacities are calculated in Table B of the Stiffened Seated Beam Connections Table (bolted) found in the AISC Manual from single shear values of the fastener group. The capacity of the connection is based on the lesser of these two values, together with consideration of the web crippling value of the supporting beam.

The effective length of stiffener bearing is assumed to be $\frac{1}{2}$ in. less than the outstanding angle leg dimension. The maximum gage in the legs of the stiffeners connected to the columns is $4\frac{3}{4}$ in. ASTM A307 bolts may be used in these connections (as in unstiffened seated connections), providing that the AISC Specification rules for HSB field connection requirements are met.

The paired stiffeners angles shown in contact in the AISC Manual for this type of connection can be separated to accommodate column gages. This separation $\leq 2(k - t)$, where k is defined in Section 11-13.5 and t is the thickness of the stiffener angle. If the connection parts are to be painted, this value should be a minimum of 1 in.

For economy, as stated for unstiffened seated connections, beam-end reactions should be shown on the contract drawings. For capacities in excess of those shown in the AISC Manual Stiffened Seated Beam Connections (bolted), special seated connections must be designed.

11-13.7 Stiffened Seated Connections (Welded)

Figure 11-15 shows allowable loads for stiffened welded seated connections based upon the use of E70 electrodes. As with unstiffened seated welded connections, these tables can be used with other electrodes, provided that the tabular values are multiplied by the proper ratio of the electrode strengths and the welds and base metal meet the provisions of Tables 11-11 and 11-12. Additional tables are given in the AISC Manual for larger W and L dimensions for the seat.

The minimum stiffener plate (based on A36 bracket material) thickness, t, must be $\geq$ the supported beam web thickness, t_w, or 1.4 times this value, for beams with $F_y = 36$ or 50 ksi, respectively, for supported beams with unstiffened webs. For bracket material of $F_y \geq 50$ ksi, $t \geq t_w[(F_y)_{beam}/(F_y)_{bracket}]$.

When using E70 electrodes, $t \geq 2a$ or $1.5a$ when the bracket material is $F_y = 36$ ksi or 50 ksi, respectively. a is the required weld size for E70 electrodes.

The AISC Manual suggests that the thickness of the horizontal seat plate or the thickness of the vertical stiffener plate $\geq \frac{3}{8}$ in. This is a rather arbitrary rule, and conceivably the use of a horizontal plate of less thickness than that of the stiffener plate would be satisfactory based upon sound judgment keyed to the condition on hand.

When the horizontal seat plate and the vertical stiffener plate are two separate plates (they can be a tee shape), the stiffener should be fitted to bear against the seat, and the welds connecting the two plates should have a minimum strength of that of the horizontal welds to the support under the seat plate. The welds connecting the beam to the seat (horizontal) plate may be replaced by rivets or A307 bolts, provided that the AISC Specification releative to the required usage of HSB for field connections is obeyed.

Combinations of material thickness and weld size selected from Fig. 11-15 should meet the limitations on minimum size of weld (Table 11-8) and maximum size of fillet welds (Section 11-12.2). If these limitations are not met, the weld size or material thickness should be increased as required.

When stiffener seats occur in line on opposite sides of the column web, for column webs of $F_y = 36$ ksi, when $a \leq t_c/2$ (a is weld size and

STIFFENED SEATED BEAM CONNECTIONS

Welded—E70XX Electrodes

Allowable Loads in Kips

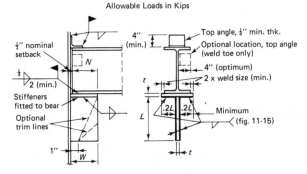

L	Width of Seat, *W*, inches											
In.	4				5				6			
	Weld Size, inches				Weld Size, inches				Weld Size, inches			
	$\frac{1}{4}$	$\frac{5}{16}$	$\frac{3}{8}$	$\frac{7}{16}$	$\frac{5}{16}$	$\frac{3}{8}$	$\frac{7}{16}$	$\frac{1}{2}$	$\frac{5}{16}$	$\frac{3}{8}$	$\frac{7}{16}$	$\frac{1}{2}$
6	22.7	28.4	34.0	39.7	23.5	28.1	32.8	37.5	19.9	23.9	27.9	31.9
7	29.9	37.4	44.9	52.4	31.2	37.5	43.7	50.0	26.7	32.0	37.3	42.7
8	37.8	47.2	56.7	66.1	39.8	47.8	55.8	63.7	34.3	41.1	48.0	54.8
9	46.1	57.6	69.2	80.7	49.1	59.0	68.8	78.6	42.5	51.1	59.6	68.1
10	54.9	68.6	82.3	96.0	59.0	70.8	82.6	94.4	51.4	61.7	72.0	82.3
11	63.9	79.8	95.8	112	69.4	83.3	97.1	111	60.9	73.1	85.2	97.4
12	73.1	91.4	110	128	80.2	96.2	112	128	70.8	85.0	99.2	113
13	82.5	103	124	144	91.3	110	128	146	81.1	97.4	114	130
14	92.0	115	138	161	103	123	144	164	91.9	110	129	147
15	101	127	152	178	114	137	160	183	103	123	144	165
16	111	139	167	195	126	151	176	202	115	138	160	183
17	121	151	181	212	138	165	193	221	126	151	176	201
18	131	163	196	229	150	180	210	240	137	164	192	219
19	140	175	211	246	162	194	227	259	149	179	208	238
20	150	188	225	263	174	209	243	278	161	193	225	257
21	160	200	240	280	189	223	260	298	173	207	242	276
22	169	212	254	296	198	238	277	317	185	221	258	295
23	179	224	269	313	210	252	294	336	197	236	275	315
24	189	236	283	330	222	267	311	356	209	250	292	334
25	198	248	297	347	234	281	328	375	221	265	309	353
26	208	260	312	364	247	296	345	394	233	279	326	373
27	217	272	326	380	259	310	362	414	245	294	343	392

Note: Loads shown are for E70XX electrodes. For E60XX electrodes, multiply tabular loads by 0.86, or enter table with 1.17 times the given reaction. For E80XX electrodes, multiply tabular loads by 1.14, or enter table with 0.875 times the given reaction.

Courtesy of American Institute of Steel Construction.

FIG. 11-15

t_c is the thickness of the column web), a may be considered fully effective when E70 electrodes are used. When the column web is of $F_y =$ 50-ksi material, and if $a \leq 0.67t_c$, a may be considered fully effective when E70 electrodes are used. These stipulations assure that when the weld is fully stressed the connected material is not overstressed.

Once again, for economy, the reaction value should be shown on the contract drawings.

11-13.8 End-Plate Shear Connections

Figure 11-1d shows an end-plate connection. This connection type uses a plate shorter in length than the beam depth placed normal to the direction of the beam span. The end plate is shop-welded to the beam web, with fillet welds on each side of the beam web. Within the ranges shown in the End Plate Shear Connections Tables found in the AISC Manual, this type of connection will furnish adequate end rotation and strength to be considered a type 2 connection. As stated before, it compares favorably with a double-angle connection (Fig. 11-1a) of like thicknesses, gages, and lengths of connection.

Closer tolerances in fabrication (beam length, square-cut beam ends, etc.) must be adhered to. To compensate for both mill and shop tolerances, shims will probably be required and should be used between the end plate and the face of the column flanges. End-plate ranges of $\frac{1}{4}$ to $\frac{3}{8}$ in. are suggested to assure adequate end rotation.

To achieve the full tabulated values of the fasteners and welds found in the AISC Manual for End Plate Shear Connections, the end-plate and web thicknesses must be equal to or greater than the tabulated thickness. If either supplied thickness is less than the tabulated value, the fastener or weld capacity must be reduced by multiplying by the ratio thickness supplied/thickness required.

As mentioned in Section 11-3, g should be $3\frac{1}{2}$ to $5\frac{1}{2}$ in. for average plate thickness, with an edge distance of $1\frac{1}{4}$ in. being used to maximize end rotational capacity and minimize prying action (edge distance values less than $1\frac{1}{4}$ in. tend to increase the effect of prying action).

End-plate thicknesses listed in the AISC Manual are for $F_y =$ 36-ksi material. End-plate thicknesses of $\frac{1}{4}$ in. for A36 material with a gage of 3 in. should assure adequate end rotation in the connection. Higher F_y values for the end plate demand engineering investigation to ascertain adequate rotational capacity. Tabulated weld values are for two fillet welds using E70 electrodes and are based on an effective weld length equal to $(L - 2a)$. The indicated weld should not be returned across the web at the top or bottom of the end plates.

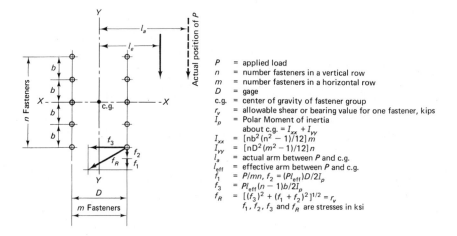

FIG. 11-16

11-13.9 Eccentric Loads on Fastener Groups

Figure 11-16 shows a group of fasteners that support an eccentric concentrated load, P. Each fastener supports an equal part of P plus an additional force due to moment, which is proportional to the fasteners distance from the center of gravity. The total force on any one fastener is the resultant of the components acting on that fastener.

The computation of l_{eff} was developed as the result of several tests on eccentric riveted connections. It was then applied to bolted joints with normal-sized holes.

Figure 11-17 shows one of four tables for obtaining coefficients, C, for various fastener arrangements (number of rows and dimensions). Shown are tabulated coefficients, C, for two vertical rows of fasteners when $D = 3$ in.

l_{eff} for any number of vertical rows of fasteners, except a single row, with any D dimension is given as

$$l_{eff} = l_a - \frac{1 + n}{2} \qquad (11\text{-}2a)$$

l_{eff} for a single vertical row of fasteners is given as

$$l_{eff} = l_a - \frac{1 + 2n}{4} \qquad (11\text{-}2b)$$

These tables are based upon an effective moment arm, l_{eff}, rather than on the actual arm, l_a. The tests performed have indicated that upon load-

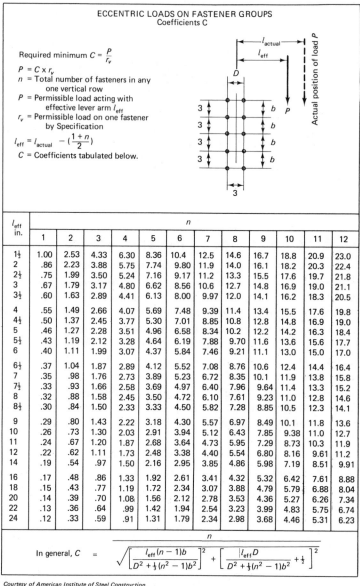

ECCENTRIC LOADS ON FASTENER GROUPS
Coefficients C

Required minimum $C = \dfrac{P}{r_v}$

$P = C \times r_v$
n = Total number of fasteners in any one vertical row
P = Permissible load acting with effective lever arm l_{eff}
r_v = Permissible load on one fastener by Specification

$l_{eff} = l_{actual} - (\dfrac{1+n}{2})$

C = Coefficients tabulated below.

l_{eff} in.	n											
	1	2	3	4	5	6	7	8	9	10	11	12
1½	1.00	2.53	4.33	6.30	8.36	10.4	12.5	14.6	16.7	18.8	20.9	23.0
2	.86	2.23	3.88	5.75	7.74	9.80	11.9	14.0	16.1	18.2	20.3	22.4
2½	.75	1.99	3.50	5.24	7.16	9.17	11.2	13.3	15.5	17.6	19.7	21.8
3	.67	1.79	3.17	4.80	6.62	8.56	10.6	12.7	14.8	16.9	19.0	21.1
3½	.60	1.63	2.89	4.41	6.13	8.00	9.97	12.0	14.1	16.2	18.3	20.5
4	.55	1.49	2.66	4.07	5.69	7.48	9.39	11.4	13.4	15.5	17.6	19.8
4½	.50	1.37	2.45	3.77	5.30	7.01	8.85	10.8	12.8	14.8	16.9	19.0
5	.46	1.27	2.28	3.51	4.96	6.58	8.34	10.2	12.2	14.2	16.3	18.4
5½	.43	1.19	2.12	3.28	4.64	6.19	7.88	9.70	11.6	13.6	15.6	17.7
6	.40	1.11	1.99	3.07	4.37	5.84	7.46	9.21	11.1	13.0	15.0	17.0
6½	.37	1.04	1.87	2.89	4.12	5.52	7.08	8.76	10.6	12.4	14.4	16.4
7	.35	.98	1.76	2.73	3.89	5.23	6.72	8.35	10.1	11.9	13.8	15.8
7½	.33	.93	1.66	2.58	3.69	4.97	6.40	7.96	9.64	11.4	13.3	15.2
8	.32	.88	1.58	2.45	3.50	4.72	6.10	7.61	9.23	11.0	12.8	14.6
8½	.30	.84	1.50	2.33	3.33	4.50	5.82	7.28	8.85	10.5	12.3	14.1
9	.29	.80	1.43	2.22	3.18	4.30	5.57	6.97	8.49	10.1	11.8	13.6
10	.26	.73	1.30	2.03	2.91	3.94	5.12	6.43	7.85	9.38	11.0	12.7
11	.24	.67	1.20	1.87	2.68	3.64	4.73	5.95	7.29	8.73	10.3	11.9
12	.22	.62	1.11	1.73	2.48	3.38	4.40	5.54	6.80	8.16	9.61	11.2
14	.19	.54	.97	1.50	2.16	2.95	3.85	4.86	5.98	7.19	8.51	9.91
16	.17	.48	.86	1.33	1.92	2.61	3.41	4.32	5.32	6.42	7.61	8.88
18	.15	.43	.77	1.19	1.72	2.34	3.07	3.88	4.79	5.79	6.88	8.04
20	.14	.39	.70	1.08	1.56	2.12	2.78	3.53	4.36	5.27	6.26	7.34
22	.13	.36	.64	.99	1.42	1.94	2.54	3.23	3.99	4.83	5.75	6.74
24	.12	.33	.59	.91	1.31	1.79	2.34	2.98	3.68	4.46	5.31	6.23

In general, $C = \sqrt{\left[\dfrac{l_{eff}(n-1)b}{D^2 + \frac{1}{3}(n^2-1)b^2}\right]^2 + \left[\dfrac{l_{eff}D}{D^2 + \frac{1}{3}(n^2-1)b^2} + \frac{1}{2}\right]^2}$

Courtesy of American Institute of Steel Construction.

FIG. 11-17

ing the fasteners the fastener the greatest distance from the centroid of the fastener group will yield and then distribute additional loads to the innermost fasteners. By using l_{eff} (somewhat less than l_a), a closer approximation of the aforementioned actual ultimate strength behavior is achieved.

The coefficient, C, may be used to compute the allowable eccentric load, P, on a fastener group (e.g., $P = Cr_v$). Knowing P and the permissible load on a fastener, the minimum value of C may be computed by P/r_v and the required number of fasteners in a vertical row, n, and l_{eff} may be found from Fig. 11-17.

With oversized or slotted holes in an eccentrically loaded connection, available research data indicate that for friction-type connections l_{eff} should be used, and for bearing-type connections, l_a should be used. In the latter case l_a is redefined as the distance from the applied load to the extreme fastener location within the slot or oversized hole rather than the center line of the slot or oversized hole.

Values of coefficient C in Fig. 11-17 (AISC, Seventh Edition, Manual of Steel Construction) lack uniformity of factor of safety when used with the allowable stresses for fasteners provided in Table 11-2 of this text. This effective length method has been superseded by an ultimate strength method of design in the AISC Manual of Steel Construction, Eighth Edition. Where unusual fastener group arrangements occur and do not conform to the configurations presented in the AISC Manual, Eighth Edition, the aforementioned elastic vector method is an acceptable and conservative procedure.

The following method of design of eccentric loads on fastener groups is based upon an ultimate strength approach* and is presented in the AISC Manual, Eighth Edition. It is compatible with and may be used with the allowable stresses for fasteners provided in Table 11-2 of this text.

Figure 11-18 shows a group of fasteners that support an eccentric load, P, and are loaded in shear. Load P does not pass through the center of gravity of the fastener group. This eccentricity, l, creates a rotation of the fastener group (moment effect), as well as a translation of the connected material. The rotation and translation can be reduced to a pure rotation about a single point, the instantaneous center of rotation, IC (instantaneous center).

For bearing-type connections (Section 11-10) each fastener supports an equal part of P plus an additional force due to moment. The individual

---

* S. F. Crawford and G. L. Kulak, Eccentrically Loaded Bolted Connections, *Journal of the Structural Division, ASCE,* vol. 97, ST3, March 1971, pp. 765–783.

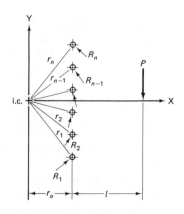

FIG. 11-18

resistance force of each fastener, R_n, is assumed to act on a line normal to the ray, r_n. This ray is measured from IC to the fastener of concern. The ultimate shear strength of fastener groups can be obtained from the load deformation relationship of a single fastener and can be expressed by

$$R = R_u(1 - e^{-\mu\Delta})^\lambda \qquad\qquad (11\text{-}2c)$$

where R = fastener shear load at any given deformation
$\quad\;\; R_u$ = ultimate shear load of a single fastener
$\quad\;\; \Delta$ = total deformation of a fastener (includes shearing, bending, and bearing deformation of the fastener and local bearing deformation of the plate)
$\;\mu,\lambda$ = regression coefficients
$\quad\;\; e$ = base of natural logarithms = 2.718

The ultimate load occurs when the fastener farthest from the IC attains the maximum deformation, $\Delta_{\max}$. The resistance of each fastener of the group can then be computed from Eq. 11-2c, in which Δ is assumed to vary linearly from the IC and R_u, μ, and λ are obtained from shear tests of individual fasteners.

The total resistance forces of all fasteners then combine to resist P, and if the correct location of the IC has been selected, the connection is in equilibrium ($\Sigma F_x = 0$, $\Sigma F_y = 0$, and $\Sigma M_{IC} = 0$ are simultaneously satisfied).

The extension of this method to friction-type connections is conservative as indicated by load tests and analytical studies.[*]

---

[*] G. L. Kulak, Eccentrically Loaded Slip-Resistant Connections, *Engineering Journal, AISC*, vol. 12, no. 2, 1975.

ECCENTRIC LOADS ON FASTENER GROUPS
Coefficients C

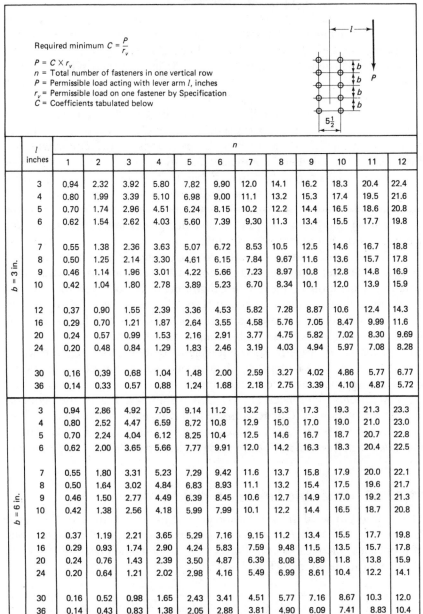

Required minimum $C = \dfrac{P}{r_v}$

$P = C \times r_v$
n = Total number of fasteners in one vertical row
P = Permissible load acting with lever arm l, inches
r_v = Permissible load on one fastener by Specification
C = Coefficients tabulated below

	l inches	n											
		1	2	3	4	5	6	7	8	9	10	11	12
b = 3 in.	3	0.94	2.32	3.92	5.80	7.82	9.90	12.0	14.1	16.2	18.3	20.4	22.4
	4	0.80	1.99	3.39	5.10	6.98	9.00	11.1	13.2	15.3	17.4	19.5	21.6
	5	0.70	1.74	2.96	4.51	6.24	8.15	10.2	12.2	14.4	16.5	18.6	20.8
	6	0.62	1.54	2.62	4.03	5.60	7.39	9.30	11.3	13.4	15.5	17.7	19.8
	7	0.55	1.38	2.36	3.63	5.07	6.72	8.53	10.5	12.5	14.6	16.7	18.8
	8	0.50	1.25	2.14	3.30	4.61	6.15	7.84	9.67	11.6	13.6	15.7	17.8
	9	0.46	1.14	1.96	3.01	4.22	5.66	7.23	8.97	10.8	12.8	14.8	16.9
	10	0.42	1.04	1.80	2.78	3.89	5.23	6.70	8.34	10.1	12.0	13.9	15.9
	12	0.37	0.90	1.55	2.39	3.36	4.53	5.82	7.28	8.87	10.6	12.4	14.3
	16	0.29	0.70	1.21	1.87	2.64	3.55	4.58	5.76	7.05	8.47	9.99	11.6
	20	0.24	0.57	0.99	1.53	2.16	2.91	3.77	4.75	5.82	7.02	8.30	9.69
	24	0.20	0.48	0.84	1.29	1.83	2.46	3.19	4.03	4.94	5.97	7.08	8.28
	30	0.16	0.39	0.68	1.04	1.48	2.00	2.59	3.27	4.02	4.86	5.77	6.77
	36	0.14	0.33	0.57	0.88	1.24	1.68	2.18	2.75	3.39	4.10	4.87	5.72
b = 6 in.	3	0.94	2.86	4.92	7.05	9.14	11.2	13.2	15.3	17.3	19.3	21.3	23.3
	4	0.80	2.52	4.47	6.59	8.72	10.8	12.9	15.0	17.0	19.0	21.0	23.0
	5	0.70	2.24	4.04	6.12	8.25	10.4	12.5	14.6	16.7	18.7	20.7	22.8
	6	0.62	2.00	3.65	5.66	7.77	9.91	12.0	14.2	16.3	18.3	20.4	22.5
	7	0.55	1.80	3.31	5.23	7.29	9.42	11.6	13.7	15.8	17.9	20.0	22.1
	8	0.50	1.64	3.02	4.84	6.83	8.93	11.1	13.2	15.4	17.5	19.6	21.7
	9	0.46	1.50	2.77	4.49	6.39	8.45	10.6	12.7	14.9	17.0	19.2	21.3
	10	0.42	1.38	2.56	4.18	5.99	7.99	10.1	12.2	14.4	16.5	18.7	20.8
	12	0.37	1.19	2.21	3.65	5.29	7.16	9.15	11.2	13.4	15.5	17.7	19.8
	16	0.29	0.93	1.74	2.90	4.24	5.83	7.59	9.48	11.5	13.5	15.7	17.8
	20	0.24	0.76	1.43	2.39	3.50	4.87	6.39	8.08	9.89	11.8	13.8	15.9
	24	0.20	0.64	1.21	2.02	2.98	4.16	5.49	6.99	8.61	10.4	12.2	14.1
	30	0.16	0.52	0.98	1.65	2.43	3.41	4.51	5.77	7.16	8.67	10.3	12.0
	36	0.14	0.43	0.83	1.38	2.05	2.88	3.81	4.90	6.09	7.41	8.83	10.4

Courtesy of American Institute of Steel Construction.

FIG. 11-19

Tables have been prepared based on the solution of the IC problem for each fastener pattern and each eccentric condition. Figure 11-19 shows one of the many tables for obtaining coefficients, C, for various fastener arrangements (number of rows, gages, pitch, number of fasteners per row, etc.). Shown are tabulated coefficients, C, for two vertical rows of fasteners when their gage is $5\frac{1}{2}$ in. and their pitch b is 3 or 6 in.

Values of $R_u = 74$ kips, $\mu = 10.0$, $\lambda = 0.55$, and $\Delta = \Delta_{\max} = 0.34$ in., which were experimentally determined by Crawford and Kulak[†] for $\frac{3}{4}$-in.-diameter A325 bolts, are used in Eq. 11-2c to develop the tables found in the AISC Manual. P/R_u gives values of the coefficient, C. Multiplying the tabulated coefficient, C, by r_v, the shear resistance or permissible load on one fastener (Table 11-2, allowable working stress for fasteners multiplied by A_b) gives the resistance of the connection as a permissible load, P, in kips. Or, knowing P and dividing by r_v gives the required coefficient, C, and a fastener group must be selected for which C is of that magnitude as a minimum.

These nondimensional coefficients, C, can safely be used with any fastener diameter and are slightly conservative when used with A490 bolts.[*] As mentioned, these coefficients may also be used for friction-type connections. Values for this type of connection are 5% to 10% on the conservative side. These tables may be used with any fastener (A307 bolts or rivets, as well as high-strength bolts). Utilizing these tables, margins of safety are provided equivalent to those bolts used in joints less than 50 in. long, subject to shear produced by concentric load only in friction and bearing-type connections.

Additional references are (1) C. L. Shermer, Plastic Behavior of Eccentrically Loaded Connections, *Engineering Journal, AISC,* vol. 8, no. 2, April 1971, and (2) P. F. Adams, H. A. Krentz, and G. L. Kulak, *Canadian Structural Steel Design,* Canadian Institute of Steel Construction.

11-13.10 *Eccentric Loads on Weld Groups*

The solution of weld groups with eccentric loads is similar to that used for eccentrically loaded fastener groups except that, in the calculations of properties, the weld is considered to be a line coincident with the edge of material to be welded (Fig. 11-20).

[†] Crawford and Kulak, Eccentrically Loaded Bolted Connections, p. 205.
[*] J. W. Fisher and J. H. A. Struik, *Guide to Design Criteria for Bolted and Riveted Joints,* John Wiley & Sons, Inc., New York, 1974.

ECCENTRIC LOADS ON WELD GROUPS

The solution of eccentric loading of weld groups is similar to the method employed for fastener groups, except that for computation of properties, the weld is considered a line coincident with the edge to be fillet welded.

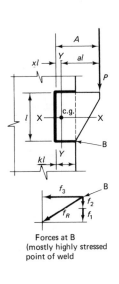

Forces at B
(mostly highly stressed
point of weld)

P	= Permissible load, kips
A	= Distance from vertical weld to P, inches = $l(a + x)$
l	= Length of vertical weld, inches
kl	= Length of horizontal weld, inches
L	= Total length of weld, inches = $l(1 + 2k)$
xl	= Distance from vertical weld to center of gravity of weld group, in inches

$$= \frac{(kl)^2}{L} \text{ or } x = \frac{k^2}{(1 + 2k)}$$

I_p = Polar moment of inertia, inches^4

$$= l^3 \left[\frac{(1 + 2k)^3}{12} - \frac{k^2(1 + k)^2}{(1 + 2k)} \right]$$

D = Number of sixteenths of an inch in weld size

$0.928D$ = Value of E70XX weld per sixteenth inch of weld per lineal inch, kips
= 0.3(70)(0.707a/16)

f_1 = Stress on weld at B due to vertical load
= $P/(l + 2kl) = P/l(1 + 2k)$

f_2 = Vertical stress on weld at B due to moment

$$= \frac{Pal(kl - xl)}{I_p} = \frac{Pal^2(k - x)}{I_p}$$

f_3 = Horizontal stress on weld at B due to moment

$$= \frac{Pal(l/2)}{I_p} = \frac{Pal^2}{2I_p}$$

f_R = Resultant of stresses on weld at B

$$= \sqrt{(f_1 + f_2)^2 + (f_3)^2}$$

and

f_R = $0.928D$ (for E70XX electrodes)

Courtesy of American Institute of Steel Construction.

FIG. 11-20

The AISC Manual includes Eccentric Loads On Weld Groups Tables for the weld configurations and loading conditions shown in Fig. 11-21. Figure 11-22 shows the table for the condition shown in Fig. 11-21d. From Fig. 11-22 the permissible eccentric load in kips, P, can be determined for a given weld configuration and loading condition. The coefficient, C, is determined from the table for various values of a and k for E70 electrodes. $P = CC_1Dl$, where all terms are defined as in Fig. 11-22 and C_1 is the ratio of the electrode strength used to that of E70 (e.g., for an E80 electrode, $C_1 = \frac{80}{70} = 1.14$). The coefficient C_1 permits the AISC Tables to be used for any electrode. $C_1 = 1.0$ for E70 electrodes.

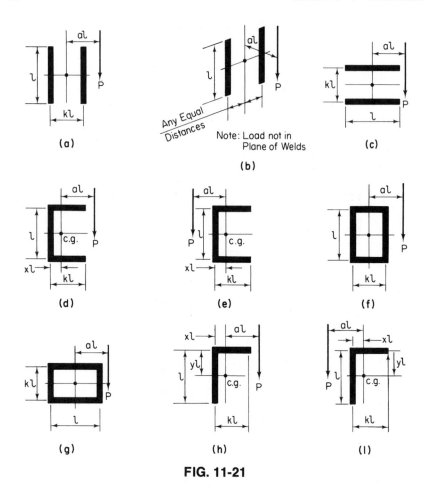

FIG. 11-21

The aforementioned elastic design method provides a simplified and conservative approach to the design of eccentric loads on weld groups and is presented in greater detail in the AISC Manual, Seventh Edition. However, the method does not give a consistent factor of safety and in some instances results in an excessively conservative design. Where conditions of unusual weld group geometry are not covered by the tables in the eighth edition of the Manual, or where the nature of load conditions suggests that inelastic deformations in the weld group should be avoided, this elastic design approach should be used. In these cases, the elastic design method is an acceptable and conservative procedure of design.

The following method of design of eccentric loads on weld groups is based upon an ultimate strength approach and is presented in the AISC Manual, Eighth Edition.

ECCENTRIC LOADS ON WELD GROUPS

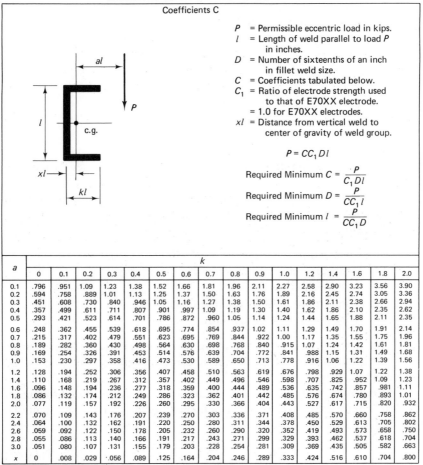

Coefficients C

P = Permissible eccentric load in kips.
l = Length of weld parallel to load P
  in inches.
D = Number of sixteenths of an inch
  in fillet weld size.
C = Coefficients tabulated below.
C_1 = Ratio of electrode strength used
  to that of E70XX electrode.
  = 1.0 for E70XX electrodes.
xl = Distance from vertical weld to
  center of gravity of weld group.

$$P = CC_1Dl$$

Required Minimum $C = \dfrac{P}{C_1Dl}$

Required Minimum $D = \dfrac{P}{CC_1l}$

Required Minimum $l = \dfrac{P}{CC_1D}$

a	\multicolumn{16}{c}{k}
	0	0.1	0.2	0.3	0.4	0.5	0.6	0.7	0.8	0.9	1.0	1.2	1.4	1.6	1.8	2.0
0.1	.796	.951	1.09	1.23	1.38	1.52	1.66	1.81	1.96	2.11	2.27	2.58	2.90	3.23	3.56	3.90
0.2	.594	.758	.889	1.01	1.13	1.25	1.37	1.50	1.63	1.76	1.89	2.16	2.45	2.74	3.05	3.36
0.3	.451	.608	.730	.840	.946	1.05	1.16	1.27	1.38	1.50	1.61	1.86	2.11	2.38	2.66	2.94
0.4	.357	.499	.611	.711	.807	.901	.997	1.09	1.19	1.30	1.40	1.62	1.86	2.10	2.35	2.62
0.5	.293	.421	.523	.614	.701	.786	.872	.960	1.05	1.14	1.24	1.44	1.65	1.88	2.11	2.35
0.6	.248	.362	.455	.539	.618	.695	.774	.854	.937	1.02	1.11	1.29	1.49	1.70	1.91	2.14
0.7	.215	.317	.402	.479	.551	.623	.695	.769	.844	.922	1.00	1.17	1.35	1.55	1.75	1.96
0.8	.189	.282	.360	.430	.498	.564	.630	.698	.768	.840	.915	1.07	1.24	1.42	1.61	1.81
0.9	.169	.254	.326	.391	.453	.514	.576	.639	.704	.772	.841	.988	1.15	1.31	1.49	1.68
1.0	.153	.230	.297	.358	.416	.473	.530	.589	.650	.713	.778	.916	1.06	1.22	1.39	1.56
1.2	.128	.194	.252	.306	.356	.407	.458	.510	.563	.619	.676	.798	.929	1.07	1.22	1.38
1.4	.110	.168	.219	.267	.312	.357	.402	.449	.496	.546	.598	.707	.825	.952	1.09	1.23
1.6	.096	.148	.194	.236	.277	.318	.359	.400	.444	.489	.536	.635	.742	.857	.981	1.11
1.8	.086	.132	.174	.212	.249	.286	.323	.362	.401	.442	.485	.576	.674	.780	.893	1.01
2.0	.077	.119	.157	.192	.226	.260	.295	.330	.366	.404	.443	.527	.617	.715	.820	.932
2.2	.070	.109	.143	.176	.207	.239	.270	.303	.336	.371	.408	.485	.570	.660	.758	.862
2.4	.064	.100	.132	.162	.191	.220	.250	.280	.311	.344	.378	.450	.529	.613	.705	.802
2.6	.059	.092	.122	.150	.178	.205	.232	.260	.290	.320	.352	.419	.493	.573	.658	.750
2.8	.055	.086	.113	.140	.166	.191	.217	.243	.271	.299	.329	.393	.462	.537	.618	.704
3.0	.051	.080	.107	.131	.155	.179	.203	.228	.254	.281	.309	.369	.435	.505	.582	.663
x	0	.008	.029	.056	.089	.125	.164	.204	.246	.289	.333	.424	.516	.610	.704	.800

FIG. 11-22

The ultimate strength approach for eccentric loads on weld groups is somewhat similar to that of bolt groups in that the external eccentric load causes a relative rotation and translation between the elements connected by the weld. The point at which rotation tends to take place is again called the IC (instantaneous center) of rotation. Its location varies with the eccentricity of the external load, the weld group geometry, and the deformation of the weld at different angles of the resultant elemental force relative to the weld axis. As for bolt groups, the individual resistance force of each weld element, R, is assumed to act on a line

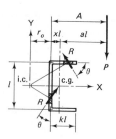

FIG. 11-23

normal to a ray, r, passing through the IC and that element's location (Fig. 11-23).

The ultimate shear strength of weld groups* can be obtained from the load deformation relationship of a single unit weld element. This can be expressed as in Eq. 11-2c,

where R = shear force in a single weld element at any deformation
R_u = ultimate shear load of a single weld unit
μ, λ = regression coefficients
Δ = deformation of a weld element
e = base of natural logarithms = 2.718

The strength and deformation performance in welds depend upon the angle θ that the resultant elemental force makes with the axis of the weld element (Fig. 11-23).

The critical weld element is usually (not always) the weld element farthest from the IC. The critical deformation, Δ_{max}, can be calculated from Eq. 11-2d.

$$\Delta_{max} = 0.225(\theta + 5)^{-0.47} \qquad (11\text{-}2d)$$

where θ is in degrees.

The deformation of other weld elements may be computed from Eq. 11-2e.

$$\Delta = \frac{r}{r_{max}} (\Delta_{max}) \qquad (11\text{-}2e)$$

Values of R_u, μ, and λ are dependent on the value of θ and can be calculated from Eqs. 11-2f, g, and h:

---

* Butler, Pal and Kulak, Eccentrically Loaded Weld Connections, *Journal of the Structural Division, ASCE,* vol. 98, ST5, May 1972, pp. 989–1005.

$$R_u = \frac{10 + \theta}{0.92 + 0.0603\theta} \qquad (11\text{-}2f)$$

$$\mu = 75e^{0.01140\theta} \qquad (11\text{-}2g)$$

$$\lambda = 0.4e^{0.01460\theta} \qquad (11\text{-}2h)$$

The total resistance force of all the weld elements combine to resist the external load, P, and if the correct location of the IC has been selected, the connection is in equilibrium ($\Sigma F_x = 0$, $\Sigma F_y = 0$, and $\Sigma M_{\mathrm{IC}} = 0$ are simultaneously satisfied).

The AISC Manual, Eighth Edition, presents tables giving coefficients that replace the elastic method C coefficients used in the Eccentric Load Tables of the seventh edition of the Manual.

Tests were performed on eccentrically loaded $\frac{1}{4}$-in. fillet weld groups made with E60 electrodes. Coefficient tables for E70 electrodes for $\frac{1}{16}$-in. fillet welds were produced by adjusting the ultimate capacities by $(\frac{1}{4})\frac{70}{60}$. The resulting values were then multiplied by 0.30 to convert to an allowable stress value. This provides a minimum factor of safety of 3.33. Finally, this value was reduced by a factor that prevents the stress in any element of weld from exceeding the allowable stress for fillet weld metal (Table 11-11). The tables are based upon an equalization of limiting allowable shear stresses between base metal (A36 material) and an E70 electrode weld [e.g., $0.4(36)(\frac{1}{16}) = 0.900$ for A36 base metal and $0.3(70)0.707(\frac{1}{16}) = 0.928$ for an E70 electrode weld]. The higher value is used as the limiting stress. The value of the coefficient, C, as used in the tables of the eighth edition of the Manual, is reduced by dividing by the weld length, l.

In much the same manner, the capacity of the weld group subjected to an eccentric load is $P = CC_1Dl$, where all terms are as defined in Fig. 11-24.

For welded framed beam connections (Figs. 11-12 and 11-13), Weld A uses IC solutions and Weld B capacity is computed using vector analysis techniques.

For welded seated beam connections (Table C of Figs. 11-14 and 11-15), the allowable weld capacities are computed using vector analysis.

11-14   ONE-SIDED CONNECTIONS

The AISC Manual, Part 4, also suggests a design procedure for one-sided connections (e.g., a single web angle or the like). Consideration is given to vertical shear or bearing shared in all the fasteners, as well as the eccentric effect of the beam loaded on the fasteners in the outstanding leg of the clip angle (Fig. 11-25). The framing angle's shear capacity

ECCENTRIC LOADS ON WELD GROUPS
Coefficients C

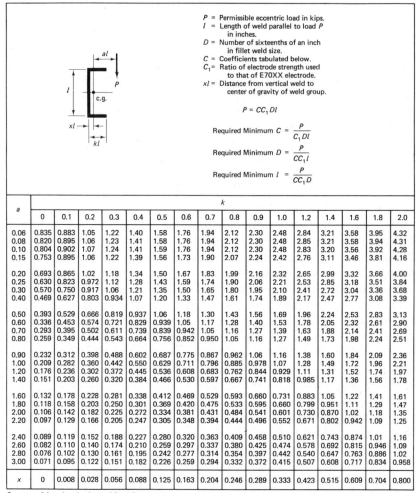

P = Permissible eccentric load in kips.
l = Length of weld parallel to load P in inches.
D = Number of sixteenths of an inch in fillet weld size.
C = Coefficients tabulated below.
C_1 = Ratio of electrode strength used to that of E70XX electrode.
xl = Distance from vertical weld to center of gravity of weld group.

$$P = CC_1 Dl$$

Required Minimum $C = \dfrac{P}{C_1 Dl}$

Required Minimum $D = \dfrac{P}{CC_1 l}$

Required Minimum $l = \dfrac{P}{CC_1 D}$

a	\multicolumn{16}{c}{k}															
	0	0.1	0.2	0.3	0.4	0.5	0.6	0.7	0.8	0.9	1.0	1.2	1.4	1.6	1.8	2.0
0.06	0.835	0.883	1.05	1.22	1.40	1.58	1.76	1.94	2.12	2.30	2.48	2.84	3.21	3.58	3.95	4.32
0.08	0.820	0.895	1.06	1.23	1.41	1.58	1.76	1.94	2.12	2.30	2.48	2.85	3.21	3.58	3.94	4.31
0.10	0.804	0.902	1.07	1.24	1.41	1.59	1.76	1.94	2.12	2.30	2.48	2.83	3.20	3.56	3.92	4.28
0.15	0.753	0.895	1.06	1.22	1.39	1.56	1.73	1.90	2.07	2.24	2.42	2.76	3.11	3.46	3.81	4.16
0.20	0.693	0.865	1.02	1.18	1.34	1.50	1.67	1.83	1.99	2.16	2.32	2.65	2.99	3.32	3.66	4.00
0.25	0.630	0.823	0.972	1.12	1.28	1.43	1.59	1.74	1.90	2.06	2.21	2.53	2.85	3.18	3.51	3.84
0.30	0.570	0.750	0.917	1.06	1.21	1.35	1.50	1.65	1.80	1.95	2.10	2.41	2.72	3.04	3.36	3.68
0.40	0.469	0.627	0.803	0.934	1.07	1.20	1.33	1.47	1.61	1.74	1.89	2.17	2.47	2.77	3.08	3.39
0.50	0.393	0.529	0.666	0.819	0.937	1.06	1.18	1.30	1.43	1.56	1.69	1.96	2.24	2.53	2.83	3.13
0.60	0.336	0.453	0.574	0.721	0.829	0.939	1.05	1.17	1.28	1.40	1.53	1.78	2.05	2.32	2.61	2.90
0.70	0.293	0.395	0.502	0.611	0.739	0.839	0.942	1.05	1.16	1.27	1.39	1.63	1.88	2.14	2.41	2.69
0.80	0.259	0.349	0.444	0.543	0.664	0.756	0.852	0.950	1.05	1.16	1.27	1.49	1.73	1.98	2.24	2.51
0.90	0.232	0.312	0.398	0.488	0.602	0.687	0.775	0.867	0.962	1.06	1.16	1.38	1.60	1.84	2.09	2.36
1.00	0.209	0.282	0.360	0.442	0.550	0.629	0.711	0.796	0.885	0.978	1.07	1.28	1.49	1.72	1.96	2.21
1.20	0.176	0.236	0.302	0.372	0.445	0.536	0.608	0.683	0.762	0.844	0.929	1.11	1.31	1.52	1.74	1.97
1.40	0.151	0.203	0.260	0.320	0.384	0.466	0.530	0.597	0.667	0.741	0.818	0.985	1.17	1.36	1.56	1.78
1.60	0.132	0.178	0.228	0.281	0.338	0.412	0.469	0.529	0.593	0.660	0.731	0.883	1.05	1.22	1.41	1.61
1.80	0.118	0.158	0.203	0.250	0.301	0.369	0.420	0.475	0.533	0.595	0.660	0.799	0.951	1.11	1.29	1.47
2.00	0.106	0.142	0.182	0.225	0.272	0.334	0.381	0.431	0.484	0.541	0.601	0.730	0.870	1.02	1.18	1.35
2.20	0.097	0.129	0.166	0.205	0.247	0.305	0.348	0.394	0.444	0.496	0.552	0.671	0.802	0.942	1.09	1.25
2.40	0.089	0.119	0.152	0.188	0.227	0.280	0.320	0.363	0.409	0.458	0.510	0.621	0.743	0.874	1.01	1.16
2.60	0.082	0.110	0.140	0.174	0.210	0.259	0.297	0.337	0.380	0.425	0.474	0.578	0.692	0.815	0.946	1.09
2.80	0.076	0.102	0.130	0.161	0.195	0.242	0.277	0.314	0.354	0.397	0.442	0.540	0.647	0.763	0.886	1.02
3.00	0.071	0.095	0.122	0.151	0.182	0.226	0.259	0.294	0.332	0.372	0.415	0.507	0.608	0.717	0.834	0.958
x	0	0.008	0.028	0.056	0.088	0.125	0.163	0.204	0.246	0.289	0.333	0.423	0.515	0.609	0.704	0.800

Courtesy of American Institute of Steel Construction.

FIG. 11-24

should be checked against the net area (refer to Section 11-13.2) using a hole diameter equal to the nominal bolt diameter plus $\frac{1}{16}$ in.

As discussed in Section 11-13.9, Eccentric Loads on Fastener Groups, the AISC Manual computes and tabulates coefficients, C, for two cases of one-sided angles (a single gage and a double gage with $g = 3$ in.) based upon the actual arm between the center line of the beam and the center of gravity of the fasteners.

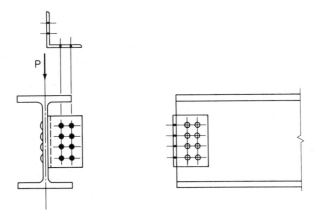

FIG. 11-25

As before, $P = Cr_v$ or $C = P/r_v$. r_v is allowable shear or bearing stress for one fastener. The smaller value of the two is the controlling value. For further details relative to maximum web leg gages, suggested angle thickness relative to fastener diameter, and other matters, refer to the AISC Manual.

11-15 ECCENTRIC CONNECTIONS

This subject is discussed in Part 4 of the AISC Manual relative to beam-to-column connections (beam on one side of column and beams on both sides of column), zee-type connections, truss members connections, and so on. Lever arms to be used are suggested to account for eccentricity in the fastener's connection element to the beam and connection element to the column.

11-16 PRYING ACTION

In the discussions of high-strength bolts, the term "prying action" was used. Prying effects occur when fasteners are in tension and transmit loads through tee flanges, angle legs, or by any other means where the external edges of the joint components act as fulcrums to pry on the fasteners.

Examples of connections with bolt groups in tension that are subject to prying forces would be tee-stub hangers with a single line of fasteners parallel and on each side of the web or a tee section that transfers the tensile component in a moment-resistant beam to a column connection (Fig. 11-28b).

The theoretical determination of the prying force, Q, is complex, but the AISC Manual, Eighth Edition, Part 4, provides design formulas and procedures from which Q values can be determined. For a more detailed discussion, the reader is referred to J. W. Fisher and J. H. A. Struik, *Guide to Design Criteria for Bolted and Riveted Joints*, John Wiley & Sons, New York, 1974, pages 260–279, and to J. H. A. Struik and J. de Back, *Tests on Bolted T-Stubs with Respect to a Bolted Beam-to-Column Connection*, Report 6-69-13, Stevin Laboratory, Delft University of Technology, Delft, the Netherlands, 1969.

The design formulas and nomenclature concerned with the final design method are as follows:

$$\delta = 1 - \frac{d'}{p} \tag{11-3a}$$

$$M = \frac{M_p}{2} = \frac{p t_f^2 F_y}{8} \tag{11-3b}$$

$$\alpha = \frac{\left(\dfrac{Tb}{M} - 1\right)}{\delta} \quad (if\ \alpha > 1.0,\ use\ \alpha = 1.0) \tag{11-3c}$$

$$B_c = T\left[1 + \frac{\delta\alpha}{(1 + \delta\alpha)}\left(\frac{b'}{a'}\right)\right] \tag{11-3d}$$

$$t_{f_{req}} = \left\{\frac{8B_c a' b'}{p F_y [a' + \delta\alpha\ (a' + b')]}\right\}^{\frac{1}{2}} \tag{11-3e}$$

$$Q = B_c - T \tag{11-3f}$$

where Q = prying force per bolt at design load, kips
  B_c = load per bolt including Q, kips
  T = applied tension per bolt (exclusive of initial tightening), kips
  M = allowable bending moment tributary to the tee-stub flange or angle leg by one bolt, kip-inches
  M_p = plastic moment, kip-inches
  F_y = yield strength of the flange material, ksi
  a = distance from center line of bolt to edge of tee flange or angle leg, inches (not to be more than $1.25b$)
  b = distance from center line of bolt (gage line) to face of tee stem or angle leg, inches
  a' = $a + d/2$, inches
  b' = $b - d/2$, inches
  d = bolt diameter, inches
  d' = width of bolt hole in flange parallel to tee stem or angle leg, inches

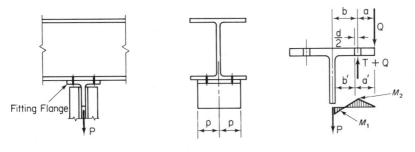

FIG. 11-26

p = length of flange, parallel to stem or leg, tributary to each bolt, inches

α = moment ratio $\dfrac{M_2}{M_1}$ $(0 > \alpha > 1.0)$

δ = ratio of net area (at bolt line) and the gross area (at the face of the stem or angle leg)

t_f = required thickness of tee-stub flange or angle leg, inches

Some of these dimensions will become more apparent by referring to Fig. 11-26.

The adequacy of the fastener subject to external tension and prying force (tension produced by deformation of the connected parts) must be investigated. The fitting flange (Fig. 11-26) must be checked for bending due to the prying-action moment.

The classic case used to illustrate prying action is the case of the hanger-type connection shown in Fig. 11-26, although the action can be illustrated by a typical top flange tee or angle used in the common wind connection. The hanger connection in Fig. 11-26 is a tee fastened to the bottom flange of a beam. P is the external applied load per inch of connection length. pP represents the downward force, which the fitting flange resists by means of four bolts loaded in tension. Force pP bends the tee flange so that its outer portions (between the bolt lines and the tee flange edges) press upward. This action is resisted by the prying force, Q. Q is assumed to be a line load applied at the edge of the tee flange (a distance, a, from the center line of the fastener) effective over the length, p, tributary to each bolt. T is the upward reaction applied by the bolts in resisting the external load, $2pP$. $T = 2pP/4 = pP/2$. For equilibrium of vertical forces, T must be increased to $T + Q$, the upward reaction applied to each bolt. $T + Q \leq r_t$, the value of one bolt in tension.

Knowing the value of r_t and T, the value for Q is found by the previously mentioned equations. Q is dependent on a number of parameters, such as material, fastener type and size, and the connection geometry. The value may differ appreciably for different connections.

STRUCTURAL TEE OR DOUBLE ANGLE HANGERS

Loads in kips per linear inch for trial section

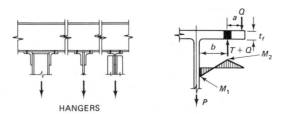

HANGERS

b in.	Thickness of angle or flange of tee, t_f, inches															
	$\frac{5}{16}$	$\frac{3}{8}$	$\frac{7}{16}$	$\frac{1}{2}$	$\frac{9}{16}$	$\frac{5}{8}$	$\frac{11}{16}$	$\frac{3}{4}$	$\frac{13}{16}$	$\frac{7}{8}$	$\frac{15}{16}$	1	$1\frac{1}{16}$	$1\frac{1}{8}$	$1\frac{3}{16}$	$1\frac{1}{4}$
1	1.76	2.53	3.45	4.50	5.70	7.03	8.51	10.13	11.88	13.78	15.82	18.00	20.32	22.78	25.38	28.13
1¼	1.41	2.03	2.76	3.60	4.56	5.63	6.81	8.10	9.51	11.03	12.66	14.40	16.26	18.23	20.31	22.50
1½	1.17	1.69	2.30	3.00	3.80	4.69	5.67	6.75	7.92	9.19	10.55	12.00	13.55	15.19	16.92	18.75
1¾	1.00	1.45	1.97	2.57	3.25	4.02	4.86	5.79	6.79	7.88	9.04	10.29	11.61	13.02	14.50	16.07
2	0.88	1.27	1.72	2.25	2.85	3.52	4.25	5.06	5.94	6.89	7.91	9.00	10.16	11.39	12.69	14.06
2¼	0.78	1.13	1.53	2.00	2.53	3.13	3.78	4.50	5.28	6.13	7.03	8.00	9.03	10.13	11.28	12.50
2½	0.70	1.01	1.38	1.80	2.28	2.81	3.40	4.05	4.75	5.51	6.33	7.20	8.13	9.11	10.15	11.25
2¾	0.64	0.92	1.25	1.64	2.07	2.56	3.09	3.68	4.32	5.01	5.75	6.55	7.39	8.28	9.23	10.23
3	0.59	0.84	1.15	1.50	1.90	2.34	2.84	3.38	3.96	4.59	5.27	6.00	6.77	7.59	8.46	9.38
3¼	0.54	0.78	1.06	1.38	1.75	2.16	2.62	3.12	3.66	4.24	4.87	5.54	6.25	7.01	7.81	8.65

Courtesy of American Institute of Steel Construction.

FIG. 11-27

Before stating the suggested design procedure steps, the following assumptions were made: tee flanges in bending and bolts in tension require a safety factor ≥ 2.0; the required ultimate tee flange bending strength tributary per bolt is equal to $M_p = pt_f^2 F_y/4$; due to a slight deformation of the flange, the force in the bolts shifts from the center line of the bolt to the stem side of the bolt by a distance of $d/2$; $M_1 \geq M_2$ (Fig. 11-27); and $a < 1.25b$.

The suggested design procedure is as follows:

(1) Determine the allowable tension on the nominal area of the type and size of bolt used ($F_t A_{bolt}$); p, the length of flange tributary to the support of one bolt; the required number of bolts ($P/F_t A_{bolt}$); and T (P/number of bolts). P is the total supported loads.

(2) Compute load per linear inch supported by a pair of bolts, located one on each side of the tee stem ($2T/p$); estimate b based on a given gage or distance required for wrench clearance. Enter the preliminary selection table for hanger-type connections (Fig. 11-27) to obtain a trial, $t_{f\,trial}$ (Section 11-27). Select a trial section and compute b, a, b', a', and d'.

(3) Compute δ, M, α, and B_c (Eqs. 11-3a, b, c, and d).

(4) Obtain $t_{f\,req}$ (Eq. 11-3e). If $t_{f\,req} < t_{f\,trial}$, the solution is valid. If $t_{f\,req} > t_{f\,trial}$, select a heavier tee, use additional bolts, or change b or p and repeat the design procedure.

(5) Where Q is required (for fatigue design or other reasons), compute Q from Eq. 11-3f (when considering fatigue or other applications where the actual prying force must be reduced to an insignificant value, $t_{f\,req}$ may be calculated from Eq. 11-3e by using $\alpha = 0$ and $B_c = T$).

This design procedure will be illustrated by a design example in Section 11-21.

11-17   YIELD-LINE ANALYSIS

Information relative to the transfer of tension to the web of a beam or column is lacking. This transfer may be accomplished by means of a plate welded to the web of a W shape or a tee bolted to the web of a W shape. Design criteria are sadly lacking. Yield-line analysis is a design tool to accomplish this end. It may be used to evaluate the strength of the web. The yield-line method, developed by K. W. Johansen, is used primarily in the design of reinforced concrete slabs. However, it is useful in steel design also. The yield-line configurations may be thought of as continuous plastic hinges. The use of this tool and suggested methods of computation to evaluate the bending effects on a web plate can be better understood by referring to the following references on the subject:

(1) A. L. Abolitz and M. E. Warner, Bending under Seated Connections, *AISC Engineering Journal,* January 1965.

(2) O. W. Blodgett, *Design of Welded Structures,* James F. Lincoln Arc Welding Foundation, Cleveland, Ohio, 1966, pp. 3.6–6 to 3.6–9.

(3) R. H. Kapp, Yield Line Analysis of a Web Connection in Direct Tension, *AISC Engineering Journal,* April 1974.

(4) F. W. Stockwell, Jr., Yield Line Analysis of Column Webs with Welded Beam Connections, *AISC Engineering Journal,* January 1974.

11-18   MOMENT CONNECTIONS
(TYPE 1 OR 3)

Figure 11-28 shows some of the more popular typical moment connections in use. Most wind connections, or connections designed to resist bending-moment, use angles, tees, or plates as connection material. Mo-

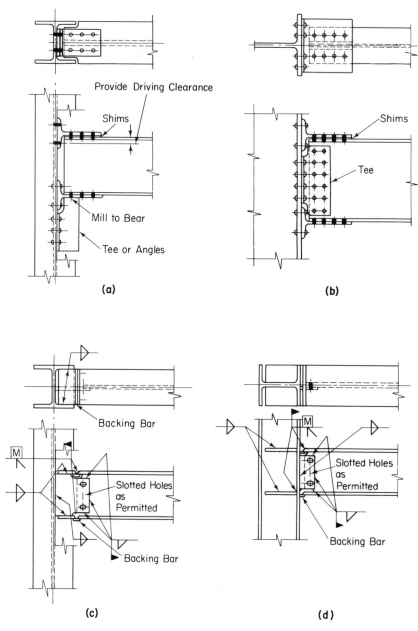

FIG. 11-28

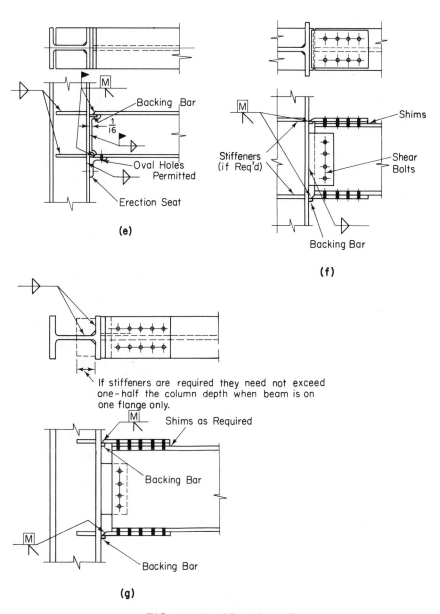

FIG. 11-28 (*Continued*)

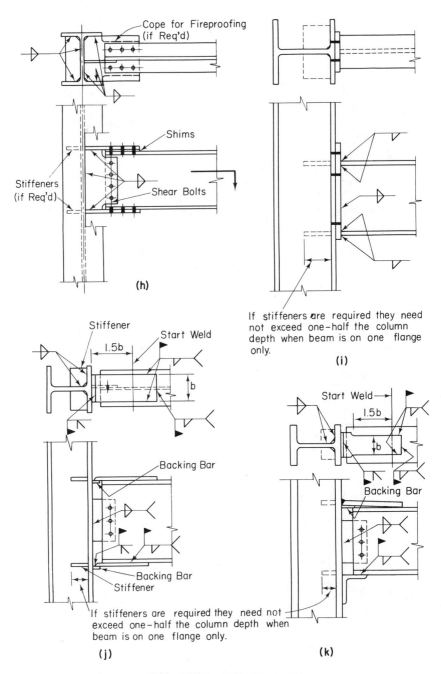

FIG. 11-28 (Continued)

ment connections may be all welded (Fig. 11-28c, d, and e), all bolted (Fig. 11-28a and b), or a combination (Fig. 11-28g, h, i, and j).

Figure 11-28j and k illustrates a typical type 3 simirigid connection that is designed to develop specific resisting moments. The moment may be assumed to be resisted by the top and bottom flange plates, which are field-welded to the top and bottom beam flange, respectively, as well as to the column as shown. The shear may be assumed to be transferred to the column by the vertical web plate, which uses field fasteners in the web and shop-welds to the column. A length of top flange plate of $1.5b$ is "weldless" to permit the necessary elongation under load to obtain the desired semirigid, type 3, connection action. A suggested design procedure is found in Part 4 of the AISC Manual.

Figure 11-28g is typical of a rigid, type 1, connection, which should be designed to develop the inherent moments. Part 4 of the AISC Manual suggests a design procedure for this shop-welded, field-bolted moment connection. The moment is assumed to be resisted by the flange plates by a horizontal couple. The flange plates are shop-welded to the column and field-bolted to the beam flanges. Shear is assumed to be transferred to the column by the web plate, which is shop-welded to the column and field-bolted to the beam web. Shims are provided between the top flange and top flange plate to facilitate field fit-up. If required, column stiffeners are shop-welded between the column flanges, in line with the top and bottom flange moment plates. A design example of this type of connection will be illustrated in Section 11-21.

Figure 11-28i shows a moment end-plate connection. The design procedure illustrated in Part 4 of the AISC Manual assumes that the centroid of the tensile and compressive flange forces acts at the center of the beam flanges. The top bolts are assumed equidistant above and below the top flange.

11-19 ECONOMY

In any practical structure, there must be an economical balance between the materials used in construction, the details of the design, and the standards of workmanship. As a practical matter, any combination of these may fall short of optimal economy. The degree of design perfection, desired materials, and construction procedures must always be considered a compromising matter to assure least cost, required margin of safety, and minimization of construction restrictions. The designer should provide for flexibility of details and fabrication.

11-20   REFERENCES

Some excellent references on connections are listed next. The AISC Manual, as well as *AISC Engineering Journal* articles pertaining to the subject, are valuable additional references.

(1)   O. W. Blodgett, *Design of Welded Structures,* James F. Lincoln Arc Welding Foundation, Cleveland, Ohio, 1966.

(2)   La Motte Grover, *Manual of Design for Arc-Welded Steel Structures,* Air Reduction Company, Inc., New York, 1964.

(3)   J. W. Fisher and J. H. A. Struik, *Guide to Design Criteria for Bolted and Riveted Joints,* John Wiley & Sons, Inc., New York, 1974.

(4)   *Fastener Standards,* Industrial Fasteners Institute, Cleveland, Ohio, 1965.

(5)   *High-Strength Bolting for Structural Joints,* Bethlehem Steel Corp., Bethlehem, Pa., Booklet 2867.

(6)   *Structural Steel Detailing,* American Institute of Steel Construction.

11-21   DESIGN EXAMPLES

An excellent reference for illustrative design examples for virtually all the connection types heretofore discussed is the AISC Manual. Numerous numerical examples are presented. In this section, three design examples will be presented: on prying action (a tee-type hanger design); the design of a one-sided type 2 angle connection (also illustrates the use of Fig. 11-12, welded framed beam connection allowable loads); moment connection design (welded/bolted Type 1).

EXAMPLE 11-1

Design a tee-section hanger using A36 steel and $\frac{7}{8}$-in.-diameter A325 high-strength bolts. A 55-kip load is suspended from the bottom flange of a W 36 × 194 shape. Assume the usual gage on the W shape.

solution:

(a)   From the AISC Manual, Part 4, Bolts and Rivets Table, tension on gross nominal area, for $\frac{7}{8}$ in. diameter A325 bolts, the allowable tension load on a bolt, $B = 26.5$ kips. Assume $p = 6$ in. (approximately $b_f/2$ of the W

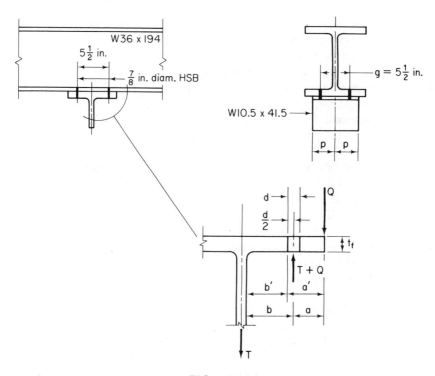

FIG. 11-29

shape; see Fig. 11-29). $P = 55$ kips; number of bolts required = $P/B = 55/26.5 = 2.08$; *try* 4 bolts.

$$T = \frac{55}{4} = 13.75 \text{ kips}$$

(b) The determination of a trial hanger section is greatly facilitated through the use of the design aid found in Fig. 11-27. This table provides a rapid selection of a trial hanger fitting using A36 material. The trial section must then be checked for bending stresses and tension in the bolts using a final design procedure (Section 11-16).

In the table, loads P, in kips per linear inch, are tabulated for corresponding values of b and t_f. b is defined as the distance from the center line of the fastener to the face of the outstanding leg of angle or the tee stem (inches) and t_f is the thickness of the hanger flange (inches); $b/2$ is the moment arm used to determine the assumed moment.

The values in the table are based upon the assumption that moments M_1 (at face of tee stem or angle leg) and M_2 (at the center line of the fastener) are equal. The allowable load on two angles or a tee, in kips per linear inch, uses a maximum bending stress of $0.75F_y = 27$ ksi (bending of a plate about its minor axis). The loads in the table are arrived at assuming $P/2 = T$ to act at each flange fastener and $f = M/S = [T(b/2)]/(t_f^2/6) = 0.75F_y$; $T = 9t_f^2/b$ or $P = 2T = 18t_f^2/b$, the hanger capacity.

Entering Fig. 11-27 with $2T/p = 27.5/6 = 4.58$ kips per linear inch and $b = (g/2) - (t_w/2) = 2\frac{3}{4} - \frac{1}{4} = 2\frac{1}{2}$ in. (assuming $t_w = \frac{1}{2}$ in.), choose $t_f = \frac{13}{16}$ in. (4.75 kips/linear inch $>$ 4.58 kips/linear inch).

Try a WT 10.5 × 41.5, which is cut from a W 21 × 83 beam, which has a flange width $b_f = 8.355$ in., a flange thickness $t_f = 0.835$ in. (slightly greater than $\frac{13}{16}$ in.). The usual gage for a W 21 × 83 accepts the assumed gage of $5\frac{1}{2}$ in. Finally, the thickness of the supporting beam flange (W 36 × 194) is 1.260 in. $>$ 0.835 in.

$$b = \frac{g}{2} - \frac{t_w}{2} = 2.75 - 0.26 = 2.49 \text{ in.} > 1\frac{3}{8} \text{ in.}$$

wrench clearance (AISC Manual, Part 4)

$$a = \frac{b_f - g}{2} = \frac{8.355 - 5.5}{2} = 1.428 \text{ in.} < 1.25b$$

$$b' = b - \frac{d}{2} = 2.49 - 0.438 = 2.052 \text{ in.}$$

$$a' = a + \frac{d}{2} = 1.428 + 0.438 = 1.866 \text{ in.}$$

$$d' = \frac{15}{16} = 0.9375 \text{ in.}$$

(c)

$$\delta = 1 - \frac{0.9375}{6} = 0.844$$

$$M = \frac{6.0(0.835)^2(36)}{8} = 18.83 \text{ kip-in.}$$

$$\alpha = \frac{[13.75(2.49)/(18.83) - 1]}{0.844} = 0.969 < 1.0$$

$$B_c = 13.75\left\{1 + \frac{0.844(0.969)}{[1 + 0.844(0.969)]}\left[\frac{2.052}{1.866}\right]\right\}$$
$$= 21.92 \text{ kips}$$

(d) $$t_{freq} = \left\{\frac{8(21.92)1.866(2.052)}{6(36)[1.866 + (0.844)(0.969)(3.918)]}\right\}^{\frac{1}{2}}$$
$$= 0.6130 \text{ in.} < 0.835 \text{ in.} \quad \text{OK}$$

(e) $$Q = 21.92 - 13.75 = 8.17 \text{ kips}$$

The trial hanger section, together with its supporting beam, is shown in Fig. 11-29.

EXAMPLE 11-2

Design a one-sided connection (Type 2) for a W 21 × 50 beam framing to the flange of a column. Consider all material to be A36 and that the connection is to be a single-angle shop-welded (E70) to the beam web and field-bolted (A325–X, $\frac{3}{4}$ in. diameter) to the column flange. The column size is W 14 × 90. The beam-end reaction is 75 kips.

solution:

(a) Outstanding angle leg: from AISC Manual, Part 4, Bolts, Threaded Parts and Rivets Allowable Shear and Bearing Load Tables: r_v (single shear) = 13.3 kips/bolt; r_v [$(t_f)_{column}$ = 0.710 $\frac{11}{16}$ in., bearing] $\gg$ 13.3 kips/bolt in single shear; r_v = 13.3 kips/bolt (governs); R = reaction = 75 kips; C = coefficent = R/r_v = 75/13.3 = 5.64 (Section 11.14).

From the AISC Manual, Part 4, reading the next largest value of C: for case I (one vertical row of fasteners, n = 7 fasteners per row), C = 6.17; for case II (two vertical rows of fasteners, n = 5 fasteners per row), C = 6.84. Use case II, because case I would necessitate too long an angle for a W 21 beam depth.

(b) Beam web leg of angle, from Fig. 11-12: since this figure is based upon double web angles and this example deals with a single web angle, use double the value of the reaction, R, or 2(75) = 150 kips.

For a $\frac{5}{16}$-in. fillet weld (E70) of 1 ft $2\frac{1}{2}$-in. length, the capacity is 156 kips > 150 kips. The minimum web thickness, t_w, listed is 0.64 in., but this is based upon having welds on both sides of the web.

Therefore, the minimum web thickness for a one-sided connection would be 0.64/2 = 0.32 in., which is < t_w = 0.380 in.

Use $\frac{5}{16}$-in. fillet weld with an angle length of 1 ft $2\frac{1}{2}$ in.

(c) Angle allowable shear stress: $F_v = 0.40F_y = 14.5$ ksi; $R = 75$ kips, $L = 14\frac{1}{2}$ in. Thickness of angle required: $t = 75/14.5(17.5) = 0.30$ in. With a $\frac{5}{16}$-in. weld, a thickness of $\frac{3}{8}$ in. is needed and $\frac{3}{8}$ in. > 0.30 in. Checking the angle for net shear capacity (Section 11-13.1):

$$[14.5 - 5(\tfrac{3}{4} + \tfrac{1}{16})]\,\tfrac{3}{8}\,(0.3)58 = 68.12 \not> 75 \text{ kips}$$

Therefore, use $\frac{1}{2}$-in. angle (90.83 kips > 75 kips). See Fig. 11-30.

Use a L6 × 3$\frac{1}{2}$ × $\frac{1}{2}$ × 1 ft-2$\frac{1}{2}$; E70 $\frac{5}{16}$-in. fillet weld.

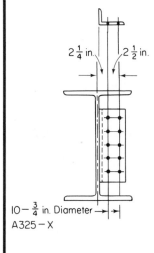

2$\frac{1}{4}$ in.   2$\frac{1}{2}$ in.

10 − $\frac{3}{4}$ in. Diameter
A325 − X

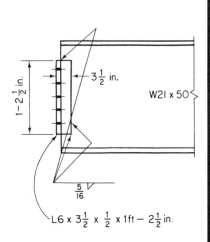

1 − 2$\frac{1}{2}$ in.

3$\frac{1}{2}$ in.

W21 × 50

$\frac{5}{16}$

L6 × 3$\frac{1}{2}$ × $\frac{1}{2}$ × 1ft − 2$\frac{1}{2}$ in.

FIG. 11-30

EXAMPLE 11-3

Design a type 1 moment connection similar to that shown in Fig. 11-28g (shop-welded and field-bolted). A W 18 × 50 beam is to be framed to the flange of a W 10 × 49 column. The end reaction is 35 kips and the design moment is 130 kip-ft. Beam, column, and connection material is A36. A325–X field bolts ($\frac{3}{4}$-in. diameter) in the flanges and A325–N in the web plate ($\frac{3}{4}$-in. diameter) and E70 shop welds are used.

solution:

(a) Flange-area reduction investigation: gross A_f = 7.495(0.570) = 4.272 in.^{2}. Assuming two rows of HSB as in Fig. 11-28g (see Section 3-3), net A_f = 4.272 − 2(0.75 + 0.125)0.570 = 3.274 in.^{2}.

Actual flange-area reduction = 100(4.272 − 3.274)/4.272 = 23.4% (Section 7-2).

$$\text{actual flange-area reduction} = \begin{array}{r} 23.4\% \\ -15.0\% \end{array}$$

design flange-area reduction = 8.4% = 0.084

For a W 18 × 50: gross I = 800 in.^{4}; d = 17.99 in.; t_f = 0.570 in.; net I = 800 − 2(4.274) [(17.99/2) − (0.570/2)]^2(0.084) = 745.5 in.^{4}. M = 130 kip-ft; $S_{\text{min req}}$ = 12(130)/24 = 65 in.^{3}. Reduced $S_{\text{net}} = I_{\text{net}}/c$ = 745.5/(17.99/2) = 82.9 in.^{3}. $S_{\text{net}} > S_{\text{min}}$; therefore, the beam is okay.

(b) Flange plates: the moment, M = 130 kip-ft can be expressed by the couple, T, acting a distance of d apart. The couple, T, is assumed to act at the outermost surface of the flanges so as to replace the moment, M.

$Td = M$ and $T = M/d$ = 12(130)/17.99 = 86.7 kips; $F_t = 0.60F_y$ = 22 ksi and net area of flange plates required = $A_p = T/F_t$ = 86.7/22 = 3.94 in.^{2}. ($A_{\text{gross min}}$ = 3.94/0.85 = 4.64 in.^{2}).

Try flange-plate thickness $t_p = \frac{3}{4}$ in. Therefore, the width of the flange plate $b = (A_p +$ fastening area deducted)$/t_p$.

Fastener area deducted two rows of $\frac{3}{4}$-in. diameter fasteners (standard holes assumed): = 2(0.75 + 0.125)0.75 = 1.31 in.^{2}; b = (3.94 + 1.31)/0.75 = 7.0 in. *Try 2 PL $\frac{3}{4}$ × 7 (one top, one bottom)*

$A_g = 5.25$ in.^{2}

$A_{P\ net} = 5.25 - 2(0.75 + 0.125)\,0.75 = 3.94$ in.$^2 < 0.85$ $(5.25) = 4.46$ in.^{2} (Section 3-3)

Use two PL $\frac{3}{4} \times 7$ (one top, one bottom).

(c) Flange-plate fasteners (A325-X): N = number of fasteners $= T/r_v$.

Single shear r_v (bottom flange plate) = 13.3 kips/bolt (from AISC Manual, Part 4, Bolts, Threaded Parts and Rivets, Allowable Shear and Bearing Load Tables); bearing r_v: ($t_f = 0.570 = \frac{9}{16}$ in.) $\$c4\frac{1}{2}$ 13.35 kips/bolt in single shear; therefore, $r_v = 13.3$ kips/bolt (governs). $N = 86.7/13.3 = 6.52$.

Use eight A325-X bolts (four in each of two rows).

(d) Web connection: A325-N, $\frac{3}{4}$-in. diameter; E70 shop welds to column flange ($t_f = 0.560$ in.).

(i) Field bolts: assuming $t_p = \frac{1}{4}$-in. web plate; single shear $r_v = 9.3$ kips/bolt (from AISC Manual, Part 4, Bolts, Threaded Parts and Rivets Allowable Shear and Bearing Load Tables); bearing $\frac{1}{4}$-in. plate $r_v \gg$ 9.3 kips/bolt; $r_v = 9.3$ kips/bolt (governs); $R = 35$ kips; $N = R/r_v = 35/9.3 = 3.76$. Try four bolts.
Bearing on beam web check:

Assume an end distance $= 1\frac{1}{2}$ in. From the AISC Manual, Part 4, Table I-E, Bolts, Threaded Parts and Rivets Allowable Bearing Loads, for $F_u = 58$ ksi (A36 material), $l_v = 1\frac{1}{2}$ in. and $t_w = 0.355$.

Allowable load $= 43.5$ (for t $= 1$ in.) $(0.355)4$

$= 61.8$ kips > 35 kips

Also, from Table I-E, for a spacing of bolts $= 3$ in.:

Allowable load $= 65.3(0.355)4 = 92.7$ kips > 35 kips

Use four $\frac{3}{4}$-in. A325-N in a bearing-type connection (standard holes).
Shear plate design:

Try a plate with 3-in. pitch and $1\frac{1}{2}$-in. end distance; length of plate $= 12$ in. $= l$.

$$l_{net} = 12 - 4(\tfrac{3}{4} + \tfrac{1}{16}) = 8.75 \text{ in.}$$

$$F_v = 0.30F_u = 0.30(58) = 17.4 \text{ ksi}$$

$$t = \frac{35}{(17.4)8.75} = 0.23 \text{ in.}$$

Try a $\frac{1}{4}$-in. plate.
Bearing check:
    From Table I-E, for $F_u = 58$ ksi, $l_v = 1\frac{1}{2}$ in., and $t = \frac{1}{4}$ in.:

Allowable load = 16.3 kips/bolt for 3-in. spacing
                       = 10.9 kips/bolt for $1\frac{1}{2}$- in. end distance
10.9 kips/bolt > 35/4 = 8.75 kips/bolt   OK

Weld to column flange:

l of plate = 12 in. = l of weld
For E70 electrode, two fillet welds (one each side of plate):

Strength of $\frac{1}{16}$-in. weld $= \dfrac{0.3(70)0.707}{16} = 0.928D$

where D is the number of $\frac{1}{16}$ths of fillet weld size. And

$$D_{min} = \frac{35}{2(0.928)12} = 1.57$$

Since $t_{f \text{ col}} = \frac{9}{16}$ in., use $\frac{1}{4}$-in. fillet weld (Table 11-8).
    Use PL $\frac{1}{4} \times 5 \times 1$ ft-0 with two $\frac{1}{4}$-in. fillet welds to the column flange full length each side.

(e) Column web shear check: Refer to Section 4-3 and AISC Specification, Section 1.5.1.2 and Commentary. When $t_{\text{web col}} \geq 32M/(A_{bc}F_y)$, no column web reinforcement is necessary. M is the algebraic sum of clockwise and counterclockwise moments (foot-kips) applied on opposite sides of the connection boundary of two or more members whose webs lie in a common plane; A_{bc} is the planar area of the connection web (in.^{2}).

$$t_{\text{web col}} = 0.560 \text{ in.}$$

$$\frac{32M}{A_{bc}F_y} = \frac{32(130)}{[17.99 + 2(\frac{3}{4})]9.98(36)}$$

$$= 0.594 \text{ in.} > 0.560 \text{ in.}$$

Therefore, column web must be reinforced.
    Use a web doubler plate fillet welded all around

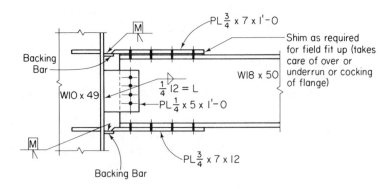

Note: (1) For flange bolts use 325-X, 2 rows
      (2) For beam web bolts use A325-N

FIG. 11-31

to make up the difference between actual and required values of web thickness or a pair of diagonal
stiffeners (refer to pages 5.7-32, 33 of Omer W. Blodgett, *Design of Welded Structures,* James F. Lincoln
Arc Welding Foundation.

  For column web stiffener requirements, reference to Section 11-21.1 should be made.

  The final connection is shown in Fig. 11-31.

11-21.1 Column Web Stiffeners*

When the flanges or moment connection plates for beam- or girder-to-
column flange connections (Figs. 11-4, 11-28, 11-31) are welded, a pair
of column web stiffeners shall be provided when A_{st}, the required combined cross-sectional area of the stiffeners, computed by Eq. 11-4a, is
positive.

$$A_{st} = \frac{P_{bf} - F_{yc}\, t_c\, (t_b + 5k)}{F_{yst}} \qquad (11\text{-}4a)$$

where F_{yc} = yield stress of column, ksi
   F_{yst} = yield stress of stiffener, ksi
   k = distance from outside face of column flange to toe of col-
     umn web fillet, or equivalent distance if column is welded,
     in.

* Ref. Beam-to-Column Flange Connections—Restrained Members—a
Design Aid. U.S. Steel Corporation ADUSS 27–7631–01, December 1979.

P_{bf} = computed force delivered by beam flange or moment connection plate multiplied by $\frac{5}{3}$ when P_{bf} is due to only live plus dead loads, or by $\frac{4}{3}$ when P_{bf} is due to live plus dead loads and and wind or earthquake force, kips.

t_c = column web thickness, in.

t_b = thickness of flange or moment connection plate delivering concentrated force, in.

In addition to the aforementioned requirements for column stiffeners, a stiffener or pair of stiffeners shall be provided opposite the appropriate beam or girder flange in accordance with Eqs. 11-4b and 11-4c.

Provide a pair of stiffeners or a single stiffener on the column web opposite the beam or girder compression flange when

$$d_c > \frac{4100 \, t_c^3 (F_{yc})^{\frac{1}{2}}}{P_{bf}} \qquad (11\text{-}4b)$$

Provide a pair of stiffeners on the the column web opposite the beam or girder tension flange when:

$$t_f < 0.4 \left[\frac{P_{bf}}{F_{yc}} \right]^{\frac{1}{2}} \qquad (11\text{-}4c)$$

where d_c = column web depth clear of fillets, in.

t_f = column flange thickness, in.

The column web stiffeners shall abide by the following requirements:

(1) Stiffener width + $t_c/2 \geq b/3$.   (11-4d)

(2) Stiffener thickness $\geq t_b/2$.   (11-4e)

(3) When P_{bf} occurs on only one column flange, the column stiffener length need not be more than the depth of column/2.

(4) Stiffener-to-column web weld sized to carry the force in the stiffener caused by the unbalanced moments on opposite sides of the column.

Where b is the width of the flange or moment connection plate delivering the concentrated force.

EXAMPLE 11-4

Determine the need for column web stiffeners and if need be design the stiffeners and their connections for Example 11-3.

solution:

(a) Requirements for stiffeners:
horizontal force delivered by moment plate

$$= \frac{12(130)}{17.99 + \frac{3}{4}}$$

$$= 83.2 \text{ kips}$$

$$P_{bf} = (\tfrac{5}{3})83.2 = 138.7 \text{ kips}$$

From Eq. 11-4a, column web stiffeners are required when the formula gives a positive answer for A_{st}.

$$A_{st} = \frac{138.7 - 36(0.560) \, [0.75 + 5(1\frac{3}{16})]}{36}$$

$$= 0.1064 \text{ in.}^2$$

Since the answer is positive, stiffeners are required at both the tension and compression flanges. If the answer were negative, a further check by Eq. 11-4b and c would still be required.
From Eq. 11-4d,

$$\text{stiffener width} = \frac{7.0}{3} - \frac{0.340}{2}$$

$$= 2.16 \text{ in. (min.)}$$

From Eq. 11-4e,

$$\text{stiffener thickness} = \frac{0.75}{2} = 0.375 \text{ in. (min.)}$$

$$\text{at top and bottom flanges}$$

For practical detailing, use $\frac{3}{8}$ in. $\times$ $4\frac{1}{2}$ in. on each side of column web at top and bottom beam flanges. Clip stiffener corners $\frac{3}{4}$ in. $\times$ $\frac{3}{4}$ in. (refer to Fig. 11-28g). Stiffener check:

$$A_{st} \text{ furnished} = 2(3)\tfrac{3}{8} = 2.25 \text{ in.}^2 > 0.1064 \text{ in.}^2 \quad \text{OK}$$

$$\frac{\text{width}}{\text{thickness}} = \frac{3}{0.375} = 8.0 \leq \frac{95}{(F_y)^{\frac{1}{2}}} = \frac{95}{6} = 15.8 \quad \text{OK}$$

$$\text{length stiffener} = \frac{d_{col}}{2} = \frac{9.98}{2} = 4.99 \text{ in.}, \quad \text{use 5 in.}$$

(b) Stiffener connection:

Use E70 fillet welds.

min. weld size to column web ($t = \frac{5}{16}$ in.) $= \frac{3}{16}$ in.

min. weld size to column flange ($t = \frac{9}{16}$ in.) $= \frac{1}{4}$ in.
(refer to Table 11-8)

Weld lengths:

From step (1), the force at the column web to be resisted by the stiffener welds is

$$P_{bf} - F_{yc}t_c(t_b + 5k) = 138.7 - [36(0.560)\,(0.75 + 5(\tfrac{13}{16})\,]$$

$$= 3.83 \text{ kips}$$

From the Column Load Tables in the AISC Manual, Part 3 (Fig. 9-5), $P_{fb} = 71$ kips for an A36, W 10 × 49 column. P_{fb} is the maximum column web resisting force at the beam tension flange (kips) and is found at the bottom of the tables. $P_{fb} = F_{yc}t_f^2/0.16$ (refer to Eq. 11-4c).

The force at the tension beam flange to be resisted by the stiffener welds is $P_{bf} - P_{fb} = 138.7 - 71 = 67.7$ kips. The tension beam flange force governs.

Weld requirements:

$$\text{stiffener load} = \frac{67.7}{2} = 33.9 \text{ kips each side}$$

l_w, length of weld at column web:

$$l_w = \frac{33.9}{2(3)0.928} = 6.1 \text{ in. (min)}$$

Make weld the full length of stiffener: 5 in. $- \frac{3}{4}$ in. $= 4\frac{1}{4}$ in. both sides.

To make $l_w < 4\frac{1}{4}$ in., a $\frac{5}{16}$-in. weld must be used ($l_w = 3.7$ in. min). Use $\frac{5}{16}$-in. weld. l_f, length of weld at column flange:

$$l_f = \frac{33.9}{2(5)0.928} = 3.7 \text{ (min)}$$

Make $\frac{5}{16}$-in. weld $4\frac{1}{2}$ in. $- \frac{3}{4}$ in. $= 3\frac{3}{4}$ in. both sides.

The minimum thickness of plate to develop $\frac{5}{16}$-in. fillet welds at maximum shear stress $= 0.51$ in. (Fig. 11-12). Shear stress check in weld metal:

$$f_v = \frac{33.9}{2(\frac{5}{16})0.707(3\frac{3}{4})} = 20.5 \text{ ksi} < 0.3(70) = 21 \text{ ksi} \quad \text{OK}$$

Shear stress check in stiffener base metal:

$$f_v = \frac{33.9}{\frac{3}{8}(3\frac{3}{4})} = 24.1 \text{ ksi} \not< 0.4(36) = 14.4 \text{ ksi}$$

Therefore, use stiffener size of $\frac{1}{2}$ in. $\times$ $4\frac{1}{2}$ in. ($f_v = 14.46$ $\approx$ 14.4 ksi).

In this problem, end moment is not reversible, and it is therefore allowed to finish the two compression flange column stiffeners to bear rather than welding them to the flange.

$F_p = 0.90F_y$ (AISC allowable bearing stress on ends of fitted bearing stiffeners)
$= 32.4$ ksi

$$f_p = \frac{33.9}{2(4\frac{1}{2} - \frac{3}{4})\frac{1}{2}} = 9.04 \text{ ksi} < 32.4 \text{ ksi} \quad \text{OK}$$

In summary, use two $\frac{1}{2}$ in. $\times$ $4\frac{1}{2}$ in. (5 in. long) fitted at compression flange and welded ($\frac{5}{16}$-in. fillet welds) at the web and at the tension flange.

In conclusion, the designer should be flexible in his or her thinking when designing connections. Generally, the fewer pieces of detail material, the more economical the connection. Designers should allow steel fabricators flexibility in the choice of connection; they may be able to produce an alternative detail more economically. The designer should judge it on its merit. The fabricator may vary the details from time to time. This is due to changes in both the amount of work and type of work that is in the shop at the particular time.

The most important advice is to keep in mind how the particular connection will behave under service loading. Remember, connections are important and are vital, serving as the link between the individual structural members. They maintain the structural integrity of the completed structure.

PROBLEMS

(11-1) Determine the following strengths using the AISC Manual.
(a) The total capacity of 12 in. of $\frac{3}{8}$ in. fillet weld using E70XX electrodes.
(b) The strength of 6 in. of $\frac{1}{2}$ in. fillet weld using E90XX electrodes.

(c) The single shear capacity of 12 high-strength bolts, A325-X of $\frac{7}{8}$ in. diameter, spaced at 3-in. increments and standard holes.

(d) The bearing capacity of the bolt specified in part (c) when acting on $\frac{5}{16}$ in. plate, $F_y = 50$ ksi, and $F_u = 70$ ksi.

(11-2) Determine the fillet weld size required to connect the web to the flange in Problem 7-3. Use E70 electrode.

(11-3) Determine the fillet weld size required to connect the web to the flange in Problem 7-6. Use F7X-EXXX submerged arc welds (allowable shear stress is 21 ksi).

(11-4) Design the fillet-welded connection in Problem 3-3. Assume E70XX electrodes.

(11-5) If the connection in Problem 3-7 is a type X A325, what is the capacity of the member? Assume only bolts connecting the back-to-back legs resist the tensile load.

(11-6) Design the connections listed in Problem 3-8. In part (a) use E70 electrodes ($\frac{5}{16}$ in. fillet welds), and in part (b) use a type F A325 connection.

(11-7) If the connection in Problem 3-9 is a type X A325, is it adequate to transfer the calculated tensile load?

(11-8) Determine the capacity of the bolted connection in Problem 3-10. The connection is a type X A325. If the number of bolts shown will not transfer the tensile load, redesign the connection to transfer that load.

(11-9) The simple tension splice shown transfers a symmetrical load P. Determine the permissible value of P, based on shear, bearing, and tension. Assume A36 steel and $\frac{3}{4}$ in. diameter A325-X bolts. Be sure to indicate the most critical section in tension.

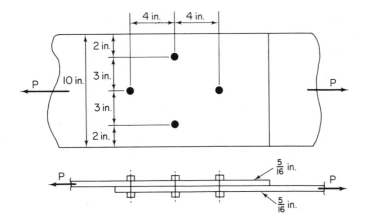

(11-10)   A single angle (A36, L6 × 4 × $\frac{1}{2}$) is fillet welded to a gusset plate
(A36, $\frac{9}{16}$ in. thick). Determine the required lengths of fillet weld
L_1 and L_2 to develop the full tensile capacity of the angle with
a "balanced" design.

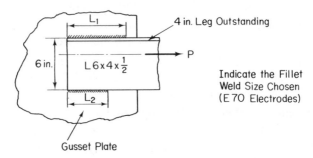

Gusset Plate

(11-11)   Revise the solution to Problem 11-10 assuming a transverse fillet
weld on the 6-in. leg in addition to the welds shown.

(11-12)   Determine the length, L, of a $\frac{1}{2}$ in. fillet weld required to transfer
the full capacity of the A588 plates. Welding runs completely
along the interface as shown. Use E70 electrodes.

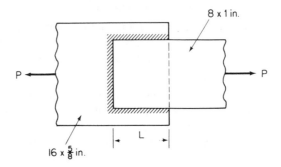

(11-13)   In Problem 11-12, if L had been limited to 8 in. by design requirements, how could you transfer the force between plates?
Design at least one alternative.

(11-14)   Design the standard framed bolted shear connections at column
C (assume column will accommodate connection angles) and
that connects beams B_1 to girder G_1 in Problem 2-2. (See Problem 6-1.) Use $\frac{3}{4}$ in. bolts in type X A325 connections (standard
holes) and A36 connection material. Check installation clearances and use standard gages, where possible, for connection
angles.

(11-15) Design a standard framed bolted connection to transfer a shear of 70 kips from a W 21 × 44 beam to the flange of a W 14 × 159 column. The members and connection material are of A588 (F_u = 70 ksi) steel and the bolts are $\frac{7}{8}$ in. diameter A325 bolts (standard holes). Design as:
(a) Friction-type connection (F).
(b) Bearing-type connection (X).

Check each for installation clearances and use standard gages, where possible, for connection angles.

(11-16) Design a standard framed (type 2) connection (Fig. 11-12) to transfer a beam end reaction of 150 kips for a W 24 × 117 beam framing into a flange of a column. Columns and beams are A36 steel. Assume column accommodates connection angles. Design the connection as:
(a) All-welded connection (Fig. 11-13).
(b) All-bolted connection ($\frac{7}{8}$ in. A325 type X, standard holes).
(c) Welded to the beam and bolted to the flange ($\frac{7}{8}$ in. A325 type X, standard holes).

(11-17) (a) Design a type 1 moment connection to transfer a shear of 200 kips and a moment of 380 kip-ft (dead and live load only). A W 36 × 170 beam is to be framed to the flange of a W 14 × 120 column. The beam (F_t = 22 ksi and F_b = 24 ksi), column, and angles are of A36 steel. Use $\frac{7}{8}$ in. diameter A325 (type X) bolts in the web plate to beam web connections and E70 field welds for flange plate to column flange connections. (Refer to Fig. 11-28.)
(b) Check and, if needed, design column web, stiffeners as well as the weld requirements of stiffeners to column web and flange.

(11-18) An all-welded unstiffened seated beam connection of the type shown in Fig. 11-14 is used to attach the web of a W 18 × 65 beam in a simple span fully laterally supported of 28 ft to the flange of a W 14 × 68 column. Assuming A36 steel and E70 electrodes, choose the appropriate seat angle and weld for this connection.

(11-19) Compute the permissible value of P using the following:
(a) Figure 11-17 (D = 3 in.).
(b) Actual equations (Figure 11-16) ($D = 5\frac{1}{2}$ in.).
(c) Figure 11-19 ($D = 5\frac{1}{2}$ in.).

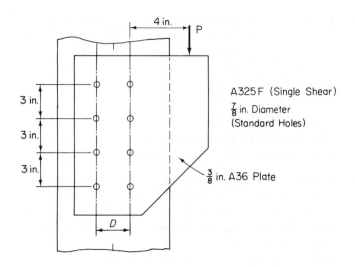

Assume that the column flange is adequately thick so that bolt shear governs.

(11-20) Compute the permissible value of *P* using the following:
   (a) Figure 11-22.
   (b) Actual equations (Figure 11-20).
   (c) Figure 11-24.

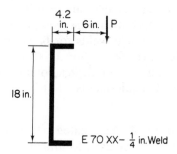

open web
steel joists

12-1   INTRODUCTION

Steel open web joists or, as they are more commonly referred to, *steel joists,* refer to standardized trusses that carry load and are suitable to directly support floors or roofs of buildings. The chord and web members of the steel joists are made from hot-rolled or cold-formed steel whose yield strength is achieved by cold working. The webs are open and permit the passage of piping, electrical conduits, and ducts within the construction depth of the floor joist readily.

The first steel joist (1923) was a Warren truss type with round bars for top and bottom chord members and one continuous bent bar for a web. Over the years, many other joist types were developed and then, as if often the case today, each manufacturer developed an independent means of design and fabrication, creating difficulties for the design professionals to compare rated capacities and to use the full economies of open web steel-joist construction.

To alleviate the problem, the Steel Joist Institute (SJI) was established in 1928, and in that year the first standard specifications were adopted, followed by the first standardized load table. A standardized series, the S-series, was established. The institute established a workable and acceptable standard, which gave a standard to what was a runaway, helter-skelter product.

Knowledge of the development of the SJI Standards is helpful to the designer of today in many instances where he or she is concerned with structures designed many years ago. With that purpose in mind, Table 12-1 lists the adoptions made by SJI.

Table 12-1

Year	Adoption
1928	S-Series Specifications
1929	S-Series Load Tables
1953	[a] L-Series (Longspan) Specifications and Load Tables (spans up to 96 ft and depths up to 48 in.)
1961	H-Series (high-strength steel, F_y = 50 ksi; spans up to 48 ft) J-Series (A36 steel, F_y = 36 ksi) replaces S-Series LA-Series (A36 steel, F_y = 36 ksi) replaces L-Series
1962	LH-Series (high-strength steel, F_y = 36 to 50 ksi; Longspan Series)
1965	[a] J- and H-Series combined specification
1966	[a] LJ- and LH-Series combined specifications and Standard Load Table for LJ-Series (replaced LA- and LH-Series) (LJ uses F_y = 36 ksi and LH uses F_y = 36 to 50 ksi)
1970	DLJ- and DLH-Series (Deep Longspan) Specifications and Load Tables (extended LJ- and LH-Spans)
1971	Number 2 chord sizes dropped and 8J3 and 8H3 added
1972	[a] J- and H-Series Specifications and Load Tables expanded to include depths up to 30 in. and spans to 30 ft [a] LJ- and LH-Series and DLJ- and DLH-Series combined into a single specification
1978	[a] J-, LJ-, and DLJ-Series eliminated as standards. Standard Specifications and Weight Tables for Joist Girders adopted by SJI

[a] Jointly approved by SJI and AISC.

The SJI membership is open to steel joist manufacturers who produce joists of the H-, LH-, or DLH-series that conform to the SJI Specifications and Load Tables. The institute's independent consulting engineer checks to see that the manufacturer's joist design conforms to the specifications and load tables. The AISC also has a procedure that checks standard series joist designs to assure that the series conforms with the

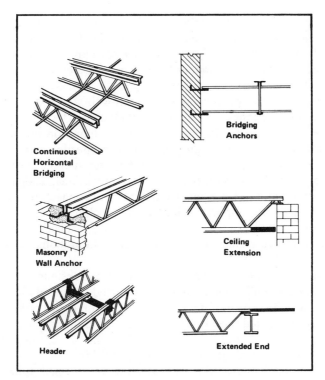

FIG. 12-1 *Copyright Steel Joist Institute. Reproduced by permission*

AISC/SJI Specifications and Load Tables. Plant facilities and some joists are investigated for the H-series. Figure 12-1 shows some of the details associated with open-web steel joists.

Several valuable references on the subject of steel open web joists are listed for the convenience of the reader:

(1) Standard Specifications and Load Tables. Steel Joists, Open Web H-Series, Longspan LH-Series, and Deep Longspan DLH-Series, SJI.

(2) *SJI Technical Digest*
No. 1: Jan. 1970, Design of Compression Chords for Open-Web Steel Joists
No. 2: Sept. 1970, Spacing of Bridging for Open-Web Steel Joists
No. 3: May 1971, Structural Design of Steel Joist Roofs to Resist Ponding Loads
No. 4: Design of Fire-Resistive Assemblies with Steel Joists
No. 5: Vibration of Steel Joist–Concrete Slab Floors
No. 6: Structural Design of Steel Roof Joists to Resist Uplift Loads

Reference (1) is an annual publication that gives, in addition to the specifications and load tables for each of the standard series of joists, a listing of manufacturers of joists along with their SJI-approved series of joist(s); Recommended Code of Standard Practice for Steel Joists; and Fire-Resistance Ratings with Steel Joists.

The Standard Specifications for open-web steel joists, for the standard H-series, the longspan LH-series and the deep longspan DLH-series are quite detailed. The Specifications cover definition of the series and steel material in the series, the mechanical properties and paint requirements, the design unit stresses and limiting parameters, design formulas and criteria for chord and web members, connections, design verification tests for chord and web elements and joints and connections; span: depth-limiting ratios, end supports, bridging, end anchorage, types and limiting dimensions of floor and roof decks, ponding, inspection, handling and erection, required camber (if any), deflection limitations (if any), and so on.

The Specifications specify that all joists be designed in accord with the particular specification as simply supported, uniformly loaded trusses that support a floor or roof deck such that the top chord of the steel joists is laterally supported (no lateral buckling). In addition, for any particular design situation not covered by the SJI Specifications, the AISC and/or AISI Specifications should be used as applicable.

The Load Tables will be discussed and described under the applicable section of this chapter. The designer need only choose one or more of the three standard series of steel joists as approved by SJI or AISC and, by the use of the load tables, choose a size that is standard and available.

12-2   STANDARD SERIES

The H-series joists are referred to as the standard joist series. These are shop-fabricated joists used for the direct support of floors and roof decks in buildings. They have *parallel* chord members. The H-series joist design is based upon a yield strength of chord of 50 ksi and a yield strength of web of either 36 or 50 ksi.

Hot-rolled or cold-formed steel, including cold-formed steel whose yield strength has been achieved by cold working, may be used. ASTM A36, 242, 441, 570, 572 (grade 50), 588, 606, 607 (grade 50) or 611 (grade D) steel material may be used for chord and web members. Any other material may be used provided such material is weldable and found to have the required mechanical properties following the specifications by tests performed by the producer or fabricator.

The usual configuration of web member is in the form of a modified Warren-type truss, although the web system may be any type, depending upon the standard of the particular manufacturer. As can be realized from the allowable steel materials for both web and chord members, they may be made from hot-rolled shapes, plates, and bars; hot-rolled or cold-rolled sheet; or hot-rolled sheet and strip, depending on the manufacturer's preference.

The joists (unless otherwise called for) have a standard shop coat of paint that meets the specified standards. They are designed following the specifications as simply supported uniformly loaded members with the top chord assumed braced against lateral buckling. The bottom chord is designed as an axially loaded tension member, while the top chord is designed either for axial compressive stress only or as a continuous member subject to combined axial and bending stress, depending upon the length of panel. All joint connections and splices use weld-type connections. The camber of joists is optional with the manufacturer.

The depth-to-clear-span ratio of a joist must not be less than $\frac{1}{24}$. The ends of joists should extend a minimum distance of $2\frac{1}{2}$ in. over steel supports or, when the bearing is less than $2\frac{1}{2}$ in., such as in the case of two opposite joists resting on a narrow steel support, special ends of joist must be provided with attachment to the support by bolting or welding. The specifications require minimum-bearing lengths measured in the direction of the joist span when the joists bear on masonry or concrete.

H-series joists frequently employ extended ends (Fig. 12-2) when required. The various joist manufacturers use different extended end de-

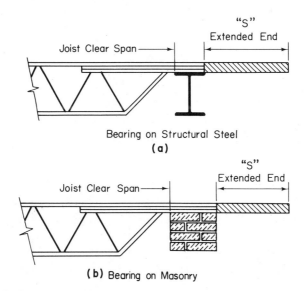

FIG. 12-2 *Copyright Steel Joist Institute. Reproduced by permission*

STANDARD LOAD TABLE OPEN WEB STEEL JOISTS, H-SERIES

Allowable Total Safe Loads in Pounds per Linear Foot Based on Allowable Stress of 30,000 psi

Joist Designation	24H6	24H7	24H8	24H9	24H10	24H11	26H8	26H9	26H10	26H11	28H8	28H9	28H10	28H11	30H8	30H9	30H10	30H11
Nominal Depth (in.)	24	24	24	24	24	24	26	26	26	26	28	28	28	28	30	30	30	30
Resist. Moment (in.-lbs)	462,000	576,000	716,000	851,000	957,000	1,106,000	784,000	925,000	1,040,000	1,203,000	846,000	1,000,000	1,124,000	1,300,000	909,000	1,075,000	1,207,000	1,397,000
Max. End React. (lbs.)	5600	5800	6000	7000	7500	8200	6700	7200	7600	8300	6700	7200	7700	8400	6800	7500	8100	8700
Approx. Wt. (lbs./ft.)	10.3	11.5	12.7	14.0	15.5	17.5	12.8	14.8	16.2	17.9	13.5	15.2	16.8	18.3	14.2	15.4	17.3	18.8
Clear Span in Feet — 24 or less	467	483	500	583	625	631	515	554	585	638	479	514	550	600	453	500	540	580
25	448	464	480	560	600	631	515	554	585	638	479	514	550	600	453	500	540	580
26	431	446	462	538	577	631	515	554	585	638	479	514	550	600	453	500	540	580
27	415 / 375	430	444	519	556	607	496	533	563	615	479	514	550	600	453	500	540	580
28	393 / 336	414 / 406	429	500	536	586	479	514	543	593	479	514	550	600	453	500	540	580
29	366 / 303	400 / 365	414	483	517	566	462	497	524	572	462	497	531	579	453	500	540	580
30	342 / 273	387 / 330	400	467 / 457	500	547	447	480	507	553	447	480	513	560	439	484	523	561
31	320 / 248	374 / 299	387 / 373	452 / 414	484 / 465	529	432 / 418	465	490	535	432	465	497	542	425	469	506	544
32	301 / 225	363 / 272	375 / 339	438 / 376	469 / 423	513 / 482	419 / 380	450 / 445	475	519	419	450	481	525	412	455	491	527
33	283 / 205	352 / 248	364 / 309	424 / 343	455 / 386	497 / 440	406 / 346	436 / 405	461 / 456	503	406 / 404	436	467	509	400	441	476	512
34	266 / 188	332 / 227	353 / 283	412 / 314	441 / 353	482 / 402	394 / 317	424 / 371	447 / 417	488 / 476	394 / 370	424	453	494	389	429	463	497
35	251 / 172	313 / 208	343 / 259	400 / 288	429 / 323	469 / 369	383 / 290	411 / 340	434 / 383	474 / 437	383 / 339	411 / 396	440	480	378 / 359	417	450	483
36	238 / 158	296 / 191	333 / 238	389 / 264	417 / 297	456 / 339	372 / 267	400 / 312	422 / 352	461 / 401	372 / 311	400 / 364	428 / 410	467	368 / 330	405 / 387	438 / 436	470
37	225 / 146	280 / 176	324 / 219	378 / 243	405 / 274	443 / 312	362 / 246	389 / 288	411 / 324	449 / 370	362 / 287	389 / 336	416 / 378	454 / 432	358 / 305	395 / 357	426 / 402	458
38	213 / 135	266 / 162	316 / 202	368 / 225	395 / 253	432 / 288	353 / 227	379 / 266	400 / 299	437 / 341	353 / 265	379 / 310	405 / 349	442 / 399	349 / 282	385 / 331	415 / 372	446 / 426
39	202 / 124	252 / 150	308 / 187	359 / 208	385 / 234	421 / 266	344 / 210	369 / 246	390 / 276	426 / 316	344 / 245	369 / 287	395 / 322	431 / 369				

STANDARD LOAD TABLE OPEN WEB STEEL JOISTS, H SERIES

Allowable Total Safe Loads in Pound per Linear Foot Based on Allowable Stress of 30,000 psi

Each cell shows two values: total safe load (top / bold) and corresponding live load (bottom). The left-hand column is the span (ft).

Span																		
40	435/395	405/345	375/306	340/262	420/342	385/299	360/266	335/227	415/292	380/256	360/228	327/194	410/247	375/217	350/193	298/174	240/139	193/115
41	424/367	395/320	366/285	332/243	410/318	376/278	351/247	327/211	405/272	371/238	351/211	311/181	400/229	366/201	337/179	284/161	228/129	183/107
42	414/341	386/298	357/265	324/226	400/295	367/258	343/229	319/196	395/253	362/221	343/197	296/168	390/213	357/187	322/166	271/150	218/120	175/100
43	405/318	377/278	349/247	316/211	391/275	358/241	335/214	305/183	386/235	353/206	334/183	283/156	381/199	345/174	307/155	258/140	208/112	167/93
44	395/297	368/259	341/230	309/196	382/257	350/225	327/200	291/171	377/220	345/193	319/171	270/146	373/186	330/163	293/145	247/130	198/105	159/87
45	387/278	360/242	333/215	299/184	373/240	342/210	320/187	279/159	369/205	338/180	305/160	258/137	364/173	315/152	280/135	236/122	190/98	152/81
46	378/260	352/227	326/202	286/172	365/225	335/197	313/175	267/149	361/192	328/168	291/150	247/128	348/162	302/142	268/127	226/114	181/92	146/76
47	370/244	345/213	319/189	274/161	357/211	328/184	302/164	255/140	353/180	314/158	279/140	237/120	334/152	289/133	257/119	216/107	174/86	139/71
48	363/229	338/200	311/177	263/151	350/198	321/173	289/154	245/131	346/169	301/148	268/132	227/112	320/143	277/125	246/111	207/100	167/81	134/67
49	355/215	331/188	298/167	252/142	343/186	312/163	278/144	235/124	334/159	289/139	257/124	218/106						
50	348/202	322/177	287/157	242/134	336/175	300/153	267/136	226/116	321/150	277/131	247/117	209/100						
51	341/191	309/166	276/148	233/126	329/165	288/144	256/128	217/110	308/141	267/124	237/110	201/94						
52	335/180	298/157	265/139	224/119	321/156	277/136	247/121	209/103	297/133	256/117	228/104	193/88						
53	328/170	286/148	255/132	216/112	309/147	267/128	237/114	201/98										
54	319/161	276/140	246/125	208/106	297/139	257/121	229/108	193/92										
55	308/152	266/133	237/118	200/101	287/132	248/115	220/102	186/87										
56	297/144	257/126	229/112	193/95	276/125	239/109	213/97	180/83										
57	287/137	248/119	221/106	187/90														
58	277/130	239/113	213/101	180/86														
59	268/123	231/108	206/95	174/81														
60	259/117	224/102	199/91	168/77														

FIG. 12-3 (continued)

tails. However, all details are designed as cantilever beams, with their reactions carried back to the first interior joist panel point as a minimum.

The standard extended end depth is $2\frac{1}{2}$ in. As a guide to the designer, allowable uniform loads in pounds per linear foot of extended end, S, are provided in the SJI annual publication (Section 12-1, reference 1) for extended ends fabricated from angle or channel sections. The capacities of these extended ends vary from manufacturer to manufacturer.

As mentioned, the H-series is completely standardized as to lengths, depths, and corresponding carrying capacities. The standard depths are 8, 10, 12, 14, 16, 18, 20, 22, 24, 26, 28, and 30 in. The clear span lengths of this standard series vary up to 60 ft. Figure 12-3 shows a typical page from the Standard Load Table for the H-series joist (based on an allowable stress of 30 ksi).

In Fig. 12-3, the **boldface** figures are the total safe uniform load-carrying capacities (plf). The dead loads, including the weight of joist, must be deducted to determine the live-load carrying capacity of the joist. The Load Tables may only be used for parallel chord joists installed to a maximum slope of $\frac{1}{2}$ in./ft. The *lightface* figures are the live loads of the joists (plf) producing an approximate deflection of $L/360$, the maximum live-load deflection permitted for floors and roofs having attached or suspended plastered ceilings following the specifications. Live load producing $L/240$, the maximum permissible live-load deflection for cases of roofs with other than plastered ceilings, may be obtained by multiplying the tabulated lightfaced figures by 1.5. Loads above the bold heavy-stepped lines indicate those loads governed by shear.

Since steel joists are relatively limber members that are dependent for stability on the metal decking or concrete decking, they require lateral bracing prior to placement of the decking for stability. This bracing is termed *bridging*. It functions to enable the steel joists to support loads during the construction phase. The bridging spans between and transverse to the steel joist spans. In addition, a secondary purpose of bridging is to hold the joists in the position shown on the plans. The specifications require that all bridging and bridging anchors be fully installed prior to the placement of construction loads on the joists. The specifications stipulate the minimum number of rows of bridging for various chord sizes [e.g., the last digit(s) of joist designation shown in the load tables] and for various ranges of clear spans of joist.

Figure 12-4a and b shows the two permissible types of bridging, horizontal and diagonal, respectively. Figure 12-4c shows how the ends of all bridging lines terminating at a wall should be anchored (a specification requirement). A similar detail should be used for bridging terminating at a beam.

Horizontal bridging (Fig. 12-4a) consists of two continuous hori-

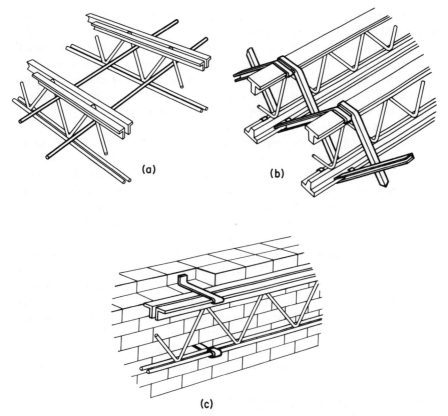

FIG. 12-4 *Copyright Steel Joist Institute. Reproduced by permission*

zontal steel members, each of which is attached to the upper *and* lower chord member by means of welding or mechanical fasteners. The welding and fasteners should be capable of resisting a minimum horizontal force of 0.700 kip. A minimum diameter of bar of $\frac{1}{2}$ in. should be used when the bridging is a round bar. The maximum l/r ratio of the bridging member shall be 300. l is the distance between bridging attachments and r is the least radius of gyration of the bridging member. The bottom horizontal bracing, even at completion of construction, should not be removed to assure stability.

Diagonal bracing (Fig. 12-4b) consists of diagonal cross members having a maximum l/r ratio of 200. l and r are as previously defined, and where the cross members connect (at their intersection), l is the distance between chord attachment and intersection attachment. A positive means of connection, mechanical fasteners or welding, shall be used.

Horizontal bridging is usually recommended for standard joists, although diagonal bridging may be used.

The end anchors for masonry supports (Fig. 12-4c) are embedded and anchored in mortar. The anchor shall be equivalent to a $\frac{3}{8}$-in. round bar with minimum length of 8 in. Every joist in a roof and every third joist on floors are required to be anchored. Two $\frac{1}{2}$-in. anchor bolts, or equal, are used in lieu of the equivalent steel bar, where the roofs have no parapet walls.

For the case where joists rest on steel supports, the connection of the end anchors to steel support shall be equivalent to two $\frac{1}{8}$-in. fillet welds, 1 in. long, or a $\frac{1}{2}$-in. bolt. Where columns are not framed in at least two directions with structural steel members, the joists shall be field-bolted at the columns to assure lateral stability. Finally, where uplift forces are a possibility, roof joists shall be anchored to resist these forces.

To assure stability when loads other than the weight of the erector are applied, all bridging, as well as joists, shall be completely fastened. Where three or more rows of bridging are to be used, instability is likely prior to the fastening of all lines of bridging. Where five bridging rows in spans more than 40 ft are used, each joist shall be properly laterally braced before the next joist is erected and before any loads are applied and the middle row shall be diagonal bridging with bolted connections at chords and intersection. Other pertinent requirements pertaining to proper and safe joist handling and erection are given in the specifications.

12-3   LONGSPAN AND DEEP LONGSPAN SERIES

The LH-series are called longspan steel joists while the DLH-series are the designations for the deep longspan steel joists. The LH-series are suitable for the direct support of both floors and roof decks in buildings, and the DLH-series are suitable for direct support of roof decks in buildings. The LH- and DLH-series extend the use of joists to span lengths in excess of the H-series. The longspan series have standardized depths of 18, 20, 24, 28, 32, 36, 40, 44, and 48 in. The clear span lengths vary up to a maximum of 96 ft. The deep longspan series have standardized depths of 52, 56, 60, 64, 68, and 72 in. for clear span lengths up to 144 ft.

Figure 12-5 shows schematically the commonly manufactured joists in these series. Although they are shown as modified Warren-type truss web configuration, any type of web configurations may be used, depending upon the particular manufacturer's standards. Notice that underslung or square ends and parallel chords or single or double-pitched ($\frac{1}{8}$ in./ft standard) top chords (to provide the required slope for roof drainage) can

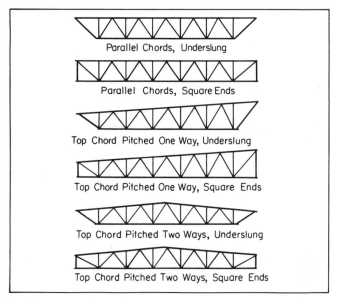

Parallel Chords, Underslung

Parallel Chords, Square Ends

Top Chord Pitched One Way, Underslung

Top Chord Pitched One Way, Square Ends

Top Chord Pitched Two Ways, Underslung

Top Chord Pitched Two Ways, Square Ends

The illustrations above indicate Longspan and Deep Longspan Steel Joists with modified WARREN type web systems. However, the web systems may be any type, whichever is standard with the manufacturer furnishing the product.

FIG. 12-5 *Copyright Steel Joist Institute. Reproduced by permission*

be furnished. Square-end joists are usually intended for bottom chord bearing. If more than $\frac{1}{8}$ in./ft of pitch of top chord is desired, the standard load tables *do not apply*. The first two digits of the joist designation indicate the joists nominal midspan depth and the last two digits are the chord-size designation.

LH- and DLH-series joists are based upon a yield strength of from 36 to 50 ksi for *both* chord and web elements. This yield strength must be achieved in the hot-rolled conditions *prior* to forming or fabrication.

All the series joists use hot-rolled or cold-formed, including cold-formed steel that has been cold-worked. ASTM A36, 242, 441, 570, 572 (grades 42, 45, and 50), 588, 607 (grades 45 and 50), or 611 (grade D) may be used for the chord or web sections of the LH- and DLH-series. Other grades of materials may be used provided that such material is weldable and is proved by tests performed by the producer or fabricator to have the required mechanical properties as per the specifications.

The depths of the bearing portion at the ends of underslung joists are 5 in. for the longspan series, 5 in. for chord sizes through 17 for the

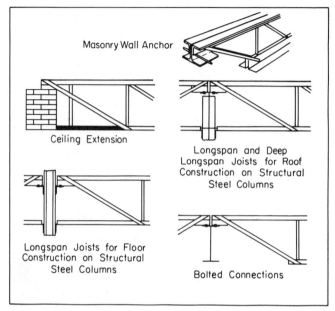

Welded or Bolted Connections
Where Longspan or Deep Longspan Joists are supported on structural
steel members, they are generally field welded. The number, size and
length of welds should be specified. Where bolted connections are
desired slotted holes are provided in the bearing plates for this purpose.

FIG. 12-6   *Copyright Steel Joist Institute. Reproduced by
permission*

deep longspan series, and $7\frac{1}{2}$ in. for chord sizes of 18 and 19 in. for the
deep longspan series. All longspan and deep longspan joists are furnished
with the recommended camber as listed in the specifications. Camber is
not optional with the manufacturer as it is in the Standard Series.

Figure 12-6 shows some of the details associated with the use of
open-web steel joists of the longspan and deep longspan series. The heavy
dots in Fig. 12-6 indicate the connectors when longspan or deep longspan
joists bear on the structural steel members. Generally, the connectors
are field welds (number, size, and length should be specified). However,
where mechanical connectors are desired, slotted holes are provided in
the bearing plates.

The specifications covering these series are similar to the standard
series with some additions. Welding electrodes are specified for the various yield-strength levels for this series of joist. The ends of joists should
extend a minimum distance of 4 in. over steel supports or, when the
bearing is less than 4 in., such as in the case of two opposite joists resting
on a narrow steel support, special ends of joist must be provided with

attachment to the support by bolting or welding. The specifications require minimum bearing lengths measured in the direction of joist span when the joists bear on masonry or concrete.

Bridging of the *diagonal* type is required for this series of joists. The maximum spacing and minimum number of rows of bridging are specified in the Specifications. Bridging must support the top and bottom chords laterally during construction and is to hold the joists in place. The required end anchorage for joists resting on masonry supports and steel supports is similar, although more stringent than that specified for the standard series. The handling and erection requirements of the longspan and deep longspan joists as per the specifications are more stringent than those stipulated for the standard series. Other than the highlights as mentioned, the specifications for the standard series are similar, if not the same, as those for the longspan and deep longspan series.

Figure 12-7 shows a typical page from the Standard Load Table for Longspan Steel Joists, LH-Series. Similar tables exist for the DLH-series. All LH- and DLH-tables are based upon an allowable stress of 30 ksi. The remarks made in Section 12-2 relative to Fig. 12-3 for the Standard Series apply to Fig. 12-7 for the Longspan and Deep Longspan Series in this section, with the following additional remarks.

These Load Tables apply to joists with *either* parallel or standard pitched top chords (see Fig. 12-5). The carrying capacities shown for pitched chords are computed by using the nominal depth of joists at midspan. If the pitch of top chord exceeds the standard pitch of $\frac{1}{8}$ in./ft, the Load Table is *not* applicable. The Load Table can be used for parallel chord joists installed to a maximum slope of $\frac{1}{2}$ in./ft. When holes are required in either chord, the joist carrying capacities must be reduced in proportion to the reduced area of chord(s).

Open-web steel joists with proper manufacture, standardization, handling, and design can offer economical floor and/or roof framing.

12-4 JOIST GIRDERS

Joist girders are open web load-carrying members. They were created to meet the need for increased spans for primary joist members. In 1978, SJI adopted standard specifications and standard Design Guide Weight Tables for joist girders. The Weight Tables will be discussed following discussion of the specifications.

The girders are designed as simple spans that support equally spaced concentrated loads from a floor or roof joist system. The concentrated loads are assumed to act at the girder panel points. The panel points are the points of intersection of the web diagonals with the top chord, and the girder panel or joist spacing is the distance between adjacent points of intersection (Fig. 12-5).

STANDARD LOAD TABLE FOR LONGSPAN STEEL JOISTS, LH-SERIES

Pounds per Linear Foot Based on Allowable Stress of 30,000 psi

CLEAR OPENING OR NET SPAN IN FEET

Joist Designation	Approx. Wt. in Lbs. per Linear Ft.	Depth in Inches	SAFE LOAD* in Lbs. Between 47–64	65	66	67	68	69	70	71	72	73	74	75	76	77	78	79	80
40LH08	20	40	16600	254 / 150	247 / 144	241 / 138	234 / 132	228 / 127	222 / 122	217 / 117	211 / 112	206 / 108	201 / 104	196 / 100	192 / 97	187 / 93	183 / 90	178 / 86	174 / 83
40LH09	23	40	21800	332 / 196	323 / 188	315 / 180	306 / 173	298 / 166	291 / 160	283 / 153	276 / 147	269 / 141	263 / 136	256 / 131	250 / 126	244 / 122	239 / 118	233 / 113	228 / 109
40LH10	25	40	24000	367 / 216	357 / 207	347 / 198	338 / 190	329 / 183	321 / 176	313 / 169	305 / 162	297 / 156	290 / 150	283 / 144	276 / 139	269 / 134	262 / 129	255 / 124	249 / 119
40LH11	27	40	26200	399 / 234	388 / 224	378 / 215	368 / 207	358 / 198	349 / 190	340 / 183	332 / 176	323 / 169	315 / 163	308 / 157	300 / 151	293 / 145	286 / 140	279 / 135	273 / 130
40LH12	32	40	31900	486 / 285	472 / 273	459 / 261	447 / 251	435 / 241	424 / 231	413 / 222	402 / 213	392 / 205	382 / 197	373 / 189	364 / 182	355 / 176	346 / 169	338 / 163	330 / 157
40LH13	36	40	37600	573 / 334	557 / 320	542 / 307	528 / 295	514 / 283	500 / 271	487 / 260	475 / 250	463 / 241	451 / 231	440 / 223	429 / 214	419 / 207	409 / 199	399 / 192	390 / 185
40LH14	37	40	43000	656 / 383	638 / 367	620 / 351	603 / 336	587 / 323	571 / 309	556 / 297	542 / 285	528 / 273	515 / 263	502 / 252	490 / 243	478 / 233	466 / 225	455 / 216	444 / 209
40LH15	41	40	48100	734 / 427	712 / 408	691 / 390	671 / 373	652 / 357	633 / 342	616 / 328	599 / 315	583 / 302	567 / 290	552 / 279	538 / 268	524 / 258	511 / 248	498 / 239	486 / 230
40LH16	47	40	53000	808 / 469	796 / 455	784 / 441	772 / 428	761 / 416	751 / 404	730 / 387	710 / 371	691 / 356	673 / 342	655 / 329	638 / 316	622 / 304	606 / 292	591 / 282	576 / 271

FIG. 12-7 *Table copyrighted by Steel Joist Institute. Reproduced by permission*

STANDARD LOAD TABLE FOR LONGSPAN STEEL JOISTS, LH–SERIES

Pound per Linear Foot Based on Allowable Stress of 30,000 psi

CLEAR OPENING OR NET SPAN IN FEET

Joist Designation	Approx. Wt. in Lbs. per Linear Ft.	Depth in Inches	SAFE LOAD in Lbs. Between 52–72	73	74	75	76	77	78	79	80	81	82	83	84	85	86	87	88
44LH09	22	44	20000	272 / 158	265 / 152	259 / 146	253 / 141	247 / 136	242 / 131	236 / 127	231 / 122	226 / 118	221 / 114	216 / 110	211 / 106	207 / 103	202 / 99	198 / 96	194 / 93
44LH10	25	44	22100	300 / 174	293 / 168	286 / 162	279 / 155	272 / 150	266 / 144	260 / 139	254 / 134	249 / 130	243 / 125	238 / 121	233 / 117	228 / 113	223 / 110	218 / 106	214 / 103
44LH11	27	44	23900	325 / 188	317 / 181	310 / 175	302 / 168	295 / 162	289 / 157	282 / 151	276 / 146	269 / 140	264 / 136	258 / 131	252 / 127	247 / 123	242 / 119	236 / 115	232 / 111
44LH12	31	44	29600	402 / 232	393 / 224	383 / 215	374 / 207	365 / 200	356 / 192	347 / 185	339 / 179	331 / 172	323 / 166	315 / 160	308 / 155	300 / 149	293 / 144	287 / 139	280 / 134
44LH13	35	44	35100	477 / 275	466 / 265	454 / 254	444 / 246	433 / 236	423 / 228	413 / 220	404 / 212	395 / 205	386 / 198	377 / 191	369 / 185	361 / 179	353 / 173	346 / 167	338 / 161
44LH14	36	44	40400	549 / 315	534 / 302	520 / 291	506 / 279	493 / 268	481 / 259	469 / 249	457 / 240	446 / 231	436 / 223	425 / 215	415 / 207	406 / 200	396 / 193	387 / 187	379 / 181
44LH15	41	44	47000	639 / 366	623 / 352	608 / 339	593 / 326	579 / 314	565 / 303	551 / 292	537 / 281	524 / 271	512 / 261	500 / 252	488 / 243	476 / 234	466 / 227	455 / 219	445 / 211
44LH16	47	44	54200	737 / 421	719 / 405	701 / 390	684 / 375	668 / 362	652 / 348	637 / 336	622 / 324	608 / 313	594 / 302	580 / 291	568 / 282	555 / 272	543 / 263	531 / 255	520 / 246
44LH17	54	44	58200	790 / 450	780 / 438	769 / 426	759 / 415	750 / 405	732 / 390	715 / 376	699 / 363	683 / 351	667 / 338	652 / 327	638 / 316	624 / 305	610 / 295	597 / 285	584 / 276

FIG. 12-7 (Continued)

STANDARD LOAD TABLE FOR LONGSPAN STEEL JOISTS, LH-SERIES

Pound per Linear Foot Based on Allowable Stress of 30,000 psi

Joist Designation	Approx. Wt. in Lbs. per Linear Ft.	Depth in Inches	SAFE LOAD* in Lbs. Between 56–80	CLEAR OPENING OR NET SPAN IN FEET															
				81	82	83	84	85	86	87	88	89	90	91	92	93	94	95	96
48LH10	25	48	20000	246 141	241 136	236 132	231 127	226 123	221 119	217 116	212 112	208 108	204 105	200 102	196 99	192 96	188 93	185 90	181 87
48LH11	27	48	2T700	266 152	260 147	255 142	249 137	224 133	239 129	234 125	229 120	225 117	220 113	216 110	212 106	208 103	204 100	200 97	196 94
48LH12	31	48	27400	336 191	329 185	322 179	315 173	308 167	301 161	295 156	289 151	283 147	277 142	272 138	266 133	261 129	256 126	251 122	246 118
48LH13	35	48	32800	402 228	393 221	384 213	376 206	368 199	360 193	353 187	345 180	338 175	332 170	325 164	318 159	312 154	306 150	300 145	294 141
48LH14	36	48	38700	475 269	464 260	454 251	444 243	434 234	425 227	416 220	407 212	399 206	390 199	383 193	375 187	367 181	360 176	353 171	346 165
48LH15	41	48	44500	545 308	533 298	521 287	510 278	499 269	488 260	478 252	468 244	458 236	448 228	439 221	430 214	422 208	413 201	405 195	397 189
48LH16	47	48	51300	629 355	615 343	601 331	588 320	576 310	563 299	551 289	540 280	528 271	518 263	507 255	497 247	487 239	477 232	468 225	459 218
48LH17	54	48	57600	706 397	690 383	675 371	660 358	646 346	632 335	619 324	606 314	593 304	581 294	569 285	558 276	547 268	536 260	525 252	515 245

*To solve for safe uniform load between spans shown, divide the Safe Load in pounds by net span in feet plus 0.67 feet. (The added 0.67 feet, eight inches, is necessary to obtain the proper span for which the load tables were developed.)

In no case shall the safe uniform load exceed the uniform load calculated for the minimum span listed.

To solve for live load between spans shown, multiply the live load of the shortest net span shown in the load table by the (shortest net span plus 0.67 feet)², and divide by the (actual net span plus 0.67 feet)². The live load shall not exceed the safe uniform load.

FIG. 12-7 (Continued)

The girders have been standardized in the Weight Tables for a range of depths from 20 to 72 in. for spans ranging from 20 to 60 ft.

The specifications cover hot-rolled or cold-formed steel members, including cold-formed steel where the yield strength is achieved by cold-working. Girder chord or web members are based on a yield strength of 36,000 psi minimum to 50,000 psi maximum. A36, 242, 441, 570, 572 (grades 42, 45, and 50), 588, 606, 607 (grades 45 and 50), and 611 (grade D) material may be used. The specifications also discuss the requirements for welding electrodes, paint, design and manufacture, allowable stresses and stress formulas, slenderness ratios, connection of joints and splices, and so on. Camber is required as a standard when girders are used in roof construction. The bottom chord is designed as an axially loaded tension member and the top chord as an axially loaded compression member. The radius of gyration of the top chord about the vertical axis should be equal to or greater than the span divided by 575. The top chord is considered to be laterally supported by the floor or roof joists when positive attachment occurs. The design of web members is also discussed.

These girders are available with underslung end and lower chord extensions. The bearing portion at joist ends has a 6-in. standard depth and is usually connected to columns using two $\frac{3}{4}$-in.-diameter A325 bolts. The lower chord to column or other support connection is recommended to be loose to stabilize the lower chord laterally to prevent overturning. If a rigid connection of the bottom chord to column or other support is used, the connection should be made after the dead load application (in this case the joist girder is no longer simply supported and the system must be analyzed for continuous frame action). Girders with unbalanced loads or when loaded from one side must be designed so that their reactions act through the joist girder centroid.

The specification goes on to discuss end supports (minimum of 4 in. over steel supports or 6 in. over masonry or concrete supports with positive attachment thereto), bracing (no bridging required with provisions for bottom chord lateral restraint and that no other loads be placed on the girder until the floor or roof joists on the girder are in place and attached), end anchorage (on masonry or concrete design of a steel bearing plate is required, and on steel supports a minimum connection of two $\frac{1}{4}$-in. fillet welds 2 in. long or two $\frac{3}{4}$-in.-diameter bolts), uplift, ponding, inspection, and handling and erection.

The Weight Tables for Joist Girders are based upon an allowable tensile stress of 30,000 psi. They list the approximate pounds per linear foot for a joist girder supporting concentrated panel point loads as listed. For a given joist girder span, the designer must first determine the number of joist spaces, calculate the panel point concentrated loads, and select a depth. The table gives weight per foot for various depths, loads, and number of joist spaces. The purpose of the tables is to assist the designer in the selection of a roof or floor support system.

Joist girders are designated by two digits and a G (indicates depth in inches and a joist girder), followed by one or two digits and an N (indicates number of joist spaces), followed by one or two digits and a K (indicates the number of kips of each concentrated panel load). For example, a designation of 48G8N9K indicates a 48-in.-deep joist girder with 8 joist spaces, having panel point concentrated loads of 9 kips.

A suggested design sequence is as follows, assuming a bay size in feet by feet, a joist spacing in feet, a live and dead load in pounds per square foot (the dead load should include an approximate joist girder weight). An interior joist girder is assumed.

(1) Joist spaces = girder span/joist spacing.

(2) Total load (plf) = (live plus dead load, psf) joist spacing

(3) Concentrated load at top chord panel points:
     P = [total load in step (2)] bay dimension normal to girder span.

(4) Select joist girder depth: a good rule for depth selection is 1 in. of depth for each foot of girder span. This is a fair compromise of limited depth and economy.

(5) Designate joist girder selection (depth, number of joist spaces, N, and concentrated load at each panel point, as discussed).

(6) Enter Weight Tables and obtain pounds per linear foot for designated girder. This weight $\leq$ the assumed dead load weight (dead load, psf, multiplied by bay dimension normal to girder span).

(7) Live load deflection:
     (a) Live load (plf) = (live load, psf) bay dimension normal to girder span.
     (b) I girder $\approx 0.027NPLd$, where P is in kips, L, the span is in feet, and d, the depth, is in inches. I is the girder moment of inertia (in.^{4}).
     (c) Allowable deflection for plastered ceilings = $l/360$, where l is the span in inches.
     (d) Actual deflection = $1.15(5wl^4/384EI)$, where w is the live load in kips per linear inch or pounds per square foot live load multiplied by the bay dimension in feet normal to the girder span divided by 12. Actual deflection should be less than allowable deflection.

Although the Weight Tables use certain depths, spans, and loads, it is not necessary to be limited by these values.

For additional details and discussions, reference to Standard Specifications Load Tables and Weight Tables for steel joists and joist girders, an annual SJI publication, should be made.

PROBLEMS

(12-1) Determine by reference to the open web steel joist load tables (Fig. 12-3) the safe uniform load that may be carried by a 28H8 spanning 42 ft between supports.

(12-2) In Problem 12-1 would the allowable live uniform load change if the deflection were limited to $L/240$? Show calculations.

(12-3) By referring to the Standard Load Tables for Longspan Steel Joists (LH series) (see Fig. 12-7), determine the live load that produces a deflection of $L/240$ (maximum permissible live load deflection for roofs with other than plastered ceilings), on a 44LH15 spanning 85 ft.

(12-4) Determine the new safe uniform load on a long-span steel joist 44LH10, spanning 76 ft, carrying 60 lb/ft estimated dead load in addition to its own weight (see Fig. 12-7).

(12-5) Determine the safe uniform load for the joist in Problem 12-4 if the deflection is limited to $L/360$.

rigid frames

13-1 INTRODUCTION

For the past thirty years or so, the single-span rigid frame has often been used for flat, gabled, or curved roofs. This type of framing is aesthetically pleasing, easily adaptable to bolted or welded fabrication, easily and rapidly erected, and economically used.

Rigid frames of rolled shapes have been used for spans of up to 200 ft with spans of up to 100 ft being used most often. Rigid frames utilizing built-up shapes have been used to span distances up to 250 ft. Economical spacing of frames for various span lengths for average roof loads has been suggested as shown in Table 13-1.*

Rigid frames have typically been used in mill-type industrial buildings, gynmasiums, tennis-court sturctures, and so on. Figure 13-1 shows the various geometric configurations used in rigid-frame construction.

---

* J. D. Griffiths, *Single-Span Rigid Frames in Steel,* AISC, 1948.

Table 13-1

Span (ft)	Spacing (ft)
30–40	16
40–60	18
60–100	20
Over 100	($\frac{1}{5}$ to $\frac{1}{6}$ of span)

The rigid frames shown in Fig. 13-1 are two-hinged rigid frames that are statically indetermine to the first degree. The analyses of this type of structure is affected by the relative sizes of the elements of the frame. The vertical reactions of the frame may be found by ordinary statics, but the horizontal reactions vary with the frame geometry. To obtain moments at any point on the frame, these horizontal reactions must be known. Several authors have published various design charts, methods, and formulas to rapidly find the horizontal reactions.*

The distribution of moments in rigid frames is affected by the h/L and f/h ratios, as well as by the relative stiffness of the elements of the

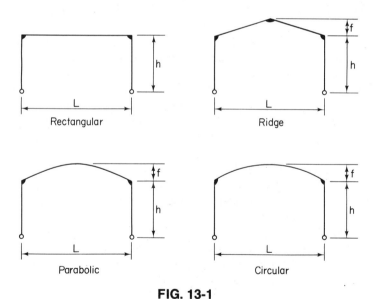

FIG. 13-1

---

* D. S. Ellifritt, Rapid Design of Tapered Rigid Frames, *Civil Engineering—ASCE,* July 1972; J. D. Griffiths, *Single-Span Rigid Frames in Steel,* AISC, 1948, pp. 14–19.

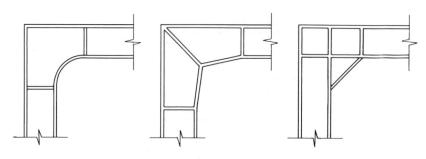

FIG. 13-2

frame, although the latter variable only has a minor influence upon the moment distribution.

All joints in the frame must be rigid. For all frames with the exception of small frames, the knee section is usually enlarged by use of a straight or curved haunch, as shown in Fig. 13-2. The knee element must be capable of transferring the end moment from the beam into the column, the vertical beam end shear into the column, and the horizontal shear of the column into the beam. The design of rigid-frame knees is treated in the AISC July 1943 publication, *Design of Rigid Frame Knees,* by Friedrich Bleich, and in *Design of Welded Structures,* by Omer W. Blodgett, James F. Lincoln Arc Welding Foundation, Cleveland, Ohio, Section 5.11. The inner flange of the knee (coopression flange) should be braced to prevent lateral movement.

The column bases of the frame are designed as if hinged (fixed against rotation). This is usually accomplished with one anchor-bolt line placed normal to the span of the frame at the column center line. The anchor bolts are usually passed through column web attached by lug angles.

Where poor foundation conditions are encountered, a horizontal tie between column bases of the frame is utilized to resist frame thrust. These ties are usually round bars, plates, or structural shapes buried below the floor line of the building.

The textbook design theories as applied to single-span rigid frames may be extended to multiple-span rigid frames. Rigid frames may be analyzed as geometrically symmetrical frames or as geometrical frames that are unsymmetrical.

13-2   TAPERED RIGID FRAMES

As discussed in Section 5-7, Appendix D to the AISC Specification covers the design provision for tapered members. Allowable compressive, bending, and combined stresses are presented for members tapered in the plane of their web. The allowable stress provisions for tapered members are similar to those provided for prismatic members with certain modifications due to the tapering of the member.

The approach in evaluating $F_{a\gamma}$, the allowable compressive stress for a tapered member, and $F_{b\gamma}$, the allowable bending stress for a tapered flexural member, is based on the concept that the critical stress for the tapered axially loaded member is equal to a prismatic column of different length but of the same cross section as the smaller end of the tapered column. This gives rise to K_γ, an equivalent effective length factor for a tapered member subject to axial compression. K_γ is used to evaluate S in Eq. 5-10. K_γ is accurately found for symmetrical rectangular rigid frames composed of prismatic beams and tapered columns by use of charts in the AISC Commentary. For the case of $\gamma = 0$, K_γ becomes K. For tapered beams, the tapered element is replaced by an equivalent prismatic element with a different length but with a cross section identical to that of the smaller end of the tapered beam. This leads to modified length factors used in allowable tension and compression-stress formulas. The design case when the restraining beams are tapered is covered by the following references:

(1)   G. C. Lee, M. L. Morrell, and R. L. Ketter, Design of Tapered Members, *WRC Bulletin 173,* June 1972.

(2)   M. L. Morrell and G. C. Lee, Allowable Stress for Web-Tapered Beams with Lateral Restraints, *WRC Bulletin 192,* Feb. 1974.

An additional reference for the design of tapered rigid frames is D. S. Ellifritt's *Rapid Design of Tapered Rigid Frames,* mentioned previously.

A general discussion of rigid frames has been the only purpose of this chapter. The reader is referred to the references listed for more detailed information on this type of structural framing.

tolerances

14-1 INTRODUCTION

Section 1-5 made mention of mill, fabrication, and erection tolerances, while Section 8-5.1 discussed erection tolerances. Certain standards and documents are acceptable to owners, architects, engineers, and others associated with construction. These standards are considered by industry to dictate rules that, when followed, yield an acceptable, trouble-free steel structure.

14-2 MILL TOLERANCES

ASTM A6, Standard Specification for General Requirements for Rolled Steel Plates, Shapes, Sheet Piling, and Bars for Structural Use, unless otherwise specified in an individual specification or in the purchase order, applies to the structural members indicated in the title of the specification. These specifications stipulate mill dimension tolerances. Permissible variations for dimensions from the published profile dimensions are expressed. These variations are caused by the rolling process due to roll wear, thermal distortions of the hot cross section after leaving the rolls, differential cooling distortions that take place on the cooling beds, and so on.

A series of tables within this specification gives the permissible dimensional variations for plates, shapes, sheet piling, bars, and bar-size shapes. These variations are expressed as under and/or over the theoretical value. Permissible variations in plate thickness and weight, plate width and length are tabulated for the varous types of plate. Permissible camber is given, and variation from flatness and variation in waviness are also given. Permissible variation in cross section (depth, flange width, out of squareness) are given for W shapes, standard beams (S shapes), H beams, channels, angles, bulb angles, tees, and zees. Permissible length variations, straightness (camber and sweep) variations, etc., are listed. Similar permissible variations for bars are given. The AISC Manual summarizes A6 and further clarifies tolerances.

Camber is the amount of deflection measured parallel to the web, while sweep is measured horizontally parallel to the plane of the flange. Figure 14-1 shows the meaning of camber and sweep for a W shape. For example, for a W30 beam shape with a span of 36 ft, A6 stipulates a length of tolerance as follows: over $= \frac{1}{2}$ in. $+ \frac{1}{16}$ in. for each additional 5 ft or fraction thereof over 30 ft $= +\frac{5}{8}$ in.; under $= \frac{1}{2}$ in. $= -\frac{1}{2}$ in. These are the variations from the specified lengths given.

ASTM A568, Standard Specification for Steel, Carbon and High-Strength Low-Alloy Hot-Rolled Sheet, Hot-Rolled Strip and Cold-Rolled Sheet, General Requirements, covers for strip and sheet (high-strength, low-alloy steel furnished as hot-rolled sheet, hot-rolled strip, and cold-rolled sheet) what A6 covers for its products.

14-3   FABRICATION TOLERANCES

Fabrication dimensional tolerances are stated in the AISC Code of Standard Practice for Steel Buildings and Bridges (Sections 6.4 and 10.3) and Section 1.23.8 of the AISC Specification. A permissible variation of $\frac{1}{32}$ in. in the overall length of members with both ends finished for contact bearing and of $\leq \frac{1}{16}$ in. for member lengths $\leq$ 30 ft and $\leq \frac{1}{8}$ in, for member lengths $>$ 30 ft for ends not finished for contact bearing (members framed to other steel members) is stated in these documents.

Permissible variations from straightness are the same as allowed for W shapes by A6, except that compression members must not deviate from straightness by more than $\frac{1}{1000}$ of the axial length between points,

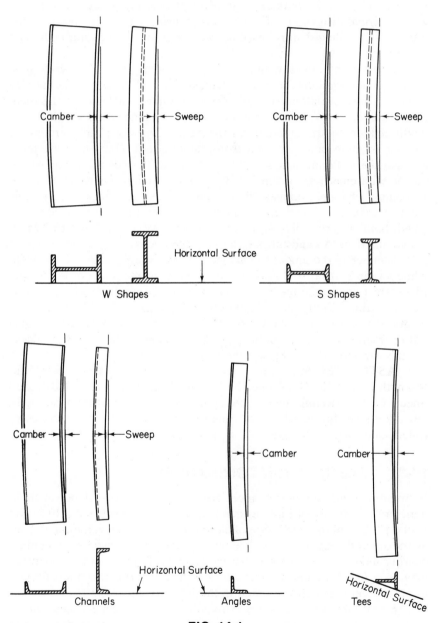

FIG. 14-1

which are to be laterally supported. For architecturally exposed structural steel (Section 10.3 of the Code), permissible tolerances for out-of-square or out-of-parallel, depth, width, and symmetry of rolled shapes are the same as specified in A6. The straightness tolerances, however, are 50% of the standard A6 camber and sweep tolerances. This holds true for rolled and built-up members.

14-4   ERECTION TOLERANCES

Section 7 of the AISC Code of Standard Practice for Steel Buildings and Bridges covers erection of structural steel. Section 7.5.1 of the Code covers the allowable dimensional variance relative to the installation of anchor and foundation bolts as set by the owner (architect, engineer, general contractor). These tolerances can be attained by using reasonably good workmanship. They will assure that the steel is erected and plumbed to the required tolerances.

Section 7.11 of the Code stipulates frame tolerances. The section states that such variations from the finished overall dimensions of structural steel frames are to be expected and are deemed to be within the limits of good workmanship when they are not greater than the cumulative effect of mill, fabricating, and erection tolerances. Allowable column plumbness deviations are stated in Section 7.11.3.1 of the Code.

It is assumed that the *owner* sets to line and grade all leveling plates and loose bearing plates that can be handled without the use of a derrick or crane; that the *erector* properly sets all other bearing devices; that the final location of bearing devices is the responsibility of the owner; and that columns are within mill and fabrication tolerances and are erected within plumbness tolerances. Therefore, the elevation of horizontal members connected to columns is considered within acceptable tolerances as expressed in Section 7.11.3.2b. These tolerances are expressed as a variation of the specified distance from the work point to the upper milled splice line of the column. The horizontal-member work point is the actual center line of the top flange or top surface at each end. The Commentary on the Code of Standard Practice helps to further clarify permissible tolerances on position and alignment of structural members.

Architecturally exposed structural steel acceptable erection tolerances are a maximum of 50% of those permitted for structural steel. These tolerances require the owner to specify adjustable connections between the architecturally exposed steel and the structural steel frame or the masonry or concrete supports. Likewise, for structural steel, the alignment of lintels, wall supports, curb angles, mullions, and similar supporting members for the use of other trades, requiring closer than the stipulated tolerances, should be specified by the owner's plans. Adjustable connection of these members to the supporting structural frame should be provided. Permissible tolerances for location of these adjustable items are given in Sections 7.11.3.4a and b of the Code.

A general discussion on mill, fabrication, and erection tolerances has been made. It was not the intent to discuss these in every detail but rather to make the reader aware of these tolerances, why they exist, and where they are stipulated. Reference to ASTM A6, the AISC Code of Standard Practice and the AISC Specification, is suggested for more detailed information.

PROBLEMS

(14-1) By referring to the ASTM A6 Specification on mill tolerances, determine the following:

    (a) Minimum possible flange width (as delivered) of a steel channel C 12 × 30.

    (b) Maximum possible depth (as delivered) of a W 14 × 145.

    (c) Permissible variation in camber on a 20-ft length of a W 10 × 60 column.

    (d) Permissible variation in sweep in a 12-ft length of beam W 8 × 15.

(14-2) What variation is permitted in the fabricated length of a 15-ft, 0 in. steel column W 12 × 79 in end bearing on both faces?

(14-3) What deviation in straightness is permitted by the AISC Code of Standard Practice to the fabricator of a W 8 × 28 column, 15 ft, 0 in. length, assuming that on installation it will have lateral support at its end points only?

appendix a

list of symbols and abbreviations

The following is a list of symbols and abbreviations used in the text.

A	Area or cross-sectional area, in.^{2} Base plate dimension, in. Distance from vertical weld to an eccentric load, in.
A'	The sum of the area for the compression flange and $\frac{1}{6}$ the web in compression for plate girder design, in.^{2}
A_b	Area of eyebar body or brace, in.^{2}
A_{bc}	Planar area of web at beam-to-column connection, in.^{2}
A_c	Area of effective concrete flange in composite design, in.^{2}
A_e	Effective net area of an axially loaded tension member, in.^{2}
$A_{eff}, A_{eff.st.}$	Effective area of bearing stiffeners when designed as columns, in.^{2}

A_f Area of flange, in.^{2} ($b_f t_f$)

Area of flange delivering concentrated load in rigid beam-column connection, in.^{2}

A_h Net area through pin hole, in.^{2}

A_{net} Net cross-sectional area, in.^{2}

A_p Area of plate, in.^{2}

A_s Area of steel section in composite design, in.^{2}

A_s' Area of longitudinal reinforcing steel located within the effective width of concrete slab used in composite design, in.^{2}

A_{sr} Total area of longitudinal reinforcing steel in the negative moment area located within the concrete slab effective width used in composite design, in.^{2}

A_{st} Gross cross-sectional area of intermediate stiffener or pair of stiffeners, in.^{2}

A_t Net area along a plane normal to the vertical line of fasteners through a row of fasteners at a beam end connection, in.^{2}

A_v Net shear area along a plane through the vertical line of web fasteners at a beam end connection, in.^{2}

A_1 Base plate bearing area, in.^{2}

A_2 Area of concrete foundation for bearing plates, in.^{2}

a   Distance from fastener line to application of prying force, in.

Coefficient used to express eccentricity of load in terms of length of weld, in.

Leg size of fillet weld, in.

Clear spacing between transverse stiffeners or between first intermediate stiffener at end of span and concentrated interior load, in.

Distance from bottom of concrete slab to neutral axis of composite encased beam, in.

Distance from edge of beam flange to edge of effective flange width used in composite design, in.

Distance from centerline of bolt to edge of tee flange or angle leg, in.

á   Required distance at ends of welded partial length cover plate to develop stress, in.

a_x, a_y	A component for bending about the X-X and Y-Y axes, respectively, of the amplification factor used to solve a modified interaction formula $= 0.149Ar^2(10)^6$
a_1	Spacing of stiffeners in end panels of plate girders, in.
AASHTO	American Association of State Highway and Transportation Officials
AISC	American Institute of Steel Construction
AISE	Association of Iron and Steel Engineers
AISI	American Iron and Steel Institute
ANSI	American National Standards Institute
AREA	American Railway Engineering Association
ASCE	American Society of Civil Engineers
ASTM	American Society for Testing and Materials
AWS	American Welding Society
ASD	Allowable stress design
B	Base plate dimension, in.
	A bending factor for determining the equivalent load
	Effective throat dimension for a submerged arc fillet weld, in.
B_c	Load per bolt including prying force, kips
B_x, B_y	Bending factor with respect to the X-X and Y-Y axes, respectively, used for determining the equivalent axial load for the bending term in the interaction formula $= A/S$, in.$^{-1}$
b	Distance from centerline of fastener (gage line) to face of leg of angle or stem of tee, in.
	Vertical spacing of fasteners, in.
	Throat dimension of fillet weld, in.
	Width of section of element, in.
	Effective width of concrete slab in composite design, in.
	Width of flange or moment connection plate delivering a concentrated force, in.
b_e	Effective net width for a tension member, in.
b_{eff}, min.	Minimum effective throat thickness for partial penetration groove welds, in.
b_f	Flange width, in.

b_g Gross width of angle, in.

b_n Net width of angle, in.

b_p Width of cover plate, in.

BOCA   The Basic Building Code (Building Officials and Code Administrators)

bott   Bottom

C   Coefficient to determine permissible load acting at an effective eccentricity on a fastener group

                      Coefficient to determine permissible eccentric load on weld groups

C_b Bending coefficient or modifying factor dependent upon moment gradient

C_c Equal to $(2\pi^2 E/F_y)^{1/2}$; defines the dividing point between elastic and inelastic buckling

C_h Coefficient for determining maximum allowable bending stress in hybrid girders.

C_m Coefficient dependent upon column curvature caused by applied moments; used in the bending term of the interaction formula—a reduction or moment factor

C_p Stiffness factor for primary members of a flat roof

C_s Stiffness factor for secondary members of a flat roof

C_t Decimal multiplier as used to express the effect of shear lag on the net area of a section

C_v Ratio of "critical" web stress according to the linear buckling theory to the shear yield stress of web material (see Sec. 7-5)

C_w Warping constant for a section, in.^{6}

C_1 Adjustment coefficient for oversized or slotted holes to compute F_v for friction-type high-strength bolts.

                      Coefficient to determine permissible eccentric load on weld groups (ratio of electrode strength used to that of E70 electrode)

                      Ratio of F_y of beam flange to F_y of column at rigid beam-column connection

                      Coefficient for web tear-out (block shear).

                      Spacing increment used in computing minimum edge distance for oversized and slotted holes

C_2	Coefficient for web tear-out (block shear). Spacing increment used in computing minimum distance for oversized and slotted holes
c	Distance from neutral axis to outermost fiber cross-section, in.
c.c.	Center to center
cos	Cosine
cwt	One hundred pounds
C	Carbon American Standard shape
Cb	Columbium
Cr	Chromium
Cu	Copper
C.E.	Carbon equivalent
CRC	Column Research Council
CSA	Canadian Standards Association
D	Diameter of fastener, thread hole, bar, or eyebar head, in. Factor depending upon type of transverse stiffeners Gage, in. Number of sixteenths of an inch of weld size
D_c	Deflection constant for uniformly loaded simply supported beams, in./ft^2
D_p	Diameter of pin, in.
D_{ph}	Diameter of pin hole, in.
d	Depth of section or element, in. Diameter, in. Diameter of fastener hole or element, in. Diameter of plug weld, in.
d'	Width of bolt hole in flange parallel to tee stem or angle leg, in.
d_b	Nominal bolt diameter, in.
d_c	Column web depth clear of fillets, in.
d_L	Larger depth of a tapered member section or element, in.
d_o	Smaller depth of a tapered member section or element, in.
d_s	Depth of steel section inclusive of steel cover plate in composite design, in.

diam.	Diameter
D.L.	Dead load
E	Electrode
E, E$_s$	Proportional limit, elastic limit, Young's modulus, or modulus of elasticity for steel, ksi
E$_c$	Modulus of elasticity for concrete, ksi
E$_t$	Tangent modulus of steel, ksi
e	Eccentricity Base of natural logarithms = 2.718
e$_x$, e$_y$	Eccentricity about X-X and Y-Y axes, respectively.
F	Externally applied load per fastener in end plate and hanger-type connection kips Force, kips Load factor equal to $(F_y/F_b)f$
F$_a$	Allowable axial stress, ksi
F$_{allow}$	Allowable force, kips
F$_{as}$	Allowable axial stress in bracing and other secondary members, ksi
F$_{a\gamma}$	Allowable axial stress for tapered member, ksi
F$_{bx}$, F$_{by}$	Allowable bending stress with respect to X-X and Y-Y axes, respectively, ksi
F$_b'$	Adjusted allowable bending stress of the compression flange due to loss of bending resistance from web buckling, ksi
F$_b$	Allowable bending stress, ksi
F$_{br}$	Force in brace or strength of brace, k
F$_{by}$	Allowable bending stress for tapered flexural member, ksi
F$_{cr}$	Critical elastic buckling stress of a beam, ksi
F$_e$	Euler stress, ksi
F$_e'$	Euler stress divided by a safety factor, ksi
F$_{horiz}$	Horizontal force, kips
F$_p$	Allowable bearing stress, ksi
F$_t$	Allowable tensile stress, ksi

F_u Specified minimum tensile strength for a particular type or grade of steel, ksi
Lowest specified minimum tensile strength of a fastener or connected part, ksi

F_v Allowable shear stress, ksi

F_w Vertical uniform load in a beam span, klf

F_y Specified minimum yield stress, yield point, or yield strength for a particular type or grade of steel, ksi

F_{yc} Specified minimum yield stress, yield point, or yield strength of column, ksi

F_{yst} Specified minimum yield stress, yield point, or yield strength of stiffener, ksi

F'_y Hypothetical maximum yield stress of the material at which the width-to-thickness ratio of half the unstiffened compression flange is satisfactory; equal to $[65/(b_f/2t_f)]^2$; beyond this value, the section is noncompact, ksi

F''_y Hypothetical maximum yield stress, based on the depth-to-thickness ratio of web, beyond which the section or shape is noncompact; based upon the conditions of pure bending only and equal to $[640/(d/t)]^2$, ksi

F'''_y Hypothetical maximum yield stress, based on the depth-to-thickness ratio of web, beyond which the section or shape is noncompact; based upon the condition of combined bending and axial stress, and equal to $[257/(d/t)]^2$, ksi

F_{yr} Specified minimum yield stress of longitudinal reinforcing steel used in composite design, ksi

f Height or rise from top of column to apex of roof rigid frame; shape factor, Z/S

f_a Calculated axial stress, ksi

f_b Calculated bending stress, ksi
Calculated bending stress of steel section in composite design, ksi

f'_c Specified 28-day compressive strength of concrete, ksi

f''_c Ultimate strength of concrete, ksi

f_p Calculated bearing stress or pressure on support, ksi

f_v	Calculated shear stress, ksi
f_{vs}	Shear between transverse stiffeners and web of single or pair of stiffeners, kli
f_1	Compressive stress on webs of plate girders resulting from concentrated loads, ksi
f_2	Compressive stress on webs of plate girders resulting from distributed loads, ksi
f_1, f_2	Vertical component of stress on fastener due to eccentric load, ksi
	Vertical force on weld at most highly stressed point of weld due to vertical load and moment, respectively, kli
f_3	Horizontal component of stress on fastener due to eccentric load, ksi
	Horizontal force on weld at most highly stressed point of weld due to moment, kli
f_R	Resultant stress on fastener due to eccentric load, ksi
	Resultant force on weld at most highly stressed point of weld, kli
ft	Foot or feet
G	Shear modulus of elasticity or modulus of rigidity, ksi
	Stiffness ratio or relative stiffness of column to beam, in.³
	Joist girder designation.
G_a, G_b or G_{top}, G_{bottom}	Stiffness ratio at the top or bottom of column under consideration, in.³
$G_{elastic}, G_{inelastic}$	Stiffness ratio in the elastic and inelastic ranges, respectively, in.³
Gr	Grade
g	Gage, in.
	Spacing between fastener gage lines, in.
H	Height of stud shear connector, in.
h	Clear distance between flanges of a section, in.
	Clear distance from bottom of steel to top of concrete of the transformed section in composite design, in.
	Column height in rigid frame

h_r	Rib height of metal deck
H	Designation for a W shape having approximately equal depth and flange width dimensions
HSB	High-strength bolt
I	Moment of inertia of cross-section, in.^{4}
I'	Sum of moment of inertia of the compression flange and $\frac{1}{6}$ the moment of inertia of the compression area of the web for plate girder design, in.^{4}
I_c	Moment of inertia of a column, in.^{4}
I_d	Moment of inertia of steel deck supported on secondary members in flat roof framing, in.^{4}/ft
I_{eff}	Effective moment of inertia used for deflection computations in composite design, in.^{4}
I_g	Moment of inertia of a girder or beam, in.^{4}
I_{gr}	Gross moment of inertia about the neutral axis of the section, in.^{4}
I_o	Moment of inertia at smaller end of a tapered member, in.^{4} Moment of inertia of a section about its own center of gravity axis, in.^{4}
I_p	Moment of inertia for a primary member in flat roof framing, in.^{4} Polar moment of inertia about the center of gravity of a fastener group, in.^{4}
I_s	Moment of inertia for a secondary member in flat roof framing, in.^{4} Moment of inertia of steel section, in.^{4}
I_{tr}	Moment of inertia of transformed section in composite design, in.^{4}
I_{xx}, I_{yy}	Moment of inertia about X-X and Y-Y axes, respectively, in.^{4}
ICBO	Uniform Building Code (International Conference of Building Officials)
I.C.	Instantaneous center.
IFI	Industrial Fasteners Institute
in.	Inch or inches

in.[2]	Square inches
in.[3]	Cubic inches
in.[4]	Inches to the fourth power
in.[6]	Inches to the sixth power
J	Torsional constant of a cross-section, in.[4]
K	Effective length factor (ratio of the length of an equivalent pinned-end column to the length of the actual column)
	Factor for the theoretical cover plate length in composite design
	A type of bracing system
	Number of kips for each concentrated panel load for a joist girder
Kγ	Effective length factor of a tapered member
k	Distance from outside face of flange to toe of web fillet of W shape or welded section, in.
	Coefficient relating linear buckling strength of a plate to its dimensions and condition of support (ref. Sec. 7-5)
	Ratio of horizontal weld length to vertical weld length, in.
kips (k)	1000 pounds
kf	Kip feet
klf	Kips per lineal foot
kli	Kips per lineal inch
ksf	Kips per square foot
L	Span or bay length, ft
	Span of rigid frame, in.
	Distance measured in line of force from center line of hole to nearest edge of connected part toward which force is directed, in.
	Length of connection angle, in.
	Length of fillet weld, in.
	Length of stiffener plate for stiffened seated beam connections, in.
L$_b$	Lateral bracing interval of the compression flange, ft
L$_c$	Length of compression member, ft
	Maximum unbraced length of the compression flange at which $F_b = 0.66F_y$ as determined by the lesser of the two values $76.0b_f/(F_y)^{\frac{1}{2}}$ or $20,000/(d/A_f)F_y$ (ref. Eq. 6-9a or b, or Eq. 6-11 as applicable), ft

L_{cp} Cover plate length, ft

L_p Length of a primary member in a flat roof, ft

L_s Length of a secondary member in a flat roof, ft

L_u Maximum unbraced length of the compression flange at which $F_b = 0.60F_y$ (the larger value of l, in feet, from Eq. 8-7c using $C_b = 1.0$ and $F_b = 0.60F_y$ or from Eq. 8-7a using $C_b = 1.0$, as applicable)

L_x, L_y Span or bay length with respect to the X-X and Y-Y axes, respectively, ft

l Unbraced length, in.
Length of weld, in.
Length of channel shear connector, in.

l_a Actual eccentricity between applied load and the center of gravity of a fastener group, in.

l_b Unbraced length, in.
Unbraced length in plane of bending, in.

l_c Unsupported length of column or column segment, in.

l_e Effective eccentricity between applied load and the center of gravity of a fastener group, in.

l_{eff} Effective length of bearing stiffeners, in.

l_e Effective arm between applied load and center of gravity of fastener group, in.

l_f Length of weld of column web stiffeners at column flange, in.

l_g Unsupported length of restraining (beam or girder) member, in.

l_h Distance from center of fastener hole to end of beam web, in.

l_v Distance from center of fastener hole to free edge of connected part in direction of the force, in.

l_w Length of weld for column stiffener at column web, in.

l_x, l_y Unsupported length with respect to X-X and Y-Y axes, respectively, in.

lb   Pounds

L   Angle shape

LL	Live load
M	Moment, kf Allowable bending moment tributary to a tee-stub flange or angle leg by one bolt, kip-in.
$\mathbf{M_{cr}}$	Moment at which a beam buckles, kf
$\mathbf{M_{DL}}$ or $\mathbf{M_D}$	Dead load moment, kf Moment produced by loads applied before the concrete hardens in composite design, kf
$\mathbf{M_{LL}}$ or $\mathbf{M_L}$	Live load moment, kf Moment produced by loads applied after the concrete hardens in composite design, kf
$\mathbf{M_o}$	End moment, kf
$\mathbf{M_p}$	Plastic moment, kf
$\mathbf{M_R}$	Beam resisting moment equal to $F_b S_x/12$, where $F_b = 0.66F_y$ for compact sections, F_b is determined by Eq. 6-11 for noncompact section (when $F_y > F_y'$), F_b is $0.60F_y$ for noncompact sections (when $F_y > F_y''$), kf
$\mathbf{M_W}$, $\mathbf{M_Z}$	Moment about W-W and Z-Z axes, respectively, for single angle columns, kf
$\mathbf{M_x}$	Moment with respect to X-X axis, kf
$\mathbf{M_y}$	Moment with respect to Y-Y axis, kf Moment at first yield, kf
$\mathbf{M_1}$	The smaller bending moment at the end of a beam, kf Moment at face of outstanding leg or web surface in hanger-type connections, kf
$\mathbf{M_2}$	The larger bending moment at the end of a beam, kf Moment at centerline of fastener in hanger-type connections, kf
m	Dimension used in base plate design, in. Stiffness modifier used to modify a restraining member's stiffness Number of fasteners in a horizontal row Factor used to convert bending moment to an approximate equivalent axial load on a column subject to combined loading
max	Maximum
min	Minimum

mph	Miles per hour
M	Miscellaneous shape
MC	Miscellaneous channel
MS	The Masons Society
MT	Structural tee cut from M shape
Mn	Manganese
Mo	Molybdenum
N	Base plate dimension, in. Length of bearing of applied load, in. Number of fasteners Number of joist spacings for a joist girder
N_c	Number of mechanical shear connectors for full composite action between the point of maximum and zero moment based on the concrete section (equal to $U_c A_c$)
N_e	Length of end bearing to develop maximum web shear, in.
N_{min}	Minimum bearing plate dimension, in.
N_s	Number of mechanical shear connectors for full composite action between the point of maximum and zero moment based on the steel section (equal to $U_s W_s$)
N_1	Required number of shear connectors required between the point of maximum and zero moments
N_2	Required number of shear connectors between any concentrated load in the positive moment area and the nearest point of zero moment
n	Dimension used in base plate design Modular ratio, E_s / E_c Number of fasteners in one vertical row Number of threads per inch of length
n′	Factor for a given range of column cross sections; an equivalent dimension in base plate thickness computations, in.
NEA	National Erectors Association
Ni	Nickel
o.c.	On center
P	Applied concentrated load, k. kips H-shaped pile Force transmitted by one fastener to critical connected part, kips

	Force transmitted by one fastener, kips
	Phosphorus
P′	Equivalent axial load due to the bending component in the interaction formula, kips
P$_{actual}$	Actual applied concentrated load, kips
P$_{allow}$	Allowable applied concentrated load, kips
P$_{bf}$	Computed force delivered by beam flange or moment connection plate multiplied by 5/3, when P_{bf} is due to only live load plus dead loads, or by 4/3, when P_{bf} is due to live plus dead loads and wind or earthquake force, kips
P$_{cr}$	Euler load divided by a factor of safety, kips
P$'_{cr}$	Buckling load used to obtain the stiffness, S, of a restraining member with axial load, kips
P$_e$	Euler or buckling load, kips
P$_{eff}$	Effective value of axial load on a column accounting for the actual value of axial load as well as moment, kips
P$_{fb}$	Force from a beam flange or connection plate that a column will resist without stiffeners, kips
P$_R$	Beam reaction/number of bolts, kips
P$_{wb}$	Force from a beam flange or moment connection plate that a column will resist without stiffeners, kips
P$_{wi}$	Force (in addition to P_{wo}) that a column will resist without stiffeners from a beam flange or moment connection plate of 1-in. thickness, kips/in.
P$_{wo}$	Force from a beam flange or moment connection plate of zero thickness that a column will resist without web stiffeners, kips
P$_y$	Plastic axial load, (AF_y), kips
PD	Piling; deep arch
	Plastic design
PL	Plate
PMA	Piling; medium arch
PS	Piling; straight
PSA	Piling; shallow arch
PSX	Piling; straight, extra strength interlock

PZ	Piling; Z shape
p	Length of flange parallel to stem or leg tributary to each bolt, in.
pcf	Pounds per cubic foot
plf	Pounds per lineal foot
psf	Pounds per square foot
Q	Statical moment (first moment) of a cross-sectional area about its neutral axis, in.^{3} Prying force per fastener, kips
q	Allowable horizontal shear value for one connector, kips
R	End reaction, kips Maximum end reaction for $3\frac{1}{2}$ in. of bearing, kips Radius of fillet between body and head of eyebar, in. Weld return, in. Shear force in a single weld element at any deformation, kips Fastener or single weld element shear load at any given deformation, kips
R$_{BS}$	Resistance (allowable reaction) to block shear, kips
R$_i$	Increase in the maximum end reaction, R, for each additional inch of beam bearing beyond $3\frac{1}{2}$ in., kips
R$_T$, R$_B$	Relative stiffness ratios at top and bottom, respectively, of a tapered column, in.^{3}
R$_u$	Ultimate shear load of a single fastener or weld unit, kips
r	Radius of gyration, in. Ray of a force that passes through the instantaneous center
r$_b$	Radius of gyration corresponding to the plane of bending, in.
r$_{min}$	Minimum radius of gyration, in.
r$_o$	Radius of gyration at smaller end of a tapered member, in.
r$_{ox}$, r$_{oy}$	Radius of gyration at smaller end of a tapered member with respect to the X-X and Y-Y axes, respectively, in.
r$_t$	Value of one fastener in tension, kips
r$_T$	Radius of gyration of a section comprising the compression flange plus one third the compression web area taken about the Y-Y axis, in.

r_v	Permissible load on one fastener by specification (shear or bearing), kips
r_x, r_{xx}	Radius of gyration with respect to X-X axis, in.
r_y, r_{yy}	Radius of gyration with respect to Y-Y axis, in.
r_z, r_{zz}	Radius of gyration with respect to Z-Z axis, in.
r_w, r_z	Radius of gyration with respect to the corresponding principal axes for single angle columns, in.
req.	Required
RCRBSJ	Research Council on Riveted and Bolted Structural Joints
S	Clear beam spacing, ft Elastic section modulus, in.^{3} Horizontal shear force, kips Length of extended end of open web joist Slenderness ratio of tapered members Spacing of secondary member in flat roof, ft Stiffness of a restraining member with axial load, in.^{3} Center-to-center spacing of plug welds, in.
S'	Additional section modulus for welded plate girders corresponding to $\frac{1}{16}$ in. increase in web thickness, in.^{3}
S_{eff}	Effective section modulus corresponding to partial composite action, in.^{3}
S_j	Section modulus for the transformed section referred to the top of the steel beam in composite design, in.^{3}
S_o	Classical stiffness of a restraining member without axial load, in.^{3}
$S_{req.}$	Required elastic section modulus, in.^{3}
S_s	Section modulus for the steel beam used in composite design referred to the bottom flange, in.^{3}
S_t	Section modulus for the transformed section referred to the top of the concrete slab in composite design, in.^{3}
S_{tr}	Section modulus for the transformed section referred to the bottom flange, in.^{3}
S_{ts}	Section modulus for the steel beam referred to the top flange in composite design, in.^{3}
S_x, S_y	Elastic section modulus with respect to the X-X and Y-Y axes, respectively, in.^{3}

S_W, S_Z	Section modulus for the W-W and Z-Z axes, respectively, for single angle columns, in.^{3}
s	Pitch or spacing between successive holes in line of stress, in.
	Spacing of secondary members in flat roof framing, ft
S	American Standard shape
	Sulfur
Si	Silicon
ST	Structural tee cut from S shape
SAE	Society of Automotive Engineers
SJI	Steel Joist Institute
SSBC	Southern Standard Building Code
SSPC	Steel Structures Painting Council
T	Horizontal force in the flanges of a beam forming a couple equal to the beam end moment, kips
	Torque or twist, kips
	Detailing dimension of web of a rolled shape clear of the web to flange fillet, in.
	Applied tension per bolt exclusive of initial tightening, kips
t	Angle or plate thickness, in.
	Thickness of concrete slab in composite design, in.
	Temperature, °F
t_b	Thickness of flange delivering concentrated load (beam flange thickness at rigid beam-to-column connection), in.
t_c	Thickness of column web, in.
t_f	Thickness of flange, in.
	Required thickness of tee-stub flange or angle leg, in.
t_p	Thickness of flange or cover plate, in.
t_s	Thickness of stiffener, in.
t_w	Thickness of web, in.
	Thickness of weld metal, in.
TS	Structural tubing
U	Factor used to convert bending moment about the minor axis to an equivalent bending moment with respect to the major axis (equal to $F_{bx}S_x/F_{by}S_y$)
U_c	Coefficient used to compute N_c in full composite design

U_s	Coefficient used to compute N_s in full composite design
UL	Underwriters Laboratories
UM	Universal plates
UN	Unified Standard Series for screw threads
UNC	Unified Standard Series (coarse) for screw threads
UNS	Unified Numbering System
V	Statical or vertical shear on a beam, kips Horizontal shear, kips
V_{allow}	Allowable shear, kips
V_c	Vertical shear at cover plate cutoff, kips
V_h	Horizontal shear to be resisted by mechanical shear connectors under full composite action, kips
V_h'	Design horizontal shear to be resisted by mechanical shear connectors under partial composite action, kips
V_{max}	Maximum shear, kips
V	Vanadium
W	Uniform load, plf Total load, kips Dimension in a lap joint, in. Width of seat plate, in.
W_c	Uniform load constant for uniformly loaded simply supported beams, kip-ft
W_s	Maximum weight of a steel section at the center of the beam span, which includes the cover plate, plf
W_u	Load defining the limit of usefulness in plastic design
w	Uniformly distributed load, psf, plf, ksf, klf Weight of concrete, pcf Width of eyebar, in. Width of slot weld, in. Length of flange tributary to each bolt, in. Width of vertical stiffener, in.
w_{eff}	Effective width of vertical stiffener, in.
w_h	Width of eyebar at head, in.
w_r	Average rib width or rib width for metal deck, in.

W	Width flange shape
WT	Structural tee cut from W shape
W-W	Principal axis of single angle
X	Stiffness distribution factor used to divide the stiffness of a column above and below a joint in a frame
x	Distance Distance from the end of a beam to the theoretical cutoff point of a cover plate, in. Coefficient to express eccentricity of a horizontal weld about the center gravity axis of the weld group, in.
$\bar{x}$	Distance from profile centroid of a shape to the shear plane of the connection, in.
X-X	Designates the X-X axis or major axis of a cross section
Y	Ratio of the yield stress of the web to that of the stiffeners
$\bar{y}$, y	Distance Distance from the center of gravity of an element to the neutral axis of the section, in. Distance from the center of gravity axis of a section to the outermost fiber of an element, in.
y_b	Distance from the neutral axis of the transformed section to the bottommost fiber of the steel section in composite design, in.
y_{bs}	Distance from the neutral axis of the steel section to the bottommost fiber of the steel section in composite design, in.
Y-Y	Designates the Y-Y axis or minor axis of a cross section
Z	Plastic section modulus, in.³ Distance from the smaller end of a tapered member, in.
$Z_{req.}$	Required plastic section modulus, in.³
Z_x, Z_y	Plastic section modulus about the X-X and Y-Y axes, respectively, in.³
Z-Z	Designation for the longitudinal axis of a shape or the principal axis for a single angle
α	Ratio of the web to flange yield stress for hybrid plate girders Moment ratio used in prying action formula

β	Angle of twist Stiffness of a brace, klf Spring constant, klf S_{tr}/S_s or S_{eff}/S_s in composite design
$\beta', \beta'', \beta'''$	First, second, and third derivatives, respectively, of the angle of twist with respect to length
ϵ	Coefficient of expansion
γ	Equal to $(d_L - d_o)/d_o$ for tapered members
ϕ	Capacity reduction factor bt/nA_s and A_c/nA_s in composite design
θ, ϕ	Angle designation
ϕ	Constant used in composite design
π	Constant equal to 3.1415927
λ	Regression coefficient used in eccentric load of fastener group design
δ	Ratio of net area (at bolt line) to the gross area at face of the stem or angle leg
Δ	Deflection, in. Horizontal translation, in. Movement in Y-Y direction, in. Total deformation of a fastener (includes shearing, bending, and bearing deformation of fastener and local deformation of the plate), in. Used in base plate design; expressed in terms of depth and flange width of the shape involved Deformation of a weld element, in.
Δ_{DL}	Deflection due to deadload, in.
Δ_{LL}	Deflection due to live load, in.
Δ_{max}	Maximum deflection, in.
Δ_o	Initial out of plumbness of a column, in. Deflection produced by end moment, in.
Δ''	Second derivative of movement in the Y-Y direction
μ	Poisson's ratio or the ratio of transverse strain to longitudinal strain under axial load (may be taken as 0.3 for steel) Regression coefficient used in eccentric load of fastener group design

°	Degrees
°F	Degrees Fahrenheit
%	Percentage
Σ	Sum
∞	Infinity
$\sqrt{}$	Square root equal to $(\ )^{1/2}$
$\approx$	Approximately equal to
$<$	Less than
$>$	Larger than
$\leq$	Equal or less than
$\geq$	Equal or larger than

list of
steel-related
organizations

The following are included in this appendix: organizations and institutes associated with the manufacture, design, fabrication and erection of steel products; related organizations concerned with other construction products; professional organizations; and code-writing organizations.

AASHTO

The American Association of State Highway and Transportation Officials
444 North Capitol Street, N.W.
Washington, D. C. 20001

A national association of state highway officials. Publications include *Standard Specifications for Highway Bridges* and many other standards for highway construction and highway structures. Interim reports, technical reports, and research reports are available for inspection at most State Highway Departments.

AISC

American Institute of Steel Construction Incorporated
National Office
400 North Michigan Ave.
Chicago, Ill. 60611

Regional Offices throughout the country.

A national organization of steel fabricators. Major publications include *Steel Construction Manual, Structural Steel Detailing Manual, Design of Orthotropic Steel Bridges, Engineering Journal* (a quarterly), and *Modern Steel Construction* (a quarterly). A list of further technical publications is available from either the Regional or National Office.

AISC

Australian Institute of Steel Construction
84 Pacific Highway
North Sidney, Australia 2060

AISE

Association of Iron and Steel Engineers
3 Gateway Center
Pittsburgh, Pa. 15222

An association of engineers employed in the steel industry. Publications include *Specifications for Design and Construction of Mill Buildings*.

AISI

American Iron and Steel Institute
1000 16th Street, N.W.
Washington, D.C. 20036

District Offices throughout the country.

A national organization of steel producing mills. Publications include *Light Gage Cold Formed Steel Design Manual, Steel Products Manual,* and *Steel Abstracts*. A list of research reports, earthquake reports, building construction reports, and many technical papers is available from the head office.

ANSI

American National Standards Institute
1430 Broadway
New York, N.Y. 10018

An organization that administers the federated national standards system, which identifies industrial and public needs for national standards. An annual catalog of standards on a variety of subjects is available.

API

American Petroleum Institute
1801 K Street, N.W.
Washington, D.C. 20060

An organization composed of individuals and companies interested in the field of petroleum engineering. Publications include specifications for oil derricks, pipe lines, and other specifics having to do with that engineering discipline.

AREA

American Railway Engineering Association
2000 L Street, N.W.
Washington, D.C. 20036

An association of individuals and companies interested in the field of railroad engineering. Publications include specifications for steel bridges and other specifics having to do with the discipline of railroad engineering.

ASCE

American Society of Civil Engineers
345 East 47th Street
New York, N.Y. 10017

A society of individuals in the field of civil engineering. There are many divisions of this organization: construction, structural, engineering mechanics, etc. Each division produces a technical journal periodical.

ASM

American Society for Metals
Metals Park, Ohio 44073

A society of companies and individuals interested in production of various metals and alloys. Publishes the Metals Handbook Series, a series comprising eleven volumes, dealing with a variety of subjects related to metals. Other engineering texts are published.

ASNT

American Society for Non-Destructive Testing
914 Chicago Avenue
Evanston, Ill. 60202

An organization of individuals interested in testing of materials. Publications include several papers on recommendations for nondestructive testing of materials.

ASTM

American Society for Testing and Materials
1916 Race Street
Philadelphia, Pa. 19103

A national organization interested in standards for materials. They publish several volumes of standards, *ASTM Standards. ASTM Standards in Building Codes* is of particular interest to those in the construction industry. Separates of any single standard can be obtained from the national headquarters.

AWS

American Welding Society
2501 N.W. 7th Street
Miami, Fla. 33125

A national organization of individuals and companies interested in the furthering of the art of welding. Publications include the five-volume series *Welding Handbook,* welding standards and specifications having to do with all forms of the art of welding, and a monthly publication, *Welding Journal.*

AWWA

American Water Works Association
National Press Building
14 and F Street, N.W.
Washington, D.C. 20045

An association composed of individuals interested in the general field of water works engineering. Publications include specifications for steel tanks and other specifics having to do with that discipline.

CISC

Canadian Institute of Steel Construction
201 Consumers Road
Willowdale, Ontario M2J 4G8
Canada

CRC (SSRC, Structural Stability Research Council)

Column Research Council
Fritz Engineering Laboratory
Lehigh University
Bethlehem, Pa. 18015

A group to promote research on the behavior of metal compression members, to assist in the development of improved design criteria, and to publish research information. The SSRC or CRC publishes the *Guide to Design Criteria for Metal Compression Members (Guide to Stability Design Criteria for Metal Structures)*.

CSI

Construction Specifications Institute
1150 17th Street N.W.
Washington, D.C. 20036

EERI

Earthquake Engineering Research Institute
424 40th Street
Oakland, Calif. 94609

A national professional society devoted to finding better ways to protect life and property from the effects of earthquake, to advancing the science and practice of earthquake engineering, and to solving earthquake engineering problems.

EIA

Electronics Industries Association
1721 DeSales Street N.W.
Washington, D.C.

An association of industries interested in the general field of radio and television. Publications include specifications for radio and television towers.

IFI

Industrial Fasteners Institute
East Ohio Bldg.
Cleveland, Ohio 44113

An association of U.S. manufacturers of technical fasteners to advance fastener technology and application. IFI produces *Fastener Standards,* a manual that lists the standards associated with most fastener sizes and designs in use today.

JFLF

The James F. Lincoln Arc Welding Foundation
P.O. Box 3035
Cleveland, Ohio 44117

A foundation created by the Lincoln Electric Company to help advance progress in welded design and construction. Award programs, educational seminars, and technical publications are some of the efforts of the foundation directed toward the growth of welding.

MBMA

Metal Building Manufacturers Association
2130 Keith Building
Cleveland, Ohio 44115

An organization of companies interested in the manufacture of prefabricated metal buildings. Publications include *Recommended Design Practices Manual* and a bimonthly publication, *Metal Building Facts.*

MLA

Metal Lath Association
221 North LaSalle Street
Chicago, Ill. 60601

An association of companies interested in the manufacture and installation of metal lath. Publications include the specifications for the installation of metal lath and general specifications for lightweight aggregate fireproofing for steel frame structures.

NCSPA

National Corrugated Steel Pipe Association
4825 West Scott Street
Schiller Park, Ill. 60176

NEA

National Erectors Association
1800 North Kent Street
Arlington, Va. 22209

A national association of steel erectors. They publish and distribute a monthly newsletter.

NFPA

National Fire Protection Association
470 Atlantic Avenue
Boston, Mass. 02210

An association that publishes and authors National Fire Codes and other authoritative information on all aspects of fire protection engineering.

NISD

National Institute of Steel Detailers
270 Madison Avenue
New York, N.Y. 10016

PFI

Pipe Fabrication Institute
1326 Freeport Road
Pittsburgh, Pa. 15238

RCRBSJ

Research Council on Riveted and Bolted
Structural Joints of the Engineering
Foundation
% Industrial Fasteners Institute

A group whose purpose is to organize, coordinate, and supervise research
on rivets and bolts and their use in structural connections.

SDI

Steel Deck Institute
9836 West Roosevelt Road
Westchester, Ill. 60153

An organization composed of companies interested in the manufacture
and installation of metal steel decks. Publications include the specifications for such work.

SEAOC

Structural Engineers Association of California
171 Second Street
San Francisco, Calif. 94105

An association that authors, through the seismology Committee and
Study Groups, *Recommended Lateral Force Requirements and Commentary*.

SJI

Steel Joist Institute
1703 Parham Road
Richmond, Va. 23229

A nonprofit organization comprised of manufacturers actively engaged in
the manufacture and distribution of open web steel joists to provide tests
and research data for public dissemination.

SPFA

Steel Plate Fabricators Association
15 Spinning Wheel Road
Hinsdale, Ill. 60521

An organization of establishments primarily engaged in the manufacture of plate products.

SSPC

Steel Structures Painting Council
4400 Fifth Avenue
Pittsburgh, Pa. 15213

An organization that performs research, recommends issues, suggests information for codes and specifications, arranges conferences, and produces literature on the subject of cleaning and painting steel structures. SSPC authors the two-volume *Steel Structures Painting Manual,* which covers good painting practices, painting systems, and specifications.

UL

Underwriters Laboratories, Inc.
333 Pfingsten Road
Northbrook, Ill. 60062

A nonprofit organization to test for public safety. It examines and tests devices, systems, and materials relating to their life, fire, casualty hazards, and crime prevention. It publishes a *Buildings Materials Directory* and a *Fire Resistance Index,* along with many other publications and motion pictures.

WRC

Welding Research Council
345 East 47th Street
New York, N.Y. 10017

An organization of engineers and scientists involved cooperatively in research in welding. The council sponsors and initiates research projects in welding, and collects and distributes authoritative information on welded materials, equipment, and processes used in the joining of metals and the design, fabrication, erection, testing, and inspection of welded structures.

WRI

Wire Reinforcement Institute
7900 West Park Drive
McLean, Va. 22101

A nonprofit service organization that publishes technical information pertaining to welded wire fabric and similar reinforcement.

WSTI

Welded Steel Tube Institute
522 Westgate Tower
Cleveland, Ohio 44116

A nonprofit corporation whose members are leading producers of welded tubing cold-formed from flat-rolled carbon, stainless, and alloy steel. They make available information, advance engineering knowledge, and user requirements, and expand uses of their products. Their Structural Tube Division produces a *Manual of Cold Formed Welded Structural Steel Tubing*.

ZI

Zinc Institute, Inc.
292 Madison Avenue
New York, N.Y. 10017

ZI makes available literature on the subject of the use of zinc coatings as a means of corrosion protection.

PROFESSIONAL ORGANIZATIONS

ACEC

American Consulting Engineers Council
Madison Building
1155 15th Street, N.W.
Washington, D.C. 20005

A national organization of engineers engaged in the practice of independent engineering. ACEC assists its members in achieving higher professional, business, and economic standards.

AIA

American Institute of Architects
1735 Massachusetts Avenue, N.W.
Washington, D.C. 20036

The national professional architectural organization devoted to the professional, ethical, economic, and social aspects of architecture.

ASME

American Society of Mechanical Engineers
315 East 47th Street
New York, N.Y. 10017

A society of individuals interested in the field of mechanical engineering. They are publishers of an extensive list of research papers, specifications, and technical papers having to do with mechanical applications of interest to designers and the relationship of machines and structures.

NSPE

National Society of Professional Engineers
2029 K. Street, N.W.
Washington, D.C. 20006

The national professional engineering organization devoted to the professional, ethical, economic, and social aspects of engineering.

SWE

Society of Women Engineers
345 East 47th Street
New York, N.Y. 10017

RELATED ORGANIZATIONS

AA

Aluminum Association
818 Connecticut Ave., N.W.
Washington, D.C. 20006

An organization composed of individuals and companies interested in the aluminum industry. Publications include the *Specifications for Design and Construction of Aluminum Frame Structures*. A list of other publications and lists of materials available from that industry can be obtained by writing to the headquarters.

ACI

American Concrete Institute
P.O. Box 19150
Redford Station
Detroit, Mich. 48217

A national organization representing the concrete industry. Major publications include the *ACI Building Code, Proceedings,* and *Journal.* A complete publication catalog is available from the national office.

BIA

Brick Institute of America
1750 Old Meadow Road
McLean, Va. 22101

CRSI

Concrete Reinforcing Steel Institute
228 North LaSalle Street
Chicago, Ill. 60601

An organization producing many publications pertaining to structural reinforced concrete and reinforcing steel.

GA

Gypsum Association
1603 Orringtone Ave.
Evanston, Ill. 60201

INCO

The International Nickel Company, Inc.
One New York Plaza
New York, N.Y. 10004

A leading producer of nickel products offering extensive literature on the product and its applications.

MS

The Masonry Society
3002 East Third Avenue
Denver, Colo. 80206

An international organization to further the technology and application of masonry in the construction industry.

NATIONAL CODE WRITING BODIES

BBC, BOCA

Basic Building Code
Building Officials and Code Administrators,
International, Inc.
17926 South Halsted
Homewood, Ill. 60430

NBC, AIA

National Building Code
American Insurance Association
230 West Monroe Street
Chicago, Ill. 60606

SSBC, SBCC

The Southern Standard Building Code
Southern Building Code Congress
3617 Eighth Avenue, South
Birmingham, Ala. 35222

UBC, ICBO

Uniform Building Code
International Conference of
Building Officials
5630 South Workman Mill Road
Whittier, Calif. 90601

Construction Review, a U.S. government publication, issues a listing every two or three years of national trade associations, professional societies, labor unions, and national associations in the finance and other areas of construction. The listing includes all involved in the construction and building materials industries. This periodic listing is a valuable and more detailed reference.

index